T0334598

Natural Water Remediation

Natural Water Remediation

Chemistry and Technology

James G. Speight

Butterworth-Heinemann
An imprint of Elsevier

Butterworth-Heinemann is an imprint of Elsevier
The Boulevard, Langford Lane, Kidlington, Oxford OX5 1GB, United Kingdom
50 Hampshire Street, 5th Floor, Cambridge, MA 02139, United States

Library of Congress Cataloging-in-Publication Data
A catalog record for this book is available from the Library of Congress

British Library Cataloguing-in-Publication Data
A catalogue record for this book is available from the British Library

ISBN: 978-0-12-803810-9

For information on all Butterworth-Heinemann publications
visit our website at https://www.elsevier.com/books-and-journals

Publisher: Matthew Deans
Acquisition Editor: Matthew Deans
Editorial Project Manager: Alexandra Romano/Lena Sparks
Production Project Manager: Sruthi Satheesh
Cover Designer: Mark Rogers

Typeset by SPi Global, India

Contents

About the author

Dr. James G. Speight

Dr. James G. Speight has doctorate degrees in Chemistry, Geological Sciences, and Petroleum Engineering and is the author of more than 80 books in petroleum science, petroleum engineering, environmental sciences, and ethics.

Dr. Speight has fifty years of experience in areas associated with (i) the properties, recovery, and refining of reservoir fluids, conventional petroleum, heavy oil, and tar sand bitumen, (ii) the properties and refining of natural gas, gaseous fuels, (iii) the production and properties of petrochemicals, (iv) the properties and refining of biomass, biofuels, biogas, and the generation of bioenergy, and (v) the environmental and toxicological effects of fuels. His work has also focused on safety issues, environmental effects, remediation, and safety issues as well as reactors associated with the production and use of fuels and biofuels.

Although he has always worked in private industry which focused on contract-based work, he has served as Adjunct Professor in the Department of Chemical and Fuels Engineering at the University of Utah and in the Departments of Chemistry and Chemical and Petroleum Engineering at the University of Wyoming. In addition, he was a Visiting Professor in the College of Science, University of Mosul, Iraq and has also been a Visiting Professor in Chemical Engineering at the following universities: University of Missouri-Columbia, the Technical University of Denmark, and the University of Trinidad and Tobago.

In 1996, Dr. Speight was elected to the Russian Academy of Sciences and awarded the Gold Medal of Honor that same year for outstanding contributions to the field of petroleum sciences. In 2001, he received the Scientists without Borders Medal of Honor of the Russian Academy of Sciences and was also awarded Dr. Speight the Einstein Medal for outstanding contributions and service in the field of Geological Sciences. In 2005, the Academy awarded Dr. Speight the Gold Medal—Scientists without Frontiers, Russian Academy of Sciences, in recognition of Continuous Encouragement of Scientists to Work Together Across International Borders. In 2007 Dr. Speight received the Methanex Distinguished Professor award at the University of Trinidad and Tobago in recognition of excellence in research. In 2018, he received the American Excellence Award for Excellence in Client Solutions from the United States Institute of Trade and Commerce. Washington, DC.

Preface

Water plays a key role in ensuring the sustainable operation of the anthrosphere and its maintenance in a manner that will enable it to operate in harmony with the environment for generations to come. Water is a complex system of chemical species that has received considerable study because of a series of environmental issues that have been raised. In fact, throughout history, the quality and quantity of water available to humans have been vital factors in determining not only the quality of life but also the existence of life. However, in many parts of the world, water pollution is a major issue. Thus, there is a need to understand water chemistry as it pertains to the various water systems and to the remediation of pollutants in these systems.

The purpose of this book is to provide an overview, with some degree of detail, of the chemistry of water in the environment within the broad framework of sustainability. This is an issue of great concern as the demands of the increasing human population of the Earth which threaten to overwhelm this important resource.

The book presents coverage of the chemistry of water and relates the science and technology of water to areas essential to sustainability science, including environmental chemistry and remediation of contaminated water systems.

Dr. James G. Speight
Laramie, WY, United States

Water systems

1

Chapter outline

1 Introduction

The Earth is a complex interrelationship between the air (the atmosphere), water (the hydrosphere), and land (lithosphere) (Speight, 1996; Speight and Lee, 2000; Weiner and Matthews, 2003: Spellman, 2008; Manahan, 2010). In addition, the geosphere is often used

Natural Water Remediation. https://doi.org/10.1016/B978-0-12-803810-9.00001-2

Table 1.1 Mineral groups of the earth

Mineral group	Example	Simple formula
Silicate	Quartz	SiO_2
	Olivine	$(Mg,Fe)_2Si_4$
	Potassium feldspar	$KAlSi_3O_g$
Oxide	Corundum	Al_2O_3
	Magnetite	Fe_3O_4
Carbonates	Calcite	$CaCO_3$
	Dolomite	$CaCO_3.MgCO_3$
Sulfides	Pyrite	FeS_2
	Galena	PbS
	Gypsum	$CaSO_4.2H_2O$
Halides	Halite	NaCl
	Fluorite	CaF_2
Native elements	Copper	Cu
	Sulfur	S

as collective name for the atmosphere, the hydrosphere, and the lithosphere and it is also used along with the atmosphere, the hydrosphere, and the biosphere to describe the systems of the Earth and the interaction of these systems. On a more specific basis, the term geosphere is often used to refer to the solid parts of the Earth—the rocks, the minerals that make up the outer core or mantle and the iron-rich material that makes up the inner core of the Earth (Table 1.1). In that context, sometimes the term lithosphere is used instead of geosphere for the solid parts of the Earth. The lithosphere, however, only refers to the uppermost layers of the solid Earth (oceanic and continental crustal rocks and uppermost mantle).

The three major constituents of air, and therefore of atmosphere of the Earth, are nitrogen, oxygen, and argon. Thus, the atmosphere of the Earth is a mixture of chemical constituents—the most abundant of them are nitrogen (N_2, (78% *v*/v) and oxygen (O_2, (21% v/v). These gases, as well as the noble gases (argon, neon, helium, krypton, xenon), possess very long lifetimes against chemical destruction and, hence, are relatively well mixed throughout the entire homosphere (below approximately 295, 000 ft altitude). Minor constituents, such as water vapor, carbon dioxide, ozone, and many others, also play an important role despite their lower concentration.

In the current context of natural water, water vapor accounts for approximately 0.25% *w*/w of the atmosphere by mass. More pertinent to the current text, the concentration of water vapor varies significantly from approximately 10 ppm by volume (ppm *v*/v) in the coldest portions of the atmosphere to as much as 5% v/v in the hot, humid air masses (typically found in the tropical zones), and concentrations of other atmospheric gases are typically quoted in terms of dry air (without water vapor). The remaining gases are often referred to as trace gases, among which are the greenhouse gases, principally carbon dioxide, methane, nitrous oxide, and ozone. The spatial and temporal distribution of chemical species in the atmosphere is determined by several processes, including surface emissions and deposition, chemical and photochemical reactions, and transport by wind and water.

Surface emissions are associated with volcanic eruptions, floral and faunal activity on the continents as well as in the ocean, as well as anthropological activity such as biomass burning, agricultural practices, and industrial activity. Chemical conversions are achieved by a multitude of reactions whose rate constants are measured in the laboratory. Transport is usually represented by large-scale advective motion (displacements of air masses in the quasi-horizontal direction), and by smaller scale processes, including convective motions (vertical motions produced by thermal instability and often associated with the presence of large cloud systems), boundary layer exchanges, and mixing associated with turbulence. Wet deposition results from precipitation of soluble species, while the rate of dry deposition is affected by the nature of the surface (such as the type of soil and the types of vegetation as well as ocean currents).

When a chemical is introduced into the environment, it becomes distributed among the four major environmental compartments: (i) air, (ii) water, (iii) land, and (iv) biota (living organisms). Each of the first three categories can be further subdivided in floral (plant) environments and faunal (animal, including human) environments. The portion of the chemical that will move into each compartment is governed by the physical and chemical properties of the chemical. In addition, the distribution of a chemical in the environment is governed by physical processes such as sedimentation, adsorption, and volatilization after which the chemical can then be degraded by chemical processes, physical processes, and/or biological processes. Chemical processes generally occur in water or the atmosphere and follow one of four reactions: oxidation, reduction, hydrolysis, and photolysis. Biological mechanisms in soil and living organisms utilize oxidation, reduction, hydrolysis and conjugation to degrade chemicals.

The degradation process for many inorganic and organic chemicals is typically controlled by the ecosystem (air, water, land, atmosphere, biota) in which the chemical is distributed and further control is exerted on the chemical by one or more of the physical processes already mentioned (i.e., sedimentation, adsorption, and volatilization). When assessing the impact of a chemicals on the environment, the most critical characteristics are: (i) the type of chemical, which depends on the type of industry and/or the process from which the chemical originated, (ii) the amount and concentration of the chemical. Each of the systems that can be affected by the entry of a chemical are presented below and it is the purpose of following sections to introduce the various environmental systems of the Earth and the interrelationships of these systems to each other.

Thus the Earth can be viewed as a combination of interrelated, interdependent, or interacting parts forming a collective whole or entity. On a macro level, the Earth system maintains its existence and functions as a whole through the interactions of the component parts which for the purposes of this text are: (i) the atmosphere, (ii) the hydrosphere, and (iii) the lithosphere. These component parts interconnected by processes and cycles, which, over time, intermittently store, transform, and/or transfer matter and energy throughout the whole Earth system in ways that are governed by the thermodynamic laws of conservation of matter and energy (Chapter 4).

These component parts—the atmosphere, the hydrosphere, and the lithosphere are described in turn in the following sections.

2 The atmosphere

The atmosphere is the layer or a set of layers of gases surrounding the Earth that is held in place by gravity. By volume, dry air contains nitrogen (78.09%), oxygen (20.95%), argon (0.93%), carbon dioxide (0.04%), and small amounts of other gases.

Chemical compounds released at the surface by natural processes and by anthropogenic processes are oxidized in the atmosphere before being removed by wet or dry deposition. Key chemical species of the troposphere include organic compounds such as methane and non-methane hydrocarbon derivatives as well as oxygenated organic species and carbon monoxide, nitrogen oxides (which are also produced by lightning discharges in thunderstorms) as well as nitric acid. Other chemical species include: hydrogen compounds (and specifically the hydroxy radical, OH, and the hydroperoxy radical, HO_2, as well as hydrogen peroxide, H_2O_2, ozone, O_3, and sulfur compounds (such as dimethyl sulfide, CH_3SCH_3, sulfur dioxide, SO_2, and sulfuric acid, H_2SO_4]. The hydroxyl radical (OH) deserves additional consideration since it has the capability of reacting with and efficiently destroying a large number of organic chemical compounds, and hence of contributing directly to the oxidation capacity (reactivity) of the atmosphere.

Finally, the release of sulfur compounds at the surface of the Earth surface and the subsequent oxidation of the sulfur compounds in the atmosphere leads to the formation of small liquid or solid particles that remain in suspension in the atmosphere. These aerosol particles affect the radiative balance of the atmosphere directly, by reflecting and absorbing solar radiation, and indirectly, by influencing cloud microphysics. The release to the atmosphere of sulfur compounds has increased dramatically, particularly in regions of Asia, Europe, and North America as a result of human activities, specifically coal combustion (Speight, 2013a,b).

Physically, the atmosphere is the thin and fragile envelope of air surrounding the Earth that is held in place around the Earth by gravitational attraction and which has a substantial effect on the environment. The atmosphere contains oxygen used by most organisms for respiration and carbon dioxide used by plants, algae, and cyanobacteria for photosynthesis. Also, the atmosphere helps protect living organisms from genetic damage by solar ultraviolet radiation, solar wind, and cosmic rays. Its current composition is the product of billions of years of biochemical modification of the paleoatmosphere by living organisms. The atmosphere can be divided (atmospheric stratification) into five main layers. Generally, the atmosphere of the Earth has four primary layers, which are (i) the troposphere, (ii) the stratosphere, (iii) the mesosphere, (iv) the thermosphere, and (v) the exosphere—these layers differ in properties such as composition, temperature and pressure.

Approximately three quarters (75% *v*/v) of the mass of the atmosphere mass resides within the troposphere, and is the layer within which the weather systems develop. The depth of this layer varies between 548,000 ft at the equator to 23,000 ft over the Polar Regions. The stratosphere, extends from the top of the troposphere to the bottom of the mesosphere, contains the ozone layer which ranges in altitude between 49,000 ft and 115,000 ft, and is where most of the ultraviolet radiation from the Sun is absorbed. The top of the mesosphere, ranges from 164,000 ft to 279,000 ft, and is the layer wherein most meteors burn up. The thermosphere extends from 279,000 ft to the base of the exosphere at approximately 2, 300,000 ft altitude and contains the ionosphere, a region where the atmosphere is ionized by incoming solar radiation.

2.1 The troposphere

The troposphere is the lowest layer of atmosphere of the Earth and the layers to which changes can greatly influence the floral and faunal environments. The troposphere extends from the surface of the Earth to a height of approximately 30,000 ft at the Polar Regions to approximately 56,000 ft at the equator, with some variation due to weather. The troposphere is bounded above by the tropopause, a boundary marked in most places by a temperature inversion (i.e. a layer of relatively warm air above a colder one), and in others by a zone which is isothermal with height.

Although variations do occur, the temperature usually declines with increasing altitude in the troposphere because the troposphere is mostly heated through energy transfer from the surface. Thus, the lowest part of the troposphere (i.e. the surface of the Earth) is typically the warmest section of the troposphere, which promotes vertical mixing. The troposphere contains approximately 80% of the mass of the atmosphere of the Earth. The troposphere is denser than all its overlying atmospheric layers because a larger atmospheric weight sits on top of the troposphere and causes it to be most severely compressed.

In the current context of water, the majority of the atmospheric water vapor or moisture is found in the troposphere.

2.2 The stratosphere

Above the troposphere, the atmosphere becomes very stable, as the vertical temperature gradient reverses in a second atmospheric region—the *stratosphere*—which extends from the top of the troposphere at approximately 39,000 ft above the surface of the Earth to the stratopause at an altitude of approximately 164,000 to 180,000 ft. The atmospheric pressure at the top of the stratosphere is approximately 1/1000 the pressure at sea level.

The stratosphere contains the ozone layer, which is the part of atmosphere that contains relatively high concentrations of that gas. In this layer ozone concentrations are approximately 2–8 ppm, which is much higher than in the lower atmosphere but still very small compared to the main components of the atmosphere. It is mainly located in the lower portion of the stratosphere from approximately 49,000 to 115,000 ft, though the thickness varies seasonally and geographically. Approximately 90% *v*/v of the ozone in the atmosphere of the Earth is contained in the stratosphere.

2.3 The mesosphere

The mesosphere is the third highest layer of atmosphere and occupies the region above the stratosphere and below the thermosphere. This layer extends from the stratopause at an altitude of approximately 160,000 ft to the mesopause at approximately 260,000 to 80,000 ft) above sea level. Temperatures drop with increasing altitude to the mesopause that marks the top of this middle layer of the atmosphere. It is the coldest place on Earth and has a temperature on the order of −85 °C (−120 °F).

2.4 The thermosphere

The thermosphere is the second-highest layer of the atmosphere and extends from the mesopause (which separates it from the mesosphere) at an altitude of approximately 260,000 ft up to the thermopause at an altitude that ranges from 1,600,000 to 3,300,000 ft. In the thermosphere, the temperature increases to reach maximum values that are strongly dependent on the level of solar activity. Vertical exchanges associated with dynamical mixing become insignificant, but molecular diffusion becomes an important process that produces gravitational separation of species according to their molecular or atomic weight.

The height of the thermopause varies considerably due to changes in solar activity. Because the thermopause lies at the lower boundary of the exosphere, it is also referred to as the exobase. The lower part of the thermosphere, from 260,000 ft to 1,800,000 ft above the surface of the Earth surface, contains the *ionosphere*.

The ionosphere is a region of the atmosphere that is ionized by solar radiation and is responsible for auroras (the *aurora borealis* in the northern hemisphere and the *aurora australis* in the southern hemisphere). The ionosphere increases in thickness and moves closer to the Earth during daylight and rises at night allowing certain frequencies of radio communication a greater range. During daytime hours, it stretches from approximately 160,000 ft to 3,280,000 ft and includes the mesosphere, thermosphere, and parts of the exosphere. However, ionization in the mesosphere largely ceases during the night, so auroras are normally seen only in the thermosphere and lower exosphere. The ionosphere forms the inner edge of the magnetosphere.

The temperature of the thermosphere gradually increases with height. Unlike the stratosphere beneath it, wherein a temperature inversion is due to the absorption of radiation by ozone, the inversion in the thermosphere occurs due to the extremely low density of its molecules. The temperature of this layer can rise as high as 1500 °C (2700 °F), though the gas molecules are so far apart that its temperature in the usual sense is not very meaningful. This layer is completely cloudless and free of water vapor. However non-hydrometeorological phenomena such as the aurora borealis and *aurora australis* are occasionally seen in the thermosphere.

2.5 The exosphere

The exosphere is the outermost layer of the atmosphere (that is, it is the upper limit of the atmosphere) and extends from the exobase, which is located at the top of the thermosphere. The exosphere begins variously from approximately 2,300,000 ft to 3,280,000 ft above the surface, where it interacts with the magnetosphere, to space. Each of the layers has a different lapse rate, defining the rate of change in temperature with height. Initial atmospheric composition is generally related to the chemistry and temperature of the local solar nebula during planetary formation and the subsequent escape of interior gases.

The exosphere layer is mainly composed of extremely low densities of hydrogen, helium and several heavier molecules including nitrogen, oxygen and carbon dioxide closer to the exobase. The atoms and molecules are so far apart that they can travel hundreds of kilometers without colliding with one another. Thus, the exosphere no longer behaves like a gas, and the particles constantly escape into space. The exosphere contains most of the satellites orbiting Earth.

3 The hydrosphere

The hydrosphere (also called the aquasphere) is the combined mass of water found on, under, and above the surface of the Earth. Although the hydrosphere has been in existence for >4 billion years, it continues to change in size. This is caused by sea floor spreading and continental drift which rearranges the land and ocean. Finally, the lithosphere is the combined land masses of the Earth the land which is the outermost shell of the Earth and is composed of the crust (which is defined on the basis of its chemistry and mineralogy) and the portion of the upper mantle that behaves elastically during geological time (a time scale of thousands of years or greater).

Another term the *cryosphere* is used to describe those portions of the surface of the Earth where to where the water is in solid form, including sea ice, lake ice, ice sheets, and glaciers. The cryosphere is an integral part of the global climate system with important linkages and through its influence on surface energy and moisture fluxes, clouds, precipitation, atmospheric circulation, and oceanic circulation.

The existence of a water system is due to a collection of factors. For example, the pore structure of the soil and the sediment are central influences on groundwater movement. Hydrologists quantify this influence primarily in terms of (i) porosity and (ii) permeability. The porosity is the proportion of total volume that is occupied by voids but it is not a direct function of the size of soil grains. Porosity tends to be larger in well sorted sediments where the grain sizes are uniform, and smaller in mixed soils where smaller grains fill the voids between larger grains. Soils are less porous at deeper levels because the weight of overlying soil packs grains closer together.

The term permeability refers to the relative ease with which a formation transmits water and is based on the size and shape of its pore spaces and the interconnectivity of the pores. Formations that have a high porosity and a high permeability produce good aquifers and include sand, gravel, sandstone, fractured rock, and basalt. Low-permeability formations that impede groundwater flow include granite, shale, and clay. Groundwater recharge enters aquifers in areas at higher elevations (typically hill slopes) than discharge areas (typically in the bottom of valleys), so the overall movement of groundwater is downhill. However, within an aquifer, water often flows upward toward a discharge area. To understand and map the complex patterns of groundwater flow, hydrogeologists use a quantity called the hydraulic head which, for a particular location within an aquifer, is the sum of the elevation of that point and the height of the column of water that would fill a well open only at that point. Thus, the hydraulic head at a point is simply the elevation of water that rises up in a well open to the aquifer at that point.

The height of water within the well is not the same as the distance to the water table. If the aquifer is under pressure, or artesian, this height may be much greater than the distance to the water table. Thus the hydraulic head is the combination of two potentials: mechanical potential due to elevation, like a ball at the top of a ramp, and pressure potential, like air compressed in a balloon. Because these are usually the only two significant potentials driving groundwater flow, groundwater will flow from high to low hydraulic head. This theory works in the same way that electrical potential (voltage) drives electrical flow and thermal potential (temperature) drives heat conduction. Like these other fluxes, groundwater flux between two points is simply proportional to the difference in potential, hydraulic head, and also to the permeability of the medium through which flow is taking place. These proportionalities are expressed in the fundamental equation for flow through porous media (Darcy's Law).

Darcy's law is an equation that describes the flow of a fluid through a porous medium and, in the absence of gravitational forces, is a proportional relationship between the instantaneous flow rate through a porous medium of permeability (, the dynamic viscosity of the fluid and the pressure drop over a given distance in a homogeneously permeable medium. Thus:

$$Q = \left[kA\left(p_{\mathrm{b}} - p_{\mathrm{a}}\right)\right] / \mu L$$

In this equation, the total discharge, Q (units of volume per unit of time) is equal to the product of the intrinsic permeability of the medium, k, the cross-sectional area to flow, A (units of area), and the total pressure drop $p_{\mathrm{b}} - p_{\mathrm{a}}$) divided by the dynamic viscosity, μ and the distance or length, L, over which the pressure drop occurs.

Hydrogeologists collect water levels measured in wells to map hydraulic potential in aquifers. These maps can then be combined with permeability maps to determine the pattern in which groundwater flows throughout the aquifer. Depending on local rainfall, land use, and geology, streams may be fed by either groundwater discharge or surface runoff and direct rainfall, or by some combination of surface and groundwater. Perennial streams and rivers are primarily supplied by groundwater, referred to as base flow. During dry periods they are completely supplied by groundwater; during storms there is direct runoff and groundwater discharge also increases. Thus, it is now possible to understand the overall structure of a water system.

Clean freshwater resources are essential for drinking, bathing, cooking, irrigation, industry, and for plant and animal survival (Dodds, 2002). Due to overuse, pollution, and ecosystem degradation the sources of most freshwater supplies—groundwater (water located below the soil surface), reservoirs, and rivers—are under severe and increasing environmental stress. The majority of the urban sewage in developing countries is discharged untreated into surface waters such as rivers and harbors. Approximately 65% v/v of the global freshwater supply is used in agriculture and 25% v/v is used in industry. Freshwater conservation therefore requires a reduction in wasteful practices like inefficient irrigation, reforms in agriculture and industry, and strict pollution controls worldwide. Aquatic regions house numerous species of plants and animals, both large and small. In fact, this is where life began billions of years ago when amino acids first started to come together. Without water, most life forms would be unable to sustain themselves and the Earth would be a barren, desert-like place. Although water temperatures can vary widely, aquatic areas tend to be more humid and the air temperature on the cooler side. The aquasphere can be broken down into two basic regions, (i) freshwater—ponds and rivers, and (ii) marine regions.

However, for the present purposes, water supply is generally considered to occur in five accessible locations: (i) groundwater, (ii) ice sheets and glaciers, (iii) lakes, (iv) rivers, and (v) oceans. Furthermore, a freshwater region is an area where the water has a low salinity (a low salt concentration, usually on the order of <1% *w*/w). Plants and animals in freshwater regions are adjusted to the low salt content and would not be able to survive in areas of high salt concentration (such as the ocean). There are different types of freshwater regions: ponds and lakes, streams and rivers, and wetlands. The following sections describe the characteristics of these three freshwater zones.

3.1 Groundwater

Groundwater is fresh water (from rain or melting ice and snow) that soaks into the soil and is stored in the tiny spaces (pores) between rocks and particles of soil (Freeze and Cherry, 1979; Fitts, 2012; Price, 2016). Groundwater is both an important direct source of supply that is tapped by wells and a significant indirect source, since surface streams **are** often supplied by subterranean water.

Contrary to the comments of some observers, a groundwater resource is not (and should not be imagined to be) an underground lake! In reality groundwater is rarely a distinct water body (large caves in limestone aquifers are one exception) but it is found in a water-bearing rock formation in which the water typically fills very small spaces (pores) within rocks and between sediment grains and the water can move within the formation due to the permeability of the rock. The packing of the rock particles dictates the amount of water that the rock can hold.

Layers of loosely arranged particles of uniform size (such as sand) tend to hold more water than layers of rock with materials of different sizes. This is because smaller rock materials settle in the spaces between larger rock materials—the decreasing the amount of open space that can hold water. Porosity (how well rock material holds water) is also affected by the shape of rock particles. Round particles will pack more tightly than particles with sharp edges. Material with angular-shaped edges has more open space and can hold more water. Groundwater can remain underground for prolonged periods (even hundreds of thousands of years) or the water can rise to the surface and in the form of rivers, streams, lakes, ponds, and wetlands. Groundwater can also come to the surface as a spring or be pumped from a well and either way provides drinking water.

Groundwater is stored in the tiny open spaces between rock and sand, soil, and gravel and is located in two zones: (i) the unsaturated zone, which is immediately below the land surface, contains water and air in the open spaces, or pores and (ii) the saturated zone, in which all of the pores and rock fractures are filled with water, underlies the unsaturated zone. The top of the saturated zone is called the water table which may be just below or hundreds of feet below the land surface.

Thus, near the surface of the earth, in the *zone of aeration,* soil pore spaces contain both air and water. This zone, which may have zero thickness in swamplands and be several hundred feet thick in mountainous regions, contains three types of moisture.

After a storm, *gravity water* is in transit through the larger soil pore spaces. *Capillary water* is drawn through small pore spaces by capillary action and is available for plant uptake. *Hygroscopic moisture* is water held in place by molecular forces during all except the driest climatic conditions. Moisture from the zone of aeration cannot be tapped as a water supply source.

In the *zone of saturation,* located below the zone of aeration, the soil pores are filled with water (known as *groundwater*). A stratum (layer of rock) that contains a substantial amount of groundwater is the aquifer. At the surface between the two zones (the *water table* or *phreatic surface*) the hydrostatic pressure in the groundwater is equal to the atmospheric pressure. An aquifer may extend to great depths, but because the weight of overburden material generally closes pore spaces, decreasing amounts of water is found at depths >2000 ft. The amount of water that can be stored in the aquifer is the volume of the void spaces between the soil grains. The fraction of voids volume to total volume of the soil (the *porosity*) is derived from the equation:

$$\text{Porosity} = (\text{Volume of void space})/(\text{Total volume})$$

However, not all of this water is available because it is so tightly tied to the rock particles in the aquifer. The amount of water that can be extracted is known as *specific yield,* defined as the percent of total volume of water in the aquifer that will drain freely from the aquifer.

If the groundwater can move rapidly though the rock (such as through gravel and sandy deposits) an aquifer can form. In the aquifer, there is enough groundwater that it can be pumped to the surface and used for drinking water, irrigation, industry, or other uses. However, if the groundwater is to move through the frock formation, the pores or fractures in the rock must be connected and the rock will have good permeability. If the pores or fractures are not connected, the rock material cannot produce water and is therefore not considered an aquifer. The amount of water an aquifer can hold depends on the volume of the underground rock materials and the size and number of pores and fractures that can fill with water. An aquifer may be a few feet to several thousand feet thick, and less than a square mile or hundreds of thousands of square miles in area.

Aquifers receive water from precipitation (rain and snow) and from water from surface waters like lakes and rivers that filters through the unsaturated zone. When the aquifer is full, and the water table meets the surface of the ground, water stored in the aquifer can appear at the land surface as a spring or seep. A recharge area is an area where the aquifer takes in water and a discharge area is the point where the groundwater flows to the land surface. Water moves from higher-elevation areas of recharge to lower-elevation areas of discharge through the saturated zone,

The groundwater and other surface water systems are part of the hydrologic cycle, which is the constant movement of water above, on, and below the surface of the Earth (Fig. 1.1). The water cyclo is the result of a collection of connected processes that distribute water and energy throughout the Earth system in cyclic patterns. Over time, on-going and repeated change in the distribution and form of water and energy around the globe is caused by processes like evaporation, condensation, freezing, melting, convection currents and infiltration.

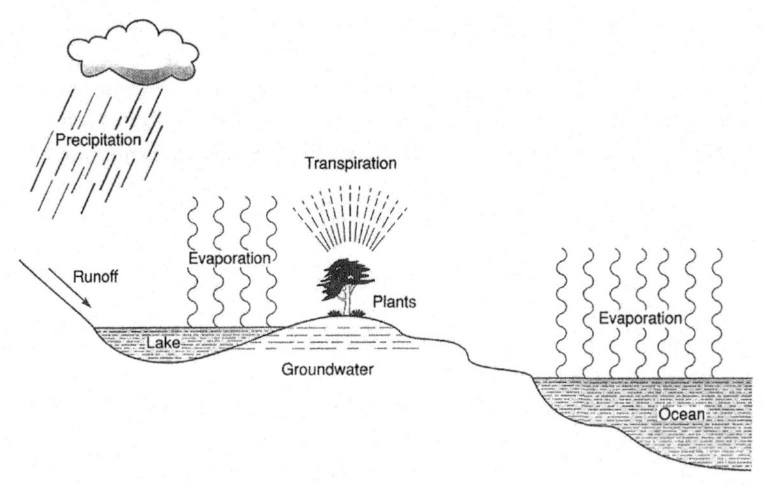

Fig. 1.1 The hydrogeological cycle (the water cycle).

A cycle has no beginning and no end, but it can be best understood but for the water cycle it is understood to commence at the precipitation stage. Selecting the beginning of the cycle (being a cycle there is a beginning or end) as the water from precipitation (i.e. rain, snow, or hail), the water soaks into the soil or water that cannot soak into the soil collects on the surface, forming runoff streams. When the soil is completely saturated, additional water moves slowly down through the unsaturated zone to the saturated zone, replenishing or recharging the groundwater. Water then moves through the saturated zone to groundwater discharge areas. Evaporation occurs when water from such surfaces as oceans, rivers, and ice is converted to vapor. Evaporation, together with transpiration from plants, rises above the surface of the Earth, condenses, and forms clouds. Water from both runoff streams and from groundwater discharge moves toward streams and rivers and may eventually reach the ocean where water can also evaporate.

Groundwater can become contaminated in many ways. If surface water that recharges an aquifer is polluted, the groundwater will also become contaminated. Contaminated groundwater can then affect the quality of surface water at discharge areas. Groundwater can also become contaminated when liquid hazardous substances soak down through the soil into groundwater. Contaminants that can dissolve in groundwater will move along with the water, potentially to wells used for drinking water. If there is a continuous source of contamination entering moving groundwater, an area of contaminated groundwater (a plume) can form. A combination of moving groundwater and a continuous source of contamination can, therefore, pollute very large volumes and areas of groundwater.

Some hazardous substances dissolve very slowly in water. When these substances seep into groundwater faster than they can dissolve, some of the contaminants will stay in liquid form. If the liquid is less dense than water, it will float on top of the water table, like oil on water. Pollutants in this form are called light non-aqueous phase liquids (LNAPLs). If the liquid is denser than water, the pollutants are called dense non-aqueous phase liquids (DNAPLs). Dense non-aqueous phase liquids sink to form pools at the bottom of an aquifer which continue to contaminate the aquifer as they slowly dissolve and are carried away by moving groundwater. As the dense non-aqueous phase liquids flow downward through an aquifer, tiny globs of liquid become trapped in the spaces between soil particles (residual contamination).

3.2 Ice sheets and glaciers

An ice sheet is a mass of glacial land ice extending—the two ice sheets on the Earth today cover most of Greenland and Antarctica. During the last ice age, ice sheets also covered much of North America and Scandinavia. Together, the Antarctic and Greenland ice sheets contain >99% *v*/v of the freshwater ice on Earth (Jansson et al., 2003; Cogley, 2012). The Antarctic Ice Sheet extends over an area of approximately 5.4 million square miles, roughly the combined areas of the contiguous United States and Mexico. The Greenland Ice Sheet extends about 656,000 mile^2 and covers most of the island of Greenland, three times the size of Texas.

Ice sheets form in areas where snow that falls in winter does not melt entirely over the summer. Over thousands of years, the layers of snow pile up into thick masses of ice, growing thicker and denser as the weight of new snow and ice layers compresses the older layers. Ice sheets are constantly in motion, slowly flowing downhill under their own weight. Near the coast, most of the ice moves through relatively fast-moving outlets (often called ice streams, glaciers, ice shelves) (Benn and Evans, 2010). As long as an ice sheet accumulates the same mass of snow as it loses to the sea, it remains stable. The ice sheets of the Earth constantly expand and contract as the climate fluctuates. During warm periods ice sheets melt and sea levels rise, with the reverse occurring when temperatures fall. Water may remain locked in deep layers of polar ice sheets for hundreds of thousands of years.

A glacier is a persistent body of dense ice that is constantly moving under its own weight; it forms where the accumulation of snow exceeds its ablation (melting and sublimation) over many years, often centuries. Glaciers form above the permanent snow line due to the accumulation of water at a solid state (snow that transforms into ice). The line varies according to the latitude on which continental glaciers (that uniformly cover wide areas) and mountain glaciers (that occupy mountain valleys) form. Below the permanent snow the ice melts and the water is present in a liquid state.

Glaciers are categorized by the morphology, thermal characteristics, and behavior. *Cirque glaciers* form on the crests and slopes of mountains whereas a glacier that fills a valley is called a *valley glacier*, or alternatively an *alpine glacier* or *mountain glacier*. A large body of glacial ice astride a mountain, mountain range, or volcano is termed an *ice cap* or *ice field*.

Glaciers slowly deform and flow due to stresses induced by their weight and also abrade rock and debris from their substrate to create landforms such as moraines. Glaciers form only on land and are distinct from the much thinner sea ice and lake ice that form on the surface of bodies of water.

Ice sheets and glaciers are not always thought of as freshwater sources, but they account for a significant fraction of world reserves. Glacial ice is the largest reservoir of fresh water on Earth. Many glaciers from temperate, alpine and seasonal polar climates store water as ice during the colder seasons and release it later in the form of *meltwater* as warmer summer temperatures cause the glacier to melt, creating a water source that is especially important for plants, animals and human uses when other sources may be scant. Within high-altitude and Antarctic environments, the seasonal temperature difference is often not sufficient to release meltwater.

3.3 Ponds and lakes

Ponds and lakes vary on aerial extent (without placing hard and fast boundaries on such waters) from just a few square years to many square miles. Scattered throughout the surface of the Earth, several are remnants from the last Ice Age after which the ice melted approximately ten-to-twelve thousand years ago—the final part of the Quaternary glaciation lasted from approximately 110,000 to 12,000 years ago and occurred during the last 100,000 years of the Pleistocene epoch.

Many ponds are seasonal, lasting just a couple of months (such as sessile pools) while lakes may exist for hundreds of years or more. Ponds and lakes may have limited diversity of floral and faunal species since they are often isolated from one another and from other water sources, such as rivers and oceans which have a larger floral and faunal diversity. Lakes and ponds are divided into three different zones, which are usually determined by depth and distance from the shoreline.

The topmost zone near the shore of a lake or pond is the *littoral zone*—this zone is the warmest since it is shallow and can absorb more of the heat of the Sun—and sustains a fairly diverse community, which can include several species of algae (like diatoms—algae in which the cell walls composed of transparent, opaline silica), rooted and floating aquatic plants, grazing snails, clams, insects, crustaceans, fishes, and amphibians. In the case of the insects, such as dragonflies and midges, only the egg and larvae stages are found in this zone. The vegetation and animals living in the littoral zone are food for other creatures such as turtles, snakes, and ducks. The near-surface open water surrounded by the littoral zone is the *limnetic zone*.

The limnetic zone is well-lighted (like the littoral zone) and is dominated by plankton, both phytoplankton and zooplankton. Plankton are small organisms that play a crucial role in the food chain. Without aquatic plankton, there would be few living organisms in the world, and certainly no humans. A variety of freshwater fish also occupy this zone. Plankton have short life spans and when plankton die, the remains fall into the deep-water part of the lake/pond, the *profundal zone*. This zone is much colder and denser than the other two and little light penetrates all the way through the limnetic zone into the profundal zone. The fauna are heterotrophs—organisms which eat other dead organisms and use oxygen for cellular respiration.

A lake is an area filled with water, localized in a basin—often referred to as a structural bas, which is a large-scale structural formation of rock strata formed by tectonic warping of previously flat-lying strata—that is surrounded by land, apart from any river or other outlet that serves to feed or drain the lake.

Lakes are temporary deposits of water on continental depressions and are supplied with water by watercourses (tributaries). The water flows into the out-flowing streams,

streams or rivers that originate from the lake. Lake water has a low salinity, but has plenty of suspended material, and its temperature depends on local climate conditions. Also the water of big lakes can move and originate variations called seiches, due to differences in atmospheric pressure.

Lakes lie on land and are not part of the ocean, and therefore are distinct from lagoons, and are also larger and deeper than ponds, though there are no official or scientific definitions. Lakes can be contrasted with rivers or streams or which are usually flowing. Most lakes are fed and drained by rivers and streams. In some parts of the world there are many lakes because of chaotic drainage patterns left over from the last Ice Age. All lakes are temporary over geologic time scales, as they will slowly fill in with sediments or spill out of the basin.

Natural lakes are generally found in mountainous areas, rift zones, and areas with ongoing glaciation. Other lakes are found in endorheic basins or along the courses of mature rivers- an endorheic basin is a limited drainage basis that normally retains water and allows no outflow to other external bodies of water. In some parts of the world, a water contained in an endorheic basin (with no outflow to other external bodies of water) often referred to as a *pond*.

A phenomenon that needs to be addressed because it can affect the survival and distribution of pollutants is the stratification of a lake which is the separation of the lake into three layers: (i) the epilimnion, which is the top of the lake, (ii) the metalimnion or thermocline, which is the middle layer and which may change depth throughout the day, and (iii) the hypolimnion, which is the bottom layer (Fig. 1.2).

The thermal stratification of lakes refers to a change in the temperature at different depths in the lake, and is due to the change in the density of the water with temperature. Cold water is denser than warm water and the epilimnion generally consists of water that is not as dense as the water in the hypolimnion but the temperature of maximum density for freshwater is 4 °C (39 °F). In temperate regions where the lake water warms up and cools through the seasons, a cyclical pattern of overturn occurs that is repeated from year to year as the cold dense water at the top of the lake sinks. For example, in dimictic lakes, the lake water turns over during the spring and the autumn. This process occurs more slowly in deeper water and as a result, a thermal bar may

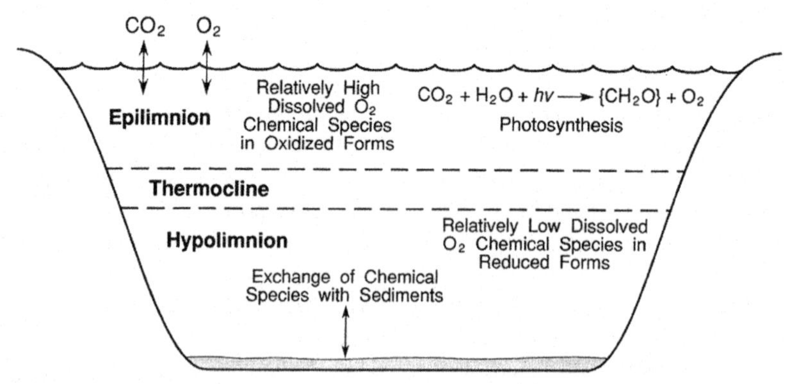

Fig. 1.2 Stratification of a lake.

form. If the stratification of water lasts for extended periods, the lake is meromictic. Conversely, for most of the time, the relatively shallower meres are unstratified; that is, the mere is considered all epilimnion. If a lake is triggered into limnic eruption (a natural disaster in which dissolved carbon dioxide suddenly erupts from deep lake waters, forming a gas cloud) the carbon dioxide can quickly leave the lake and displace the oxygen needed for life by people and animals in the surrounding area.

Also with severe thermal stratification in a lake, the quality of drinking water can also be adversely affected. In addtion, the spatial distribution of fish within a lake is often adversely affected by thermal stratification and in some cases may indirectly cause large die-offs of recreationally important fish. One commonly used tool to reduce the severity of these lake management problems is to eliminate or lessen thermal stratification through aeration which has met with some success, but is not always the cure-all for such a problem.

It is appropriate at this point to mention another surface water system that is often ignored many presentations of water system and that is the system known as the wetlands systems.

A wetland is a place where the land is covered by water, either salt, fresh or somewhere in between (Mitsch and Gosselink, 2007). Wetlands vary widely and are difficult to classify because of regional and local differences in soils, topography, climate, hydrology, water chemistry, vegetation and other factors, including human disturbance. Wetlands are often located alongside waterways and in floodplains and it is not surprising that there is a classification that is based on location in which wetlands are subdivided into five general types: (i) marine or ocean wetlands, (ii), estuary wetlands, (iii) river wetlands, (iv) lake wetlands also called lacustrine wetlands, and (v) marsh wetlands, also called palustrine wetlands. Common names for wetlands include names such as marshes, estuaries, mangroves, mudflats, mires, ponds, fens, swamps, deltas, coral reefs, billabongs, lagoons, shallow seas, bogs, lakes, and floodplains. A large wetland area may be comprised of several smaller wetland types.

Thus, marshes and ponds, the edge of a lake or ocean, the delta at the mouth of a river, low-lying areas that frequently flood all fall into the definition of wetlands. Wetlands are some of the most productive habitats on the Earth and often support high concentrations of animals such as mammals, birds, fish and invertebrates, and serve as nurseries for many of these species. Wetlands also support the cultivation of rice, a staple in the diet of the population in many countries. Also, wetlands provide a range of ecosystem services that benefit human populations, including water filtration, storm protection, flood control and recreation.

Wetlands act as natural water filters, but they can only do so much to clean up the fertilizers and pesticides from agricultural runoff, mercury from industrial sources and other types of pollution. There is growing concern about the effect of this pollution on the supply of drinking water and the biological diversity of wetlands. Climate change brings a variety of alterations to patterns of water and climate. In some places, rising sea levels are swamping shallow wetlands and drowning some species of mangrove trees while in other locations, droughts are destroying estuaries, floodplains and marshes. Also, dams alter the natural flow of water through a landscape. There are possibilities for building dams or locating them in more sustainable ways that limit impact on existing ecosystems, but many have been very destructive to wetlands.

Temperature varies in ponds and lakes seasonally. For example, during the summer, the temperature can range from 4 °C (39 °F) near the bottom of the pond or lake to 22 °C (72 °F) at the top. During the winter, the temperature at the bottom of the pond or lake can be 4 °C (39 °F) while the top is 0 °C (39 °F). In between the two layers, there is a narrow zone (the *thermocline*) where the temperature of the water changes rapidly. Outside of the summer and winter season (during the spring and autumn/fall seasons), there is a tendency for the top layers and the bottom layers to mix, usually due to winds, which results in a uniform water temperature of around 4 °C (39 °F) but this is very dependent upon the climatology of the region. This mixing also circulates oxygen throughout the lake but if the lake or pond did not freeze during the winter, thus the top layer can be expected to be warmer than the bottom of the pond or lake.

3.4 Streams and rivers

Streams and rivers are bodies of flowing water that are ubiquitous throughout the surface of the Earth and which move predominantly move naturally in one direction (unless there are upsets such as earthquakes). These water courses start at the headwaters which arise from springs, from snowmelt, or even from lakes after which the water flow to the mouth (or delta) of the river into another water channel, typically the ocean.

The characteristics of a river or stream change during the journey from the source to the mouth. For example, the temperature of the water is cooler at the source than it is at the mouth. The water is also clearer, has higher oxygen levels, and freshwater fish such as trout and heterotrophs can be found there. Toward the middle part of the stream/river, the width increases, as does species diversity—numerous aquatic green plants and algae can be found. Toward the mouth of the river/stream, the water becomes murky from all the sediments that it has picked up upstream, decreasing the amount of light that can penetrate through the water. Since there is less light, there is less diversity of flora, and because of the lower oxygen levels, fish that require less oxygen, such as catfish and carp, can be found.

A river is a natural flowing watercourse, usually freshwater, flowing toward another river, a lake, a sea, or an ocean. In some cases a river flows into the ground and becomes dry at the end of its course without reaching another body of water. Small rivers are often referred to using names such as stream, creek, brook, rivulet, rill (a shallow channel cut into soil by the erosive action of flowing water), and *beck* in north-eastern England where remnants of a Saxon-related dialect (with the insertion of some Old Norse words) remain.

Rivers are part of the hydrological cycle (Fig. 1.1)—water generally collects in a river from precipitation through a drainage basin from surface runoff and other sources such as groundwater recharge, springs, and the release of stored water in natural ice and snow packs (e.g., from glaciers). Every river is part of a larger system—a watershed, which is the land drained by a river and its tributaries. Rivers are large natural streams of water flowing in channels and emptying into larger bodies of water. This diagram shows some common characteristics of a river system. Every river is different, however, so not all rivers may look exactly like this illustration.

The river source, also called the headwaters, is the beginning of a river and is often located in mountains, the source may be fed by an underground spring, or by runoff from rain, snowmelt, or glacial melt. Other terminology related to overs includes: (i) the main river, which is the primary channel and course of a river, (ii) a **tributary**, which is a smaller stream or river that joins a larger or main river, (iii) a **meander**, which is a loop in a river channel—a meandering river winds back and forth, rather than following a straight course—and the meander may be referred to as an ox-bow, (iv) **upstream, which is** the opposite direction from that in which a stream or river flows, nearer to the source—the converse is downstream, and (v) the floodplain, which is relatively flat land stretching from either side of a river, which may flood during heavy rain or snowmelt—the floodplain consists of materials deposited by a river and is often rich in nutrients for plants.

Rivers contain a relatively small share of fresh water, but the flux of water down rivers is a large part of the global hydrologic cycle and they are centrally important in shaping landscapes. The flow of rivers erodes solid sediment and carries it toward the sea, along with dissolved minerals. These processes shape land into valleys and ridges and deposit thick layers of sediment in flood plains. Over geologic time the erosion caused by rivers balances the uplift driven by plate tectonics.

3.5 Wetlands

Wetlands are areas where water covers the soil, or is present either at or near the surface of the soil all year or for varying periods of time during the year, including during the growing season. Water saturation (hydrology) largely determines how the soil develops and the types of plant and animal communities living in and on the soil. Wetlands may support both aquatic and terrestrial species. The prolonged presence of water creates conditions that favor the growth of specially adapted plants (hydrophytes) and promote the development of characteristic wetland (hydric) soils. Wetlands vary widely because of regional and local differences in soils, topography, climate, hydrology, water chemistry, vegetation and other factors, including human disturbance. Indeed, wetlands are found from the tundra to the tropics and on every continent except Antarctica. Two general categories of wetlands are recognized: coastal or tidal wetlands and inland or non-tidal wetlands.

Tidal wetlands are, as the name suggests, located found along coast lines and are often linked to river estuaries where sea water mixes with fresh water to form an environment of varying salinity. The salt water and the fluctuating water level (due to tidal action) combine to create a rather difficult environment for most plants. Consequently, many shallow coastal areas are non-vegetated mud flats or sand flats. Some plants, however, have successfully adapted to this environment.—certain types of grasses and grass-like plants that are able to adapt to the saline conditions form the tidal salt marshes that are found along a coast line. Moreover, mangrove swamps, with salt-loving shrubs or trees, are common in tropical climate areas. Some tidal freshwater wetlands form beyond the upper edges of tidal salt marshes where the influence of salt water ends.

Non-tidal wetlands are most common on floodplains along rivers and streams (riparian wetlands), in isolated depressions surrounded by dry land, along the margins of lakes and ponds, and in other low-lying areas where the groundwater intercepts the soil surface or where precipitation sufficiently saturates the soil (vernal pools and bogs). Inland wetlands include marshes and wet meadows dominated by herbaceous plants, swamps dominated by shrubs, and wooded swamps dominated by trees.

Many of wetlands are seasonal (they are dry one or more seasons every year), and, particularly in the arid and semiarid areas, may be wet only periodically (i.e., less than seasonal). The quantity of water present and the timing of its presence in part determine the functions of a wetland and its role in the environment. Even wetlands that appear dry at times for significant parts of the year—such as vernal pools (temporary pools of water that provide habitat for distinctive plants and animals)—often provide critical habitat for wildlife adapted to breeding exclusively in these areas.

Thus, wetlands are areas of standing water that support aquatic plants. Marshes, swamps, and bogs are all considered wetlands. Plant species adapted to the very moist and humid conditions (hydrophytes) include pond lilies, cattails, sedges, tamarack, and black spruce as well as marsh flora which include species such as cypress and gum. Wetlands have the highest species diversity of all ecosystems and are attractive to many species of amphibians, reptiles, birds (such as ducks and waders), and furbearers can be found in the wetlands. Wetlands are not considered freshwater ecosystems as there are some, such as salt marshes, that have high salt concentrations and which support different species of animals, such as shrimp, shellfish, and various grasses.

Wetlands—in which high rainfall, appropriate topography, or low evapotranspiration lead to permanent or seasonal flooding—are an exception to the relationship between emissions and climatic conditions. Wetlands are disproportionately found in cooler latitudes, and northern wetlands are thought to share in importance with methane emissions over tropical wetlands simply because of the greater area (albeit with lower fluxes) of northern wetlands. These areas become anoxic (deficient in oxygen) because microbial activity consumes all the oxygen—anoxic conditions lead to reduced productivity, reduced organic matter decomposition, and greatly enhanced methane production. Wetlands may also contribute significant amounts of reduced sulfur-containing gases to the atmosphere.

3.6 The oceans

The oceans (sometime referred to as *marine regions*) cover approximately 75% of the surface of the Earth and also (under the general term *marine regions*) includes coral reefs, and estuaries—estuaries are areas where freshwater streams or rivers merge with the ocean. This mixing of waters with such different salt concentrations creates a very interesting and unique ecosystem. Microflora (such as algae) and macroflora (such as seaweeds, marsh grasses, and mangrove trees), can be found here. Estuaries support a diverse fauna, including a variety of worms, oysters, crabs, and waterfowl. Marine algae supply much of the oxygen supply of the Earth and, at the same time, remove a substantial amount of carbon dioxide from the atmosphere. In addition, the evaporation of the seawater provides rainwater for the land.

The largest of all of the marine region ecosystems sand are very large bodies of water that dominate the surface of the Earth. Like ponds and lakes, the ocean regions are separated into separate zones: intertidal, pelagic, abyssal, and benthic. All four zones have a great diversity of species—often claimed (with some justification) to be richest diversity of species even though the oceans may contain fewer species than exist on land-based ecosystems.

The ocean, which covers approximately 75% of the surface of the Earth, is coupled to the atmosphere from both a physical and a biogeochemical perspective. The basic structure of the ocean is set by the geographic patterns of surface heating and freshwater input (precipitation—rainfall, snowfall—minus water lost due to evaporation), which influences the salinity distribution in the ocean. In general, there is net warming of the ocean surface in the tropics and subtropics and net cooling at of the oceans in the temperate and polar latitudes.

In addition, the ocean can be divided into two general regions: (i) a warm, surface pool—typically 18 °C (64 °F—that is approximately 3280 ft thick and (ii) the deep water—typically: 3 °C (37 °F)—that outcrops to the surface at high latitudes and forms the bulk of the ocean volume. Unlike the atmosphere, heating of the ocean surface stabilizes the water and also prevents rapid exchange (or mixing) between the surface and deep water. In fact, contact between the surface and deep waters is limited to localized polar regions where losses of heat and freshwater lead to sinking and deep water formation. The resulting thermohaline circulation (part of the large-scale ocean circulation that is driven by global density gradients that are created by surface heat and fresh-water flow) is especially important over long timescales (such as glacial cycles).

The other component of the ocean circulation is produced by the drag of the surface winds on the ocean. The zonal wind patterns over the ocean result in rotating circulation patterns in the subtropical and subpolar regions and are also responsible for the circumpolar current in the southern oceans (i.e., oceans to the south of the equator). Typically, the wind-driven surface currents move heat and trace species (such as organic chemical compounds) from the tropics to the poles, and approximately an equal amount of solar energy received in the tropics is transported toward the pole by oceanic and atmospheric circulations. Wind forcing also causes divergence of the surface water and upwelling along both the equator and coastal regions on the eastern margin of ocean basins. The upwelling of cooler, nutrient-rich waters in these areas greatly enhances ocean productivity. The ocean circulation, which exhibits variations on different timescales (including perturbations such as the El Niño events that occur in the equatorial Pacific on an average of four years and produce massive warming of the coastal waters off Peru and Ecuador with torrential rainfall in the region), greatly affects biogeochemical cycles as well as the global climate.

An ocean is a body of saline water that composes much of a hydrosphere of the Earth. The word *ocean* is often used interchangeably with *sea* but, more specifically, an ocean is a large area of salt water between continents while a sea is a body of saline water partly or fully enclosed by land. The main characteristics of ocean water and sea water are:

1. Salinity, which is refers to the total salt content in 1000 g of seawater and it has a value of approximately 35 g. The percentage of the different substances that are present in the solution depends on the river contribution, on chemical reactions that occur in sea sediments, on volcanic activity and on the decomposition of organisms. In fact, the quantity of salt is stable only at a certain depth, while on the surface and coastal areas it is also subject to seasonal variations.

2. Melted gases: oxygen and carbon dioxide are necessary for the life of water organisms. Oxygen is largely present on the surface, since water is in contact with the atmosphere and where photosynthetic organisms live, and deep underwater, where water temperature is very low. Carbon dioxide is a very soluble gas that easily spreads from the atmosphere into sea water, transported by river water to the sea, and which derives from decomposing organic materials.
3. The temperature: as well as mitigating the climate of coastal regions, temperature influences the chemical and physical characteristics that are responsible for the vertical movement of water masses. On the most superficial layer (50–200 m) the temperature is similar to the superficial one; on the thermocline layer (200–1000 m) the temperature rapidly diminishes; on the deep layers it keeps on diminishing, but very slowly. The thermocline is an important surface as far as the spreading of organisms in the oceans is concerned. It represents an obstacle for many animals, plants, and tropical algae that need a temperature of 15 to 20 °C 59 to 68 °F).
4. Brightness: it depends on the ability of the light to penetrate in the water and light up only the superficial part even though the water is clear. This area (the photic zone, approximately 0 yo 550 ft of depth) is where most of marine life and phytoplancton are concentrated.

The ocean is a continuous body of saltwater that covers >70% of the surface of the Earth. On a geographical basis, the oceans can be divided into four major sections: (i) the Pacific Ocean, (ii) the Atlantic Ocean, (iii) the Indian Ocean, and (iv) the Arctic Ocean. Smaller ocean regions typically fall under the names: seas, gulfs, and bays, such as the Mediterranean Sea, Gulf of Mexico, and the Bay of Bengal but stand-alone bodies of saltwater such as the Caspian Sea and the Great Salt Lake are distinct from the oceans. The ocean water is approximately 3.5% *w*/w salt and contains traces of all of the chemical elements found on Earth (Table 1.2). The oceans absorb the heat from the Sun and transfer this heat to the atmosphere and distribute it around the world via the ever-moving ocean currents as well as carrying various forms of pollutants from one continent to another, have a major effect on the climate of the Earth by transferring heat from the tropics to the Polar Regions.

Huge masses of water displace for long distances due to the wind action. The direction of the movement is determined by the earth rotation (Coriolis force), which

Table 1.2 Elements in the crust of the earth

Element (symbol)	% w/w[a]
Oxygen (O)	46.6
Silicon (Si)	27.7
Aluminum (Al)	8.1
Iron (Fe)	5
Calcium (Ca)	3.6
Sodium (Na)	2.8
Potassium (K)	2.6
Magnesium (Mg)	2.1
Titanium (Ti)	0.4

[a] Approximation.

creates circular movements. In the Atlantic Ocean, regular and constant winds, the trade winds, move superficial water masses toward the Equator where they are diverted to the west by the Coriolis force (North-Equatorial current); when they reach the American continent they are pushed to the north and accumulate in the Gulf of Mexico. The water continues to flow toward the Atlantic ocean and form the Gulf current along the coast of the United States, and then divide into two: (i) one current goes toward the Canary islands and starts the tour again, and (ii) the other current moves to the north-east, reaches the north-western coasts of Europe and mitigates their climate.

In polar areas the water cools down, becomes denser, falls deep down and moves to the Equator. As it gets warmer, it becomes less dense and lighter and tends to rise to the surface. This movement, that forms deep sea currents, is very slow: it takes even a thousand years to a water mass to go back to the surface.

Transferring warm or cold air and precipitation to coastal regions, winds may carry them inland. Surface heat and freshwater fluxes create global density gradients that drive the thermohaline part of large-scale ocean circulation. Changes in the thermohaline circulation governs the rate at which deep waters reach the surface, it may also significantly influence the concentration of carbon dioxide in the atmosphere. These vast bodies of water surrounding the continents are critical to human life but overfishing and global warming threaten to leave this vital habitat barren.

Life began in the ocean, and the ocean remains home to the majority of floral and faunal species of the Earth, extending from tiny single-celled organisms to the animals as large as the blue whale which is the largest living animal on the Earth. However, most of the plant life of the ocean consists of microscopic algae (phytoplankton) that float at the surface and through photosynthesis produce about half of the oxygen that humans and all other terrestrial creatures breathe. Seaweed and kelp are big algae easily visible to the naked eye. Marine plants with roots, like seagrasses, can only survive as deep as the rays of the Sun can support photosynthesis—approximately 650 ft (200 m). Ocean depths vary but can go down to 9800 ft.

The Pacific Ocean the largest of the oceans, also reaches northward from the Southern Ocean to the Arctic Ocean. It spans the gap between Australia, Asia, and the Americas and meets the Atlantic Ocean south of South America at Cape Horn. On the other hand, the Atlantic Ocean, the second largest ocean, extends from the Southern Ocean between the Americas, Africa, and Europe to the Arctic Ocean. The Atlantic Ocean meets the Indian Ocean south of Africa at Cape Agulhas.

The Indian Ocean, the third largest, extends northward from the Southern Ocean to India, the Arabian Peninsula, and Southeast Asia and, in Asia, between Africa in the west to Australia in the East. The Indian Ocean joins the Pacific Ocean to the east, near Australia. The Arctic Ocean is the smallest of the five oceans and joins the Atlantic Ocean near to Greenland and Iceland and joins the Pacific Ocean at the Bering Strait. The Arctic Ocean overlies the North Pole and touches North America in the Western Hemisphere and Scandinavia and Siberia in the Eastern Hemisphere. The Arctic Ocean is partially covered by sea ice, the extent of which varies according to the season.

The Southern Ocean is the general name given to the waters the surround Antarctica and is partially covered by sea ice, the extent of which varies according to the season. The Southern Ocean is the second smallest of the five oceans.

From a chemical point of view, the ocean influences the atmosphere through the exchanges of trace gases across the air-sea interface. The transfer of carbon dioxide from the atmosphere to the ocean is controlled by the two competing factors of temperature: warming of surface waters, which releases carbon dioxide to the atmosphere, and biological productivity. Photosynthesis by marine phytoplankton converts dissolved carbon dioxide into organic carbon, leading to a reduction in surface carbon dioxide values and a carbon dioxide flow into the ocean. The amount of carbon dioxide dissolved in seawater is quite large due to its high solubility and its reactivity with water to form carbonic acid and its dissociation products. The ocean, therefore, serves as a major reservoir for carbon dioxide, approximately 65 times larger than the atmosphere and oceans have played an important role in the evolution of atmospheric carbon dioxide over the geological history of the Earth and is a primary sink for anthropogenic carbon dioxide.

Other chemical species are released by the ocean, such as reduced sulfur, certain hydrocarbon derivatives, and carbon monoxide. The largest oceanic source of sulfur is provided by dimethyl sulfide (CH_3SCH_3), which is produced by various, but specific, types of phytoplankton. These emissions appear to be most intense in regions where the net primary productivity of the ocean is highest, modified somewhat by poorly understood large-scale patterns in the distribution of phytoplankton species.

The *intertidal zone* is where the ocean meets the land—sometimes this zone is a submerged zone and at other times exposed, as the ocean ebbs (outgoing tide) and flows (incoming tide). Because of this, the floral and faunal communities are constantly changing. For example, on rocky coasts, the zone is stratified vertically and, where only the highest tides reach, there are only a few species of algae and mollusks. In those areas usually submerged during high tide, there is a more diverse array of algae and small animals, such as herbivorous snails, crabs, sea stars, and small fishes. At the bottom of the intertidal zone, which is only exposed during the lowest tides, many invertebrates, fishes, and seaweed can be found. The intertidal zone on sandier shores is not as stratified as in the rocky areas—waves action maintain the mud and sand in a state of constant motion and, thus, very few algae and plants can establish themselves—the fauna includes worms, clams, predatory crustaceans, crabs, and shorebirds.

The *pelagic zone* includes those waters further from the land and the open ocean. The pelagic zone is generally cold though it is hard to give a general temperature range since, just like ponds and lakes, there is thermal stratification with a constant mixing of warm and cold ocean currents. The flora in the pelagic zone include surface seaweeds. The faunal animals include many species of fish and some mammals, such as whales and dolphins. Many feed on the abundant plankton.

The *benthic zone* is the area below the pelagic zone, but does not include the very deepest parts of the ocean. The bottom of the zone consists of sand, slit, and/or dead organisms. Here temperature decreases as depth increases toward the abyssal zone, since light cannot penetrate through the deeper water. Flora are represented primarily by seaweed while the fauna, since it is very nutrient-rich, include all sorts of bacteria, fungi, sponges, sea anemones, worms, sea stars, and fishes.

The deep ocean is the *abyssal zone* and in this region is very cold (typically: approximately 3 °C, 347 °F), highly pressured, high in oxygen content, but low in nutritional value. The abyssal zone supports many species of invertebrate species and fishes. The mid-ocean ridges (spreading zones between tectonic plates), often with hydrothermal vents, are found in the abyssal zone along the ocean floor. Chemosynthetic bacteria thrive near these vents because of the large amounts of hydrogen sulfide and other minerals produced from the vents. These bacteria are thus the start of the food web as they are eaten by invertebrates and fishes.

Coral reefs are widely distributed in warm shallow waters and can occur as barriers along continents, such as the Great Barrier Reef located in the Coral Sea off the coast of Queensland, Australia. The Great Barrier Reef is the largest coral reef system in the world and is composed of over 2900 individual reefs and 900 islands stretching for >1400 miles and over an area of approximately 133,000 mile^2. The dominant organisms in coral reefs are corals. Corals are interesting since they consist of both algae and tissues of animal polyp. Since reef waters tend to be nutritionally poor, corals obtain nutrients through the algae via photosynthesis and also by extending tentacles to obtain plankton from the water. Besides corals, the faunal animals include several species of microorganisms, invertebrates, fishes, sea urchins, octopuses, and sea stars.

4 The lithosphere

The lithosphere (the land, sometime referred to as is the *terrestrial biosphere*) solid, outer part of the Earth, including the brittle upper portion of the mantle and the crust of the Earth. As part of the biosphere, the anthrosphere is that part of the environment made or modified by humans and used for their activities. The biosphere is key to the existence of life on the Earth. The life forms (including human life forms) that exist in the biosphere are adapted to make a variety of complex and specialized chemical products for which it is essential to have adequate supplies of water, which emphasizes the important connection between the biosphere and the hydrosphere.

4.1 Types

There are two types of lithosphere: (i) the continental lithosphere, which is associated with the continental crust of the Earth and (ii) the oceanic lithosphere, which is associated with the oceanic crust and exists in the ocean basins.

The thickness of the lithosphere is considered to be the depth to the isotherm associated with the transition between brittle and viscous behavior. The oceanic lithosphere is typically about 30 to 80 miles thick crust. The oceanic lithosphere thickens as it ages and moves away from the mid-ocean ridge. This thickening occurs by conductive cooling, which converts hot asthenosphere into lithospheric mantle and causes the oceanic lithosphere to become increasingly thick and dense with age. The thickness of the mantle part of the oceanic lithosphere can be approximated as a thermal boundary layer that thickens as the square root of time.

Oceanic lithosphere consists mainly of mafic (rich in magnesium and iron) crust and ultramafic (in excess of 90% *w*/w mafic) mantle and is denser than continental lithosphere. It thickens as it ages and moves away from the mid-ocean ridge. This thickening occurs by conductive cooling, which converts hot asthenosphere into lithospheric mantle. It was less dense than the asthenosphere for tens of millions of years, but after this becomes increasingly denser. The gravitational instability of mature oceanic lithosphere has the effect that when tectonic plates come together, oceanic lithosphere invariably sinks underneath the overriding lithosphere. New oceanic lithosphere is constantly being produced at mid-ocean ridges and is recycled back to the mantle at subduction zones, so oceanic lithosphere is much younger than its continental counterpart. The oldest oceanic lithosphere is approximately 170 million years old compared to parts of the continental lithosphere which are billions of years old.

The terrestrial lithosphere is important to environmental chemistry as a source and sink for many compounds—a major activity within atmospheric chemistry has been (and remains) the determination of such flows. The structure of the biosphere is controlled by the interaction of climate with the patterns of soils and topography resulting from geological processes on a range of time scales, and further modified by the biogeographic distribution of organisms. Climate patterns are reflected in productivity (annual carbon fixation through photosynthesis), with warmer and wetter regions having higher productivity—the rate of nitrogen cycling follows similar trends—and as a result, trace gas emissions are usually higher (in some cases higher by one or more orders of magnitude) in the tropics than in the mid-to-high-latitude regions. This is clearly true for all soil trace gas fluxes, and may be true for plant-mediated fluxes.

For example, large quantities of hydrocarbon derivatives such as isoprene (C_5H_8) are produced by the foliage of the abundant vegetation in productive ecosystems. Biomass burning fluxes are highest in tropical savanna ecosystems, which are warm and have sufficient rainfall during the wet seasons to accumulate significant biomass, which burns readily during the dry seasons. Large quantities of atmospheric carbon dioxide, carbon monoxide, hydrocarbon derivatives, and nitrogen oxides (NO_x) are produced as a result of the combustion of biomass. Soil and nitrogen oxide relationships are enhanced when the soil is rapidly wetted, dried, and wetted again in succession, and so may be higher in regions of sporadic rainfall, despite higher overall rates of nitrogen cycling, and nitrogen oxide emissions in moist areas.

Soil, a mixture of mineral, plant, and animal materials, is essential for most plant growth and is the basic resource for agricultural production. In the process of developing the land and clearing away the vegetation that holds water and soil in place, erosion has caused devastation on a worldwide scale. The rapid deforestation taking place in the tropics is especially damaging because the thin layer of soil that remains is extremely fragile and quickly washes away when exposed to the heavy tropical rain storms.

Technically, soil is a mixture of mineral constituents—the inorganic components of soil are principally produced by the weathering of rocks and minerals—plant materials, and animal materials, that forms during a long process that may take thousands of years and it is an unconsolidated, or loose, combination of inorganic and organic materials. Soil is necessary for most plant growth and is essential for all agricultural production. The organic materials are composed of debris from plants and from the decomposition of animals as well as the many tiny (microscopic) life forms that inhabit the soil. The chemical composition and physical structure of soils is determined by a number of factors such as: the kinds of rocks, minerals, and other geologic materials from which the soil is originally formed. The vegetation that grow in the soil are also important.

Food sources grown on soils are predominately composed of carbon, hydrogen, oxygen, phosphorous, nitrogen, potassium, sodium and calcium. Plants take up these elements from the soil and configure them into the plants that are recognized as food-plants. Each plant has unique nutritional requirements that are obtained through the roots from the soil. Nutrients are stored in soil on "exchange sites" of the organic and clay components. Calcium, magnesium, ammonium, potassium and the vast majority of the micronutrients are present as cations in soils of varying acidity and alkalinity (varying under most soil pH).

4.2 Composition of soil

Soil is a mixture of organic matter and minerals that support the floral and faunal species of the Earth. Soil, also called the pedosphere, has four important functions: (i) as a medium for plant growth, (ii) as a means of water storage, water supply, and water purification, and (iii) as a habitat for various life forms and all of these functions, in turn, modify the soil.

Soil consists of a solid phase of minerals and organic matter (the soil matrix), as well as a porous phase that holds gases (the soil atmosphere) and water (the soil solution). Accordingly, soil scientists can envisage soils as a three-state system of solids, liquids, and gases. In addition, soil continually undergoes development by way of numerous physical, chemical and biological processes, which include weathering and the associated erosion.

Soil is a major component of the ecosystem of the Earth and, with respect to the carbon cycle of the Earth, soil is an important carbon reservoir, and it is potentially one of the most reactive to the effects of humans. A typical soil is approximately 50% *w*/w solids (45% w/w mineral and 5% w/w organic matter), and 50% *v*/v voids (or pores) of which half is occupied by water and half by gas. Networks of pores hold water within the soil and also provide a means of water transport. Oxygen and other gases move through pore spaces in soil and the pores also serve as passageways for small animals and provide room for the growth of plant roots.

The percent soil mineral and organic content can be treated as a constant (in the short term), while the percent soil water and gas content is considered highly variable whereby a rise in one is simultaneously balanced by a reduction in the other. The pore space allows for the infiltration and movement of air and water, both of which are critical for life existing in soil. Compaction, a common problem with soils, reduces this space, preventing air and water from reaching plant roots and soil organisms.

The mineral component of soil consists of an arrangement of particles that are <2.0 mm in diameter. Technically, soil is composed of particles that fall into three main mineral groups (i) sand, (ii) silt, and (iii) clay each of which is determined by particle size: sand, 0.05 to 2.00 mm. silt 0.002 to 0.05 mm, and clay, <0.002 mm. Depending upon the parent rock materials from which these mineral were derived, the assorted mineral particles ultimately release the chemicals on which plants depend for survival, such as potassium, calcium, magnesium, phosphorus, sulfur, iron, and manganese.

Organic materials constitute another essential component of soils. Some of the organic material arises from the residue of plants, such as the remains of the roots of plants deep within the soil, or materials that fall on the ground, such as leaves on a forest floor or even a dead animal. These materials become part of a cycle of decomposition and decay, a cycle that provides important nutrients to the soil. In general, soil fertility depends on a high content of organic materials.

Soils are also characterized according to how effectively they retain and transport water. Once water enters the soil from rain or irrigation, gravity comes into play, causing water to trickle downward. Soil differs in the capacity to retain moisture against the pull exerted by gravity and by plant roots. Coarse soil, such as soil consisting of mostly of sand, tend to hold less water than do soils with finer textures, such as those with a greater proportion of clays. Water also moves through soil pores by capillary action, which is the type of movement in which the water molecules move because they are more attracted to the pore walls (adhesion) than to one another (cohesion). Such movement tends to occur from wetter to drier areas of the soil. The attraction of water molecules to each other is an example of cohesion.

Water is a critical agent in soil development due to its involvement in the dissolution, precipitation, erosion, transport, and deposition of the materials of which a soil is composed. Soil water is never pure water, but contains dissolved organic and mineral substances. In fact, water is a critical agent in soil development due to its involvement in the dissolution, precipitation, erosion, transport, and deposition of the materials of which a soil is composed. Water is central to the dissolution, precipitation, and leaching of minerals from the soil profile. Finally, water affects the type of vegetation that grows in a soil, which in turn affects the development of the soil.

4.3 Soil pollution

Unhealthy soil management methods have seriously degraded soil quality, caused soil pollution, and enhanced erosion. In addition to other human practices, the use of chemical fertilizers, pesticides, and fungicides has disrupted the natural processes occurring within the soil resulting in soil pollution. Soil pollution is a buildup of toxic chemical compounds, salts, pathogens, or radioactive materials that can affect plant and animal life. The concern over soil contamination stems primarily from health risks, both of direct contact and from secondary contamination of water supplies. All kinds of soil pollutants originate from a source. The source is particularly important because it is generally the logical place to eliminate pollution. After a pollutant is released from a source, it may act upon a receptor. The receptor is anything that is affected by the pollutant. The following sub-unit describes some of the most common sources of soil pollution.

Some of the most common toxic soil pollutants include organic chemicals, oils, tars, pesticides, biologically active materials, combustible materials, asbestos and other hazardous materials. These substances commonly arise from the rupture of underground storage tanks; application of chemical fertilizers, pesticides, and fungicides; percolation of contaminated surface water to subsurface strata; leaching of wastes from landfills or direct discharge of industrial wastes to the soil. Pesticides that are used in agricultural practices pollute the soil directly by affecting the organisms that reside in it. Pesticides include many types of chemicals that are spread around in the environment to kill some specific sort of pest, usually insects (insecticides), weeds (herbicides), or fungi (fungicides).

Organic pollutants enter the soil via atmospheric deposition, direct spreading onto land, contamination by wastewater and waste disposal. Organic contaminants include pesticides and many other components, such as oils, tars, chlorinated hydrocarbon derivatives, such as polychlorobiphenyl derivatives (PCBs) and dioxins. The use of pesticides may lead to: (i) destruction of the micro-flora and fauna of the soil, leading to both physical and chemical deterioration, (ii) severe yield reduction in crops; and (iii) leaching of toxic chemicals into groundwater and potentially threatening drinking water resources.

Existence of the ecosystems requires existence of plants. Humans and animals cannot survive without plants. Soil is not only a source of nutrition but also a place for plants to stand. Pollution of agricultural soils is known to reduce agricultural yield and increase levels of these toxic heavy metals in agricultural products, and thus to their introduction into the food chain. Vegetables and crop plants grown in such soils take up these toxic elements and pose health risk to humans and animals feeding on these plants. The major concern approximately soil pollution is that there are many sensitive land uses where people are in direct contact with soils such as residences, parks, schools and playgrounds. Other contact mechanisms include contamination of drinking water or inhalation of soil contaminants which have vaporized. There is a very large set of health consequences from exposure to soil contamination depending on pollutant type, pathway of attack and vulnerability of the exposed population.

Organic pollutants which are directly applied into soils or deposited from the atmosphere may be taken up by plants or leached into water bodies. Ultimately they affect human and animal health when taken up through the food they eat and the water they drink. More recently research has revealed that many chemical pollutants, such as DDT and polychlorobiphenyls (PCBs), mimic sex hormones and interfere with the reproductive and developmental functions of the human body—the substances are known as *endocrine disrupters*. Although, soil might be affected less by pollution compared to water or air but cleaning polluted soil is more difficult, complex, and expensive than cramming water and air.

As part of the biosphere, forests are very important for maintaining ecological balance and provide many environmental benefits. In addition to timber and paper products, forests provide wildlife habitat, prevent flooding and soil erosion, help provide clean air and water, and contain tremendous biodiversity. Forests are also an important defense against global climate change. Forests produce life-giving oxygen and consume carbon dioxide, the compound that is claimed to be the most responsible for global warming through photosynthesis, thereby reducing the effects of global warming.

5 Interrelationships

In terms of the inter-relationships of the Earth, there is a strong connection between the atmosphere, the hydrosphere, and the lithosphere because of the human activities that affect both the hydrosphere and the lithosphere. For example, disturbance of land by conversion of grasslands or forests to agricultural land or intensification of agricultural production may reduce vegetation cover, decreasing transpiration (loss of water vapor by plants) and affecting the microclimate and the result is increased rain runoff, erosion, and accumulation of silt in bodies of water. The nutrient cycles may be accelerated, leading to nutrient enrichment of surface waters. This, in turn, can profoundly affect the chemical and biological characteristics of bodies of water.

In fact, the characteristics of the natural water that is available for human use and consumption are the outcome of the interaction of the hydrosphere with the geosphere. For example, the hardness of water, is generally ascribed to the soluble calcium ion (Ca^{2+} ion) while the alkalinity of water is generally ascribed to dissolved bicarbonate ion (HCO_3^- ion). Both of these ions are present in water that is in content with limestone ($CaCO_3$) formations and/or dolomite ($CaCO_3.MgCO_3$) formation. In the latter case (i.e. dolomite formations, magnesium ions (Mg^{2+} ions) will also be present in the water. Other elements (particularly trace elements) are also present in groundwater having been extracted from mineral formations.

Thus, water is a complex system of chemical species that has received considerable study because of a series of environmental issues that have been raised (Eglinton, 1975; Jenne, 1979; Melchior and Bassett, 1990; Easterbrook, 1995). In fact, throughout history, the quality and quantity of water available to humans have been vital factors in determining not only the quality of life but also the existence of life (Boyd, 2015; Laws, 2017). However, in many parts of the world, water pollution is a major issue.

By means of clarification, water pollution refers to any change in natural waters which may impair their further use, caused by the introduction of organic or inorganic substances with undesirable properties for the ecosystem (Table 1.3), or a change in

Table 1.3 Key properties of wastes that will be harmful to water systems

Property	Comments
Ignitable	Waste that will ignite; includes liquids with a flash point below 60 °C (140 °F), non-liquids that cause fire through specific conditions, ignitable compressed gases and oxidizers.
Corrosive	Waste that is corrosive; includes corrosive waste which include aqueous wastes with a pH of less than or equal to 2, a pH greater than or equal to 12.5 or based on the liquids ability to corrode steel.
Reactive	Waste that is reactive and which may react with water, may give off toxic gases and may be capable of detonation or explosion under normal conditions or when heated.
Toxic	Waste that is toxic when ingested or absorbed; may be able to leach from deposited waste and pollute groundwater.

temperature of the water (ASTM, 2019). Furthermore, water—preferably clean and unpolluted water—is essential to all living organisms. Precipitation and evaporation rates, humidity, and available soil moisture are factors governing water availability for various life forms. Precipitation varies in relation to the position and movement of air masses and weather system, location relative to mountain ranges (rain shadow effect), and altitude. Seasonal distribution of rainfall is as important as the total amount; rainfall evenly distributed throughout the year usually results in greater availability. The water that humans use is primarily fresh (but treated) surface water and groundwater. In arid regions, a small fraction of the water supply comes from the ocean, a source that is likely to become more important as the supply of fresh water available to the Earth dwindles relative to demand. Saline or brackish groundwater may also be utilized in some areas.

Obviously, humans have also been polluting water since the early days of civilization. The development of towns and cities in close proximity to rivers also caused the rivers to become polluted by human waste and *effluents*. Indeed, whole civilizations have disappeared not only because of water shortages resulting from changes in the climate but also because of water-borne diseases, such as cholera and typhoid (Cartwright and Biddiss, 1972). In fact, in medieval times when a castle was protected from invaders by the moat (which tended to be the dumping area for all type of castle waste), one can wonder whether or not the castle moat actually protected the inmates from the attackers, although it is more likely that the disease- ridden moat was fatal to both sides!

Serious epidemics of waterborne diseases such as cholera, dysentery, and typhoid fever were caused by underground seepage from privy vaults into town wells (James and Thorpe, 1994). Such direct bacterial infections through water systems can be traced back for several centuries, even though the germ or bacterium as the cause of disease was not proved for nearly another century. In the modern world, the indirect reuse of wastewater that has been discharged into rivers is common around the world, and it is acceptable as long as the discharges are treated and contamination from chemical waste is avoided (Lacy, 1983). Less acceptable is the direct reuse of wastewater for potable use, even after a high level of treatment and the major concern in such cases is the possibility of an outbreak of disease initiated by the use of incorrectly treated or untreated water.

In considering water pollution, it is useful to keep in mind an overall picture of possible pollutant cycles which involve the major routes of pollutant interchange among the biotic, terrestrial, atmospheric, and aquatic environments. Problems with water supply quantity and quality remain and in some respects are becoming more serious. These problems include increased water use due to population growth, contamination of drinking water by improperly discarded waste materials, and destruction of wildlife by water pollution.

Water pollutants can be divided into a series of general chemical types (Table 1.4) and water pollution control is closely allied with the water supplies of communities and industries because both generally share the same water resources (Noyes, 1991). There is great similarity in the pipe systems that bring water to each home or business property, and the systems of sewers or drains that subsequently collect the wastewater

Table 1.4 Water pollutants and the effects on water systems

Pollutant type	Effect[a]
Trace elements	Health, aquatic biota
Metal-organic combinations	Metal transport
Inorganic pollutants	Toxicity, aquatic biota
Algal nutrients	Eutrophication
Radionuclides	Toxicity
Acidity, alkalinity, salinity (in excess)	Water quality, aquatic life
Sewage	Water quality, oxygen levels
Biochemical oxygen demand	Water quality, oxygen levels
Trace organic pollutants	Toxicity
Pesticides	Toxicity, aquatic biota
Crude oil, crude oi products, crude oil wastes	Effect on water biota
Detergents	Eutrophication
Sediments	Water quality, aquatic biota

[a] Will also effect land-based wildlife (including humans) that consume the water.

and conduct it to a treatment facility. Treatment should prepare the flow for return to the environment so that the receiving watercourse will be suitable for beneficial uses such as general recreation, and safe for subsequent use by downstream communities or industries.

Water supply can be considered to be one part of a hydrogeological cycle (also called the water cycle and the hydrological cycle) in which a major portion of the water occurs in the oceans (Pickering and Owen, 1994; Hiscock, 2005; Weingärtner et al., 2016). In the present context, the hydrogeologic cycle is a useful starting point for the study of water supply. This cycle Fig. 1.1) includes precipitation of water from clouds, infiltration into the ground or runoff into surface water, followed by evaporation and transpiration of the water back into the atmosphere. The rates of precipitation and evaporation and/or transpiration help define the baseline quantity of water available for human consumption.

Precipitation is the term applied to all forms of moisture (rain, ice, snow) falling to the ground. *Evaporation* and *transpiration* are the movement of water back to the atmosphere from open water surfaces and from plant respiration. The same meteorological factors that influence evaporation are at work in the transpiration process: solar radiation, ambient air temperature, humidity, and wind speed. The amount of soil moisture available to plants also affects the transpiration rate. Evaporation is measured by measuring water loss from a pan. Transpiration can be measured with a *phytometer,* a large vessel filled with soil and potted with selected plants. The soil surface is hermetically sealed to prevent evaporation; thus moisture can escape only through transpiration. The rate of moisture escape is determined by weighing the entire system at intervals up to the life of the plant. The data from a phytometer can be used as an index of water demand by a crop under field conditions, and thus relate to calculations that help an engineer determine water supply requirements for that crop. Because it is often not necessary to distinguish between evaporation and transpiration, the two processes are often linked as *evapotranspiration,* or the total water loss to the atmosphere.

Water (as water vapor) is also present in the atmosphere and other water is contained as ice and snow in snowpack, glaciers, and the polar ice caps. Surface water is found in lakes, streams, and reservoirs. Groundwater is located in aquifers (water bearing rock strata) underground.

An aquifer is a geologic formation consisting of porous and permeable rocks, material such as sand, gravel, or sandstone, through which water flows and is stored. An artesian aquifer is a confined aquifer containing groundwater under positive pressure and is, essentially, trapped water, surrounded by layers of impermeable rocks or clay minerals, which applies positive pressure to the water contained within the aquifer. If a well is drilled into an artesian aquifer, water in the well-pipe will rise to a height corresponding to the point where hydrostatic equilbrium is reached i.e. the hydrostatic pressure is equal to atmospheric pressure. A well drilled into such an aquifer is called an *artesian well*, and if water reaches the ground surface under the natural pressure of the aquifer (a *flowing artesian well*).

Groundwater is a vital resource that plays a crucial role in geochemical processes, such as the formation of secondary minerals. The nature, quality and mobility of groundwater are all strongly dependent upon the rock formations in which the water is held. Groundwater is the part of the hydrosphere (also called the aquasphere, or the region of the Earth that consists of the various water systems) most vulnerable to damage from chemical waste materials. Although surface water supplies are subject to contamination, groundwater can become almost irreversibly contaminated by the improper land disposal of chemicals.

Once there is penetration into aquatic systems, chemical species are subject to a number of chemical and biochemical processes. These include acid-base, oxidation -reduction, precipitation-dissolution, and hydrolysis reactions, as well as biodegradation. Under many circumstances, biochemical processes largely determine the fates of chemical species in water. The most important such processes are those mediated by microorganisms. In particular, the oxidation of biodegradable organic wastes in water generally occurs by means of microorganism-mediated biochemical reactions. Bacteria produce organic acids and chelating agents, such as citrate, which have the effect of solubilizing heavy metal ions. Some mobile methylated forms, such as compounds of methylated arsenic and mercury, are produced by bacterial action.

Physically, an important characteristic of such formations is the porosity, which determines the percentage of rock volume available to contain water. A second important physical characteristic is permeability, which describes the ease of flow of the water through the rock. High permeability is usually, but not always, associated with high porosity—a serious fracture in a rock formation may give the impression that the formation in highly permeable but the remainder of the formation may be impermeable to the water and other fluids. However, clay minerals tend to have low permeability even when a large percentage of the volume is filled with water. Most groundwater originates from precipitation in the form of rain or snow.

Water from precipitation that is not lost by evaporation, transpiration, or to steam runoff, may infiltrate into the ground. Initial amounts of water from precipitation on to dry soil held very tightly as a film on the surfaces and in the micropores of soil particles. At intermediate levels, the soil particles are covered with films of water, but air is still present in larger voids in the soil. The region in which such water is held is called the unsaturated zone or zone of aeration (the vadose zone) and the water present in it is vadose water.

At lower depths, in the presence of adequate amounts of water, all voids are filled to produce a zone of saturation, the upper level of which is the water table. The water table is crucial in explaining and predicting the flow of wells and springs and the levels of streams and lakes. It is also an important factor in determining the extent to which pollutant and hazardous chemicals are likely to be transported by water. The water table tends to follow the general contours of the surface topography and varies with differences in permeability and water infiltration. The water table is at surface level in the vicinity of swamps and frequently above the surface where lakes and streams are encountered. The water level in such bodies may be maintained by the water table. In-flowing streams or reservoirs that are located above the water table; these water systems lose water to the underlying aquifer and cause an upward bulge in the water table beneath the surface water.

Groundwater flow is an important consideration in determining the accessibility of the water for use and transport of pollutants from underground waste sites. Various parts of a body of groundwater are in hydraulic contact so that a change in pressure at one point will tend to affect the pressure and level at another point. For example, infiltration from a heavy, localized rainfall may affect water level at a point remote from the infiltration. Groundwater flow occurs as the result of the natural tendency of the water table to assume even levels by the action of gravity. Also, the flow of groundwater is strongly influenced by the permeability of the stratum (or strata, if there are several such formations) through which the water flows rock permeability. Porous or extensively fractured rock is relatively highly pervious. Because water can be extracted from such a formation, it is commonly referred to as an aquifer.

Briefly, an aquifer is a subsurface zone that yields economically important amounts of water to wells. The term is synonymous with water-bearing formation. An aquifer may be porous rock, unconsolidated gravel, fractured rock, or cavernous limestone. Economically important amounts of water may vary from less than a gallon per minute for cattle water in the desert to thousands of gallons per minute for industrial, irrigation, or municipal use. Also an artesian is a confined aquifer containing groundwater that will flow upwards out of a well without the need for pumping. Two other terms that are worthy of definition at this point are (i) permeability and (ii) porosity.

The permeability of a formation is a measure of the ease with which water and other fluids migrate through geological strata or landfill liners. For example, a sandstone formation may have very high permeability (i.e., water or other fluids, such as crude oil, flow relatively easily though the formation) while a shale formation or a clay formation will typically have very low permeability (i.e., water or other fluids, such as crude oil, flow with difficulty or not at all through the formation) (Speight, 2014).

Aquifers are important reservoirs storing large amounts of water relatively free from evaporation loss or pollution. If the annual withdrawal from an aquifer regularly exceeds the replenishment from rainfall or seepage from streams, the water stored in the aquifer will be depleted. Lowering the pressure in an aquifer by over pumping may cause the aquifer and confining layers of silt or clay to be compressed under the weight of the overburden. The resulting subsidence of the ground surface may cause structural damage to the aquifer and to surface buildings, damage to wells, and other problems.

By contrast, an aquiclude is a rock formation that is too impermeable to yield groundwater. Impervious rock in the unsaturated zone may retain water infiltrating from the surface to produce a perched water table that is above the main water table and from which water may be extracted. However, the amounts of water that can be extracted from such a formation are limited and the water is vulnerable to contamination.

A river is a natural, freshwater surface stream that has considerable volume compared with its smaller tributaries. The tributaries are known as brooks, creeks, branches, or forks. Rivers are usually the main stems and larger tributaries of the drainage systems that convey surface runoff from the land. Typically, a river flows from the headwater areas of small tributaries to their mouth of the river, where they may discharge into the ocean, a major lake, or a desert basin. Rivers flowing to the ocean drain about two-thirds of the land systems. The remainder of the land either is covered by ice or drains to closed basins (common in desert regions). Regions draining to the sea are termed exoreic while those regions which drain into interior closed basins are endoreic. Areic regions are those which lack surface streams because of low rainfall or lithologic conditions.

Groundwater and surface water have different characteristics: substances either dissolve in surface water or become suspended in it on its way to the ocean. Surface water in a lake or reservoir that contains the mineral nutrients essential for algal growth may support a heavy growth of algae. Surface water with a high level of biodegradable organic material, used as food by bacteria normally contains a large population of bacteria. All these factors have a profound effect upon the quality of surface water. Groundwater may dissolve minerals from the formations through which it passes. Most microorganisms originally present in groundwater are gradually filtered out as it seeps through mineral formations. Occasionally, the content of undesirable salts may become excessively high in groundwater, although it is generally superior to surface water as a domestic water source. The movement of water from waste landfills to aquifers is also an important process whereby pollutants in the landfill leachate may be adsorbed by solid material through which the water passes.

Natural waters have a broad range of total dissolved solids (TDS). Some fresh mountain streams might have a low content of total dissolved solids whilst sea water has a high content of the solids. Extreme amounts of total dissolved solids are found in highly evaporated lake water or isolated seawater basins and in the deep subsurface water (so-called formation water).

Flowing water, whether in aquifers or streams, interacts with rocks and soils and slowly dissolves some of their chemical constituents. The pH (hydrogen ion activity) of the water determines the rate of dissolution and solubility of many chemical species. Some chemical substances, particularly redox-sensitive trace metals (e.g. Fe, Mn, Pb, As, and others), are more soluble when natural waters are depleted in

Table 1.5 Chemical species commonly occurring in water

Chemical	Natural source	Other source
Sodium (Na^+)	Sea salt, soil dust	Biomass burning
Magnesium (Mg^+)	Sea salt, soil dust	Biomass burning
Potassium (K^+)	Sea salt, soil dust	Biomass burning, fertilizers
Calcium (Ca^+)	Sea salt, soil dust	Biomass burning, cement manufacture
Hydrogen (H^+)	Gas emissions	Fuel combustion
Chloride (Cl^-)	Sea salt	Industrial chemicals
Sulfate (SO_4^{2-})	Sea salt, soil dust, biological decay, volcanic activity	Biomass burning, fuel combustion
Nitrate (NO_3^{2-})	Lightning, biological decay	Biomass burning, fuel combustion, automobile emissions, fertilizers
Ammonium (NH_4+)	Biological decay	Fertilizers, biological decay
Phosphate (PO_4^{3-})	Soil dust, biological decay	Biomass burning, fertilizers
Bicarbonate (HCO_3^{2-})	Carbon dioxide in the air	Volcanic activity, biomass burning, automobile emissions, fuel combustion
SiO_2, Al, Fe	Soil dust	Fuel combustion

dissolved oxygen. Most chemical species in natural waters have both natural and pollutant sources of many types (Table 1.5).

Natural waters also contain dissolved gases, such as carbon dioxide from the atmosphere which, when dissolved in water, contributes through a series of chemical reactions, contributes to the total dissolved carbon in waters—primarily bicarbonate (HCO_3^{2-}). Also, the solubility of a gas is inversely proportional to temperature and the amount of total dissolved solids but natural waters can have low concentrations of dissolved oxygen concentrations as the result of biological activity such as the metabolism of water inhabitants, including bacteria.

Moreover, the photosynthetic activity of algae and aqueous plants can add oxygen to the water in which these primary producers grow but the breakdown of organic material by bacteria consumes dissolved oxygen. Thus, in waters below the surface wind-mixed layer (usually tens of meters or more) or in stably stratified lakes or bays, for which rates of oxygen replenishment to deeper depths are slow, deficiencies in dissolved oxygen can develop, with anoxia (total depletion of dissolved oxygen) at the extreme and excess nutrient supply can have the same impact on a water body.

6 Aquatic organisms

Microorganisms (algae, bacteria, and fungi) are living catalysts that enable a vast number of chemical processes to occur in water and soil. Aquatic organisms can be classified into four major groups, each varying in their biological characteristics, habitat, and adaptations, but linked within a complex network of ecological roles and relationships: (i) microorganisms, (ii) plants, (iii) invertebrates, and (iv) vertebrates.

In addtion, the living organisms (biota) in an aquatic ecosystem may be classified as either (i) autotrophic or (ii) heterotrophic. Autotrophic organisms utilize solar or chemical energy to fix elements from simple, nonliving inorganic material into complex life molecules that compose living organisms. Algae are typical autotrophic aquatic organisms. Generally, carbon dioxide (CO_2), nitrate derivatives (NO_3^-), and phosphate derivatives ($H_2PO_4^-$,HPO_4^{2-}) are sources of carbon, nitrogen, and phosphorus, respectively, for autotrophic organisms. Organisms that utilize solar energy to synthesize organic matter from inorganic materials are called producers.

Heterotrophic organisms utilize the organic substances produced by autotrophic organisms as energy sources and as the raw materials for the synthesis of their own biomass. Decomposers (or reducers) are a subclass of the heterotrophic organisms and consist of chiefly bacteria and fungi, which ultimately break down material of biological origin to the simple compounds originally fixed by the autotrophic organisms.

6.1 Algae and phytoplankton

The term *algae* is an informal term that is used for a large, diverse group of photosynthetic organisms that that are not necessarily closely related. Most algae are aquatic and autotrophic organisms insofar as they are capable of self-nourishment by using inorganic materials as a source of nutrients and using photosynthesis or chemosynthesis as a source of energy, as is the case with most plants and certain bacteria. The largest and most complex marine algae are the seaweeds, while the most complex freshwater forms are the Charophyta, a division of green algae which includes, for example, Spirogyra and stonewort.

Thus, algae may be considered as generally microscopic organisms that subsist on inorganic nutrients and produce organic matter from carbon dioxide by photosynthesis. The general nutrient requirements of algae are carbon (from carbon dioxide or bicarbonate, nitrogen (generally as nitrate), phosphorus (as some form of orthophosphate), sulfur (as sulfate), and trace elements including sodium, potassium, calcium, magnesium, iron, cobalt, and molybdenum.

In the absence of light, algae metabolize organic matter in the same manner as do non-photosynthetic organisms. Thus, algae may satisfy their metabolic demands by utilizing chemical energy from the degradation of stored starches or oils, or from the consumption of algal protoplasm itself. In the absence of photosynthesis, the metabolic process consumes oxygen, so during the hours of darkness an aquatic system with a heavy growth of algae may become depleted in oxygen.

Several groups of largely autotrophic protists are referred to as algae. Like the term ‘microorganisms’ it is an informal term, used for convenience to describe microorganisms that carry out photosynthesis; the cyanobacteria are often included as algae. Algae vary in size from microscopic to large colonies that can be considered macrophytes. Several types of algae (including phytoplankton) play an important role in supplying the energy at the base of many aquatic food webs.

Phytoplankton are small, microscopic plants that live suspended in the open water and are generally more abundant in lakes than in rivers, and are absent from fast-flowing streams, or where the rate at which the plants are washed downstream is greater than the rate at which they reproduce. Damming a river leads to still-water conditions more suitable for phytoplankton, and nuisance algal blooms may develop in reservoirs. Inputs of nutrients, including nitrogen and phosphorus, can also lead to algal blooms.

Phytoplankton can exist as single cells or in chains or colonies and are direct food sources for many zooplankton and some fish, and are the base of the food web in deep waters. Phytoplankton vary in their requirements for nutrients, light, and other conditions. Waterbodies support a complex mixture of phytoplankton that can change markedly with environmental conditions. In rivers containing significant amounts of phytoplankton, the concentration of algal cells (number per unit volume) is generally highest when flows are lowest, while elevated suspended sediment loads during high flows can lead to reduced light and photosynthesis. Some phytoplankton can cause taste and odor problems in water, and anoxic conditions that can kill fish. Some cyanobacteria produce toxins lethal to various fish, wildlife, and domestic species.

6.2 Bacteria

Bacteria (singular: bacterium) are a type of biological cell which are typically a few micrometers in length and have a number of shapes, including spheres rods, and spirals. Bacterial colonies were among the first life forms to appear on Earth, and are present in most of its habitats. Bacteria inhabit soil, water, acidic hot springs, radioactive waste, and the deep portions of the crust of the Earth. Bacteria also live in a symbiotic relationship (any type of a close and long-term mutual biological interaction between two different biological organisms) or parasitic relationship (a relationship is one in which one organism, the parasite, lives off of another organism, the host, harming it and possibly causing death) with plants and animals.

Some of the smallest and most ancient organisms on earth, bacteria are present in virtually every environment and are abundant in all aquatic systems. In rivers and streams, many of the bacteria wash in from the surrounding land, and their abundance can increase dramatically after a rainfall. The abundance of bacteria is typically in the millions per milliliter (mL), and in the hundreds of millions per milliliter in especially productive or polluted waters.

Bacteria occur individually or grow as groups ranging from two to millions of individual cells. Individual bacteria cells are very small and may be observed only through a microscope. Most bacteria fall into the size range of 0.5–3.0 μm (1 μm = 1 m × 10^{-6}). However, considering all species, a size range of 0.3–50 μm is observed. In general, it is assumed that a filter with 0.45 μm pores will remove all bacteria from water passing through it.

The metabolic activity of bacteria is greatly influenced by their small size. Their surface-to-volume ratio is extremely large, so that the inside of a bacterial cell is highly accessible to a chemical substance in the surrounding medium. Thus, for the same reason that a finely divided catalyst is more efficient than a more coarsely divided one, bacteria may bring about very rapid chemical reactions compared to those mediated by larger organisms. Bacteria excrete enzymes that can act outside the cell (*exo*-enzymes) that break down solid food material to soluble components which can penetrate bacterial cell walls, where the digestion process is completed.

If conditions are right, bacteria reproduce extremely rapidly by simple division to produce very large numbers in a short period of time. Typically, bacteria can be found suspended in the water, associated with decaying material (such as dead wood or leaves), or coating the surface of rocks, stones and sand grains as part of the biofilm (the slippery coating on hard surfaces in rivers). They can make up a large fraction of the living material in aquatic systems.

Bacteria display the greatest range in metabolic ability of any group of organisms. There are both autotrophic and heterotrophic bacteria. Heterotrophic bacteria are a crucial link in the decomposition of organic matter and the cycling of nutrients in aquatic systems. Autotrophic bacteria are primary producers in aquatic systems as are true algae. For this reason, autotrophic bacteria (predominantly cyanobacteria) are often categorized as 'algae', though the organisms are by no means closely related. Cyanobacteria used to be mistakenly called 'blue-green algae'. Ecologically, much of what applies to algae is relevant to autotrophic bacteria.

Bacteria obtain the energy needed for their metabolic processes and reproduction by mediating redox reactions. Also, bacteria are essential participants in many important elemental cycles in nature, including those of nitrogen, carbon, and sulfur. They are responsible for the formation of many mineral deposits, including some of iron and manganese. On a smaller scale, some of these deposits form through bacterial action in natural water systems and even in pipes used to transport water.

6.3 Fungi

Fungi occur as single cells, and in filaments (hyphae). Most aquatic fungi are microscopic; those known as hyphomycetes are the most abundant and important. Fungi are heterotrophic, and, like heterotrophic bacteria, obtain their nutrition by secreting *exo*-enzymes into their immediate environment, which break compounds down into simpler substances the fungi can absorb. Fungi are critical to the decomposition of plant matter in aquatic systems, because they are among the few organisms that can break down certain plant structural compounds such as cellulose and lignin.

Also, fungi are non-photosynthetic organisms and are the group of eukaryotic organisms (organisms whose cells have a nucleus enclosed within membranes) that includes microorganisms such as molds, yeasts, as well as mushrooms. Some fungi are as simple as the microscopic unicellular yeasts, whereas other fungi form large, intricate toadstools. The microscopic filamentous structures of fungi generally are much larger than bacteria and usually are 5–10 μm in width. Fungi are aerobic (oxygen-requiring) organisms and generally can thrive in more acidic media than can bacteria. They are also more tolerant of higher concentrations of heavy metal ions than are bacteria.

Perhaps the most important function of fungi in the environment is the breakdown of cellulose in wood and other plant materials. To accomplish this, fungal cells secrete a biological catalyst (enzyme) called cellulase. This enzyme breaks insoluble cellulose down to soluble carbohydrates that can be absorbed by the fungal cell. Because it acts outside the organism, it is called an extracellular enzyme or *exo*-enzyme.

Although fungi do not grow well in water, they play an important role in determining the composition of natural waters and waste waters because of the large amount of their decomposition products that enter water. An example of such a product is humic material, which interacts with hydrogen ions and metals.

6.4 Protozoa

Protozoa are one-celled animals found worldwide in most habitats (Imam, 2009). Most species are free living, but all higher animals are infected with one or more species of protozoa. Infections range from asymptomatic to life threatening, depending on the species and strain of the parasite and the resistance of the host. Protozoa are microscopic unicellular eukaryotes (organisms whose cells have a nucleus enclosed within membranes) that have a relatively complex internal structure and carry out complex metabolic activities. Some protozoa have structures for propulsion or other types of movement.

Free-living protozoans are common and often abundant in fresh, brackish and salt water, as well as in other moist environments, such as soils and mosses. Some species thrive in extreme environments such as hot springs and hypersaline lakes and lagoons. All protozoa require a moist habitat; however, some can survive for long periods of time in dry environments, by forming resting cysts which enable the protozoa to remain dormant until conditions improve.

The stages of parasitic protozoa that actively feed and multiply are frequently called trophozoites; in some protozoa, other terms are used for these stages. Cysts are stages with a protective membrane or thickened wall. Protozoan cysts that must survive outside the host usually have more resistant walls than cysts that form in tissues.

7 The global water cycle

Water covers about three-quarters of surface of the Earth and is necessary for sustaining life. During their constant cycling between land, the oceans, and the atmosphere, water molecules pass repeatedly through solid, liquid, and gaseous phases (ice, liquid water, and water vapor), but the total supply remains fairly constant. A water molecule can travel to many parts of the globe as it cycles.

Not all the water that comes back to the mainland through precipitations is collected by the rivers, lakes or trapped in glaciers. A part of it filters on the soil and goes down into it due to the force of gravity until it reaches a layer of waterproof rocks that stop the passage of water: a water-bearing stratum is created. When this condition does not occur, in order to reach the water-bearing stratum artesian wells are built and the water can be extracted as it reaches the surface after being subject to a high pressure. The continuous exploitation of underground water determines the emptying of water-bearing stratums and lowering of soils. Instead, when the exploitation occurs near coastal regions, the seawater filters underground to occupy the free spaces left by fresh water: this generates serious damages to agriculture and vegetation, as it is happening along Ravenna coast, where wide pinewood areas are dying.

Water absorbs a considerable amount of energy when it changes from the liquid state to the gaseous state. Even though the temperature of the water vapor may not increase by any significant amount when it evaporates from liquid water, this vapor now contains more energy, which is referred to as latent heat. Atmospheric circulation moves this latent heat around the Earth, and when water vapor condenses and produces rain, the latent heat is released.

Very little water is consumed in the sense of actually taking it out of the water cycle permanently, and unlike energy resources such as oil, water is not lost as a consequence of being used. However, anthropogenic activities often increases the flow of water out of one water storage area into another, which can lead to the depletion of the stores of water that are most usable. For example, pumping groundwater for irrigation depletes aquifers by transferring the water to evaporation or river flow. Human activities also pollute water so that it is no longer suitable for human use and is harmful to ecosystems. There are three basic steps in the global water cycle: (i) water precipitates from the atmosphere, (ii) water travels on the surface and through groundwater to the oceans, and (iii) water evaporates or transpires back to the atmosphere from land or evaporates from the oceans.

Supplies of freshwater (water without a significant salt content) exist because precipitation is greater than evaporation on land. Most of the precipitation that is not transpired by plants or evaporated, infiltrates through soils and becomes groundwater, which flows through rocks and sediments and discharges into rivers. River systems are primarily supplied by groundwater, and in turn provide most of the freshwater discharge to the sea. Within the ocean systems, evaporation is greater than precipitation and, thus, the net effect is a transfer of water back to the atmosphere. In this way freshwater resources are continually renewed by counterbalancing differences between evaporation and precipitation on land and at sea, and the transport of water vapor in the atmosphere from the sea to the land. Approximately 97% *v*/v of the water supply is held in the oceans—the other large reserves are groundwater (4% *v*/v) and icecaps and glaciers (2% v/v), with all other water systems together accounting for a fraction of 1% v/v. Residence times of pollutants in the component systems of the Earth can vary from several thousand years in the oceans (Fig. 1.3) to a few days in the atmosphere.

Solar radiation drives evaporation by heating water so that it changes to water vapor at a faster rate. This process consumes an enormous amount of energy—nearly one-third of the incoming solar energy that reaches the surface of the Earth. On land, most evaporation occurs as transpiration through plants: water is taken up through roots and evaporates through stomata in the leaves as the plant takes in carbon dioxide. A single large oak tree can transpire up to 40,000 gal per year. Much of the water moving

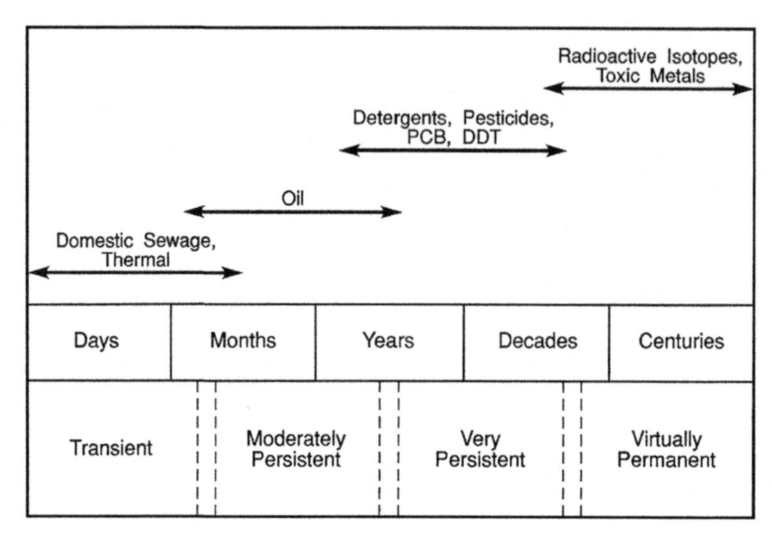

Fig. 1.3 Relative lifetimes of pollutants in the oceans.

through the hydrologic cycle thus is involved with plant growth. Since evaporation is driven by heat, it rises and falls with seasonal temperatures.

In temperate regions, water stores rise and fall with seasonal evaporation rates, so that net atmospheric input (precipitation minus evaporation) can vary from positive to negative. The temperature is more constant in tropical regions where large seasonal differences in precipitation, such as monsoon cycles, are the main cause of variations in the availability of water. In an effort to reduce these seasonal variations, many countries have built reservoirs to capture water during periods of high precipitation and release water during periods of low precipitation or drought.

The hydrologic cycle is also coupled with material cycles because rainfall erodes and weathers rock. Weathering breaks down rocks into gravel, sand, and sediments, and is an important source of key nutrients such as calcium and sulfur. Actual rates may be much higher at specific locations and may have been accelerated by human activities, such as emissions from the combustion of carbonaceous fuels that release gases such as sulfur dioxide and nitrogen oxides which make react with the moisture in the atmosphere thereby rendering the rain more acidic (Chapter 5). Thus:

$$SO_2 + H_2O \rightarrow H_2SO_3\,(\text{sulfurous acid})$$

$$SO_3 + H_2O \rightarrow H_2SO_4\,(\text{sulfuric acid})$$

$$NO + H_2O \rightarrow HNO_2\,(\text{nitrous acid})$$

$$3NO_2 + 2H_2O \rightarrow HNO_3\,(\text{nitric acid})$$

7.1 Freshwater resources

Freshwater accounts for only some 6% *v*/v of the water supply of the world but is essential for human uses such as drinking, agriculture, manufacturing, and sanitation. As discussed above, two-thirds of global freshwater is found underground and constituents the water table.

The water table is the upper surface of the saturation zone in which the pores and fractures of the ground are saturated with water. The water table is the surface where the water pressure (the pressure head) is equal to the atmospheric pressure and may be described as the surface of the subsurface formations that are saturated with groundwater (Freeze and Cherry, 1979).

The groundwater may be from precipitation (rain, snow, or hail) or from groundwater flowing into the aquifer. In areas where there is sufficient precipitation, water infiltrates through pore spaces in the soil, passing through the unsaturated zone until, at increasing depths, the water enters and fills more of the pore spaces in the soil, until a saturation is achieved (the saturated zone). Below the water table, in the zone of saturation (the phreatic zone), the groundwater is stored in the rock (the aquifer). In less permeable formations, such as tight bedrock formations and clay formations, the water table may be more difficult to identify. The water table should not be confused with the water level in a deeper well. The elevation of the water in the deep well is dependent upon the pressure in the aquifer.

The water table may lie a substantial distance (hundreds of feet) below the ground surface and a water table may shift from season to season as precipitation and transpiration levels change. For example, the water table may move up toward the surface during rainy periods or periods of little transpiration and sink during dry phases when the rate of recharge (precipitation minus evaporation and transpiration that infiltrates from the surface) drops. In temperate regions of the Earth, the water table tends to follow surface topography, rising under hills where there is little discharge to streams and falling under valleys where the water table intersects the surface in the form of streams, lakes, and springs.

Above the water table lies the unsaturated zone (the vadose zone) where the pores (spaces between grains) are not completely filled with water. Although air in the vadose zone is at atmospheric pressures, the water in the zone may be under tension due to the presence of strong adhesive forces between the water and the rock grains, and by surface tension at the small interfaces between the water and air. Thus, fine-grained formations, such as those formation consisting of clay minerals, can hold water tight (usually referred to as *suction*) under very large suctions. Water flows upward under suction through small pores from the water table toward plant roots when precipitation is less than precipitation.

Evapotranspiration (the sum of the evaporation and plant transpiration from the land and ocean surface to the atmosphere) represents a significant water loss from drainage basins and is the sum of evaporation and plant transpiration from the land surface and the ocean surface to the atmosphere of the Earth. Evaporation accounts for the movement of water to the air from sources such as the soil and water systems. Transpiration accounts for the movement of water within plants and the subsequent loss of water as vapor through the stomata and leaves. Stomata (singular: stoma) are the pores that occur in the epidermis of leaves, stems, and other organs, that facilitates gas exchange. Each pore is bordered by a pair of specialized cells (often referred to as guard cells) that are responsible for regulating the size of the opening of the associated stoma.

After a rainstorm, water may recharge the groundwater by saturating large pores and cracks in the soil and flowing very quickly downward to the water table and, often from there, to an aquifer—a permeable geologic formation through which water flows almost unrestricted. Under natural conditions many aquifers are artesian: the water they hold is under pressure, so water will flow to the surface from a well without pumping. Aquifers may be either capped by an impermeable layer (confined) or open to receive water from the surface (unconfined). Confined aquifers are often artesian because the confining layer prevents upward flow of groundwater, but unconfined aquifers are also artesian in the vicinity of discharge areas. Thus, groundwater will discharge into a river or a stream. A confined aquifers is less likely to be contaminated because the impermeable layers above them prevent surface contaminants from reaching the water and, thus, the confined aquifer can provide good-quality water.

Water has an average residence time of thousands of years to tens of thousands of years (in the case of deep groundwater even millions of years old—as old as the rocks that hold the water in their pores) in many aquifers, but the actual age of a water sample collected from a particular well will vary tremendously within an aquifer. Shallow groundwater can discharge into streams and rivers in weeks or months. Because of this distribution of residence times in aquifers, contaminants that have been introduced at the surface over the last century are only now beginning to reach well depths and contaminate drinking water in many aquifers. Indeed, much of the solute load (salt and other mineral contaminants as well as organic contaminants) that has entered aquifers due to increased agriculture and other land use changes over the last several centuries has yet to reach discharge areas where it will contaminate streams and lakes.

7.2 Depletion of freshwater resources

Although the international community is focusing on global warming as the paramount environmental threat, the scarcity of fresh water for agriculture may emerge as a more immediate problem. Almost one-fifth of the world's population lives in areas where water is physically scarce. Rapid population growth, as well as increases in per capita food consumption and changes in the composition of the diet, has led to an enormous expansion of agriculture, which is now estimated to account for 92% *v*/v percent of the use of fresh water (Hoekstra and Mekonnen, 2012).

Water resources, specifically the resources of water in aquifers, are being depleted by the withdrawal of water from the aquifers more quickly than the aquifers are replenished by recharge. In addition, excessive pumping of groundwater an aquifer can (will) pumping change the groundwater flow patterns around wells and the aquifer water may even can drain into a nearby river or stream. The pumping changes the natural equilibrium that exists in an undeveloped aquifer with discharge balancing recharge—when pumping is initiated, the groundwater in the vicinity of the well is depleted which creates a cone of depression in the hydraulic head. A cone of depression is an actual depression of the water levels. In confined aquifers (artesian aquifers), the cone of depression is a reduction in the pressure head surrounding the pumped well.

If a new water source such as a river or stream is available close by the position of the well, the well may draw (capture) water from) that source and increase the recharge rate until the inflow matches the pumping rate. However, pumping causes a reduction in pressure in a confined aquifers so that water no longer rises to the surface naturally. If no such source is available and pumping draws the water table down far enough, the aquifer will be dried up or depleted to a point where it is not physically possible to pump out the last stores of water.

In an unconfined aquifers, air fills the pores above the water table with the result that the water table falls much more slowly than in a confined aquifer. As aquifers are depleted, water has to be lifted from much greater depths and, in addtion, the overuse of groundwater can also reduce the quality of the remaining water if a well draws water from contaminated surface sources or if the water table is near the coast and is below sea level, there is a strong likelihood that salt water will flow into the aquifer.

7.3 Water pollution

Water pollution is the contamination of a water system, usually as a result of human activities which is the results of introducing contaminants into the natural environment. For example, releasing inadequately treated wastewater into a natural water system can lead to degradation of the water system. The causes of water pollution include a wide range of chemicals and pathogens as well as physical parameters, such as elevated temperatures. A common cause of thermal pollution is the use of water as a coolant by power plants and other industrial operations. Elevated water temperatures decrease the oxygen level of the water system which can kill fish and alter the composition of the food chain by reducing the and allow invasion by new thermophilic (temperature-resistant) species.

Thus, many different types of contaminants can pollute water and render it unusable. Pollutants regulated in the United States—and in many other countries—under national primary drinking water standards (legally enforceable limits for public water systems to protect public health) include: (i) microorganisms such as cryptosporidium, giardia, and fecal coliform bacteria, (ii) disinfectants and water disinfection byproducts including chlorine, bromate, and chlorite, (iii) inorganic chemicals such as arsenic, cadmium, lead, and mercury, (iv) organic chemicals such as benzene, dioxin, and vinyl chloride, and (v) radionuclides including uranium and radium.

Microorganisms are typically found in human and animal waste. Some inorganic contaminants such as arsenic and radionuclides such as uranium occur naturally in geologic deposits, but many inorganic and most major organic pollutants are emitted from industrial facilities, mining, and agricultural activities such as the application of fertilizers and pesticides. Sediments (soil particles) from erosion and activities such as excavation and construction also pollute rivers, lakes, and coastal waters. The availability of light is the primary constraint on photosynthesis in aquatic ecosystems, so adding sediments can severely affect productivity in these ecosystems by clouding the water. The clouded water (whatever the cause of the clouding) can results in the smothering of fish and shellfish spawning grounds and degrading of the habitat by filling in rivers and streams through deposition of the sediment.

Water supplies often become polluted because contaminants are introduced into the vadose zone or are present there naturally and penetrate to the water table or to groundwater, where they move into wells, lakes, rivers, and eventually (in many cases) into the ocean. Many dissolved compounds can be toxic and carcinogenic, so keeping them out of water supplies is a serious goal in terms of the protection of public health goal.

One particular issue is the manner in which some chemicals compounds behave in a water system. For example, non-aqueous phase liquids (NAPLs) form a separate phase that does not mix with water and can reside as small blobs within the pore structure of an aquifer while other compounds, such as gasoline and diesel fuel, are lighter than water and will float on top. Other chemicals, including chlorinated hydrocarbon derivatives, are denser that water and will on to and into the sediment that forms on the base of the water system. Both types are difficult to remove and will slowly dissolve into groundwater, migrating downgradient as groundwater flows. Other contaminants completely dissolve in water (Table 1.5) and, if they enter the aquifer at a single location (e.g., from a point source), they are transported with flowing groundwater as contaminant plume (Fig. 1.4) that gradually mixes with native groundwater. Over time, contaminated zones become larger but the concentration of the contaminants will decrease as the plume spreads. The paths that plumes follow can be extremely complex because of the complicated patterns of permeability within aquifers (even in a single aquifer). Nevertheless, the velocity of the groundwater is much higher through channels of high permeability (compared to the velocity of the

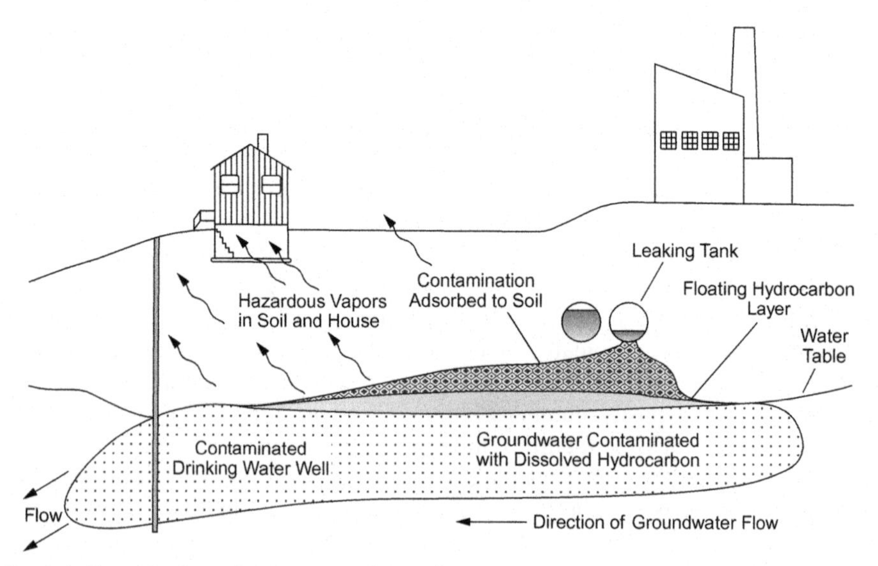

Fig. 1.4 General schematic of a contaminant plume.

water in channels of lo permeability) and so these channels (i.e. the high permeability) will transport dissolved contaminants rapidly through the subsurface.

As a plume moves through groundwater, some of the contaminants may bind (sorb) on to the mineral particles. A high content of clay minerals in a formation generally increases sorption because these minerals particles are often chemically reactive and typically a have large surface area. The sorption process can prevent contaminants from migrating to other sites—for example, in some spills containing uranium, the uranium has moved very little (often only several feet) over decades. However, contaminants such as uranium can also adsorb to very small particles suspended in the water phase that promote facile migration through an aquifer. Even if a contaminated plume is pumped out, sorbed contaminants may remain on the solid matrix and, given time, will desorb from the mineral into the groundwater. This sorption-desorption process will hinder full cleanup of the contaminant.

Along with freshwater bodies, many coastal areas and estuaries (areas where rivers meet the sea, mixing salt and fresh water) are severely impacted by water pollution and sedimentation. Ocean pollution kills fish, seabirds, and marine mammals; damages aquatic ecosystems; causes outbreaks of human illness; and causes economic damage through impacts on activities such as tourism and fishing.

Nutrient-rich runoff into ocean waters stimulates plankton to increase photosynthesis and causes blooms, or explosion (a rapid and sudden increase) of the plankton population. When excess plankton die and sink, their decomposition consumes oxygen in the water. Since the beginning of the Industrial Revolution (the transition to new manufacturing processes in Europe and the United States, in the period from approximately 1760

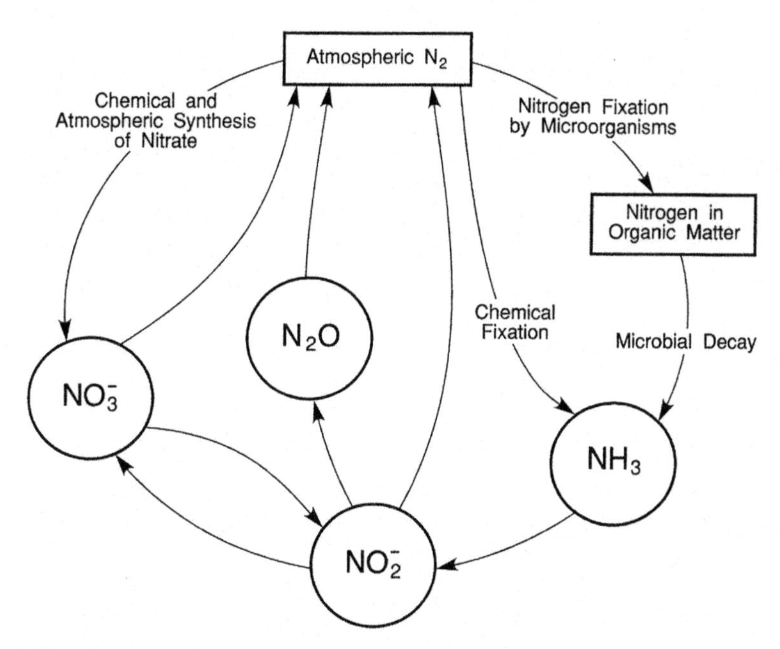

Fig. 1.5 The nitrogen cycle.

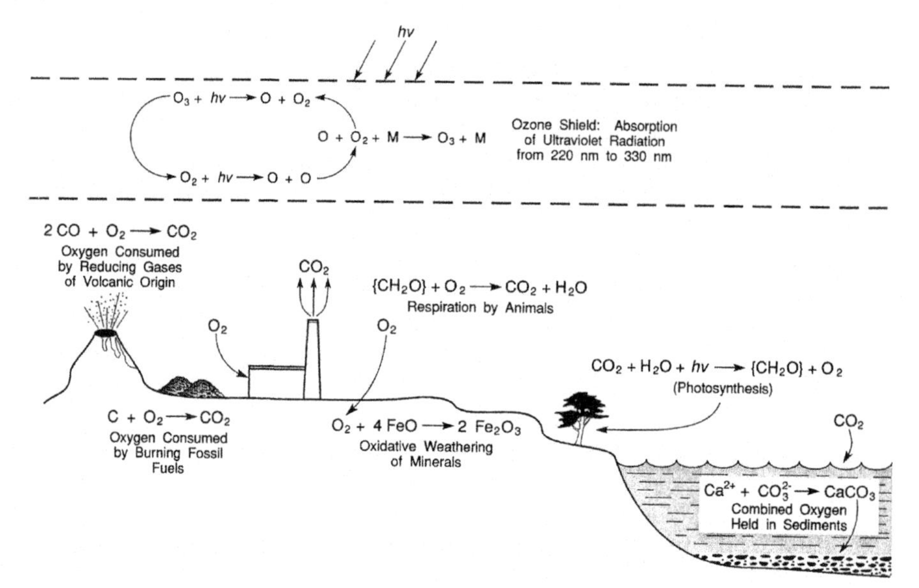

Fig. 1.6 The oxygen cycle.

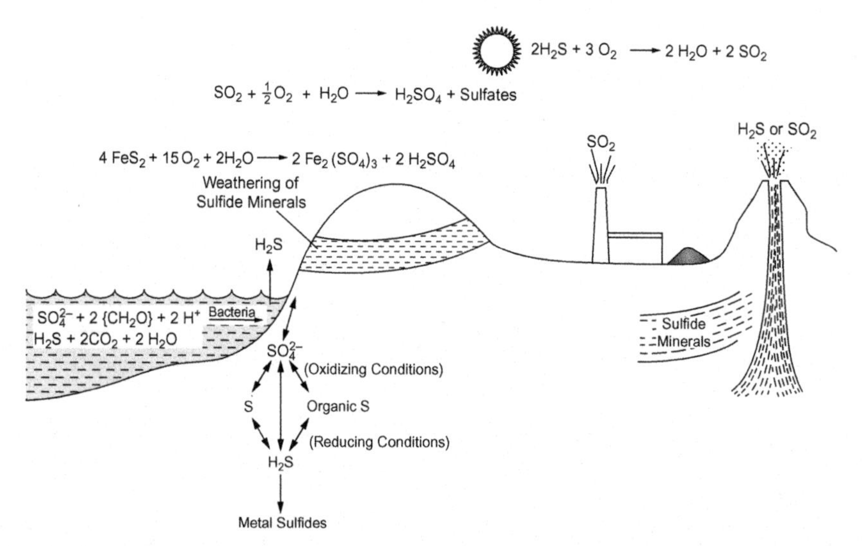

Fig. 1.7 The sulfur cycle.

to sometime between 1820 and 1840), human activities, especially the use of fertilizers in agriculture and the increased use fossil fuels for the generation of heat and energy, have roughly doubled the amount of nitrogen circulating globally, increasing the frequency and size of plankton blooms. This process (part of the nitrogen cycle, which is the biogeochemical cycle by which nitrogen is converted into multiple chemical forms as it circulates among atmospheric, terrestrial, and aqueous ecosystems) (Fig. 1.5) can create hypoxic areas (dead zones), where dissolved oxygen levels are too low to support marine life, which is typically less than two to three milligrams per liter. Oxygen and sulfur have similar cycles which can also play a role in defining the properties of water (Figs. 1.6 and 1.7).

Water pollution is relatively easier to control when it comes from a point source—a single identifiable *localized* source of (in the current context) a contaminant; a point source has negligible extent, distinguishing it from other source geometries; the actual source need not be physically small. On the other hand, nonpoint source pollution consists of discharges from many contributors, such as runoff from city streets or agricultural fields, so it is more challenging to control. Approaches for controlling nonpoint source pollution include improving urban storm water management systems; regulating land uses; limiting broad application of pesticides, herbicides, and fertilizer; and restoring wetlands to help absorb and filter runoff.

7.4 Water-related diseases

Water-related diseases (water-borne diseases) diseases are caused by pathogenic microorganisms that most commonly are transmitted in contaminated fresh water. Infection commonly results during bathing, washing, drinking, in the preparation of food, or the consumption of food thus infected. The water-related diseases generally fall into four major categories: (i) waterborne diseases, (ii) water-washed diseases, (iii) water-based diseases, and (iv) water-related insect vectors.

Thus, water-borne diseases, including cholera, typhoid, and dysentery, are caused by drinking water containing infectious viruses or bacteria, which often come from human or animal waste. Water-washed diseases, such as skin and eye infections, are caused by lack of clean water for washing. Water-based diseases, such as schistosomiasis, are spread by organisms that develop in water and then become human parasites. They are spread by contaminated water and by eating insufficiently cooked fish. Water-related insect vectors, such as mosquitoes, breed in or near water and spread diseases, including dengue and malaria. This category is not directly related to water supply or quality.

Water-borne diseases spread by contaminating drinking water systems with feces and urine of infected animals or people. The spread of contaminated water is likely to happen where private and public drinking systems get the water such as from surface waters - creeks, rivers, lakes, and rain. These sources of water may be contaminated by infected animals or people. Runoff from: (i) landfills, (ii) sewer pipes, (ii) septic fields, and (iv) industrial or residential developments. Generally, food that is contaminated is the one most common way people become infected. The germs in feces may cause the diseases by even slight contact and transfer. The contamination might happen because of floodwaters, septic fields, water runoff from landfills, and sewer pipes.

In order to sever the path of the continued transmission of water-borne diseases, it is necessary to improve the hygienic behavior of people and provide them with basic needs such as: (i) sanitation, (ii) drinking water, (iii) bathing facilities, and (iv) washing facilities. Transmission of malaria is facilitated when large numbers of people sleep outside in hot weather, or sleep in homes that have no protection against mosquitoes. Malaria mosquitoes, bilharzias snails and tropical black flies can all be controlled with efficient drainage because they (these insects) depend on water to complete their respective life cycles.

Clean water is a prerequisite for reducing the spread of water-borne diseases and it is well recognized that the prevalence of water-borne diseases may be greatly reduced by providing safe, sanitary disposal facilities, and clean drinking water. Water is disinfected to kill any pathogens that might be present in the water supply and to prevent them from growing again in distribution systems. Disinfection is then used in order to prevent the growth of pathogenic organisms and to protect human health. Without disinfection, the risk of water-borne disease increases. The two most common methods of killing microorganisms in the water supply are irradiation with ultra-violet radiation, or oxidation with chemicals like chlorine dioxide or ozone, or chlorine.

7.5 General effects

The ability of a water system) to produce living material (the productivity of the water system) results from a combination of physical and chemical factors. Water of low productivity generally is desirable for water supply or for swimming while a relatively high productivity is required for the support of fish. On the other hand, excessive productivity can result in choking by weeds and can cause odor problems. For example, the growth of algae may become quite high in very productive waters, with the result that the concurrent decomposition of dead algae reduces oxygen levels in the water to very low values (eutrophication). In inland lakes, the issue of eutrophication is due mainly to excessive, but inadvertent, introduction of domestic and industrial wastes, runoff from fertilized agricultural and urban areas, precipitation, and groundwater.

The interaction of the natural processes within the lake with the artificial disturbance caused by human activities complicates the overall problem and leads to an accelerated rate of deterioration in lakes. Since a population increase necessitates an expanded utilization of lakes and streams, cultural eutrophication has become one of the major water resource problems throughout the world. Moreover, cultural eutrophication is reflected in changes in species composition, population sizes, and productivity in groups of organisms throughout the aquatic ecosystem. Thus the biological changes which are caused by excessive fertilization are of considerable interest from both the practical and academic viewpoints.

Life forms higher than algae and bacteria (such as fish) comprise a comparatively small fraction of the biomass in most aquatic systems and the influence of these higher life forms upon aquatic chemistry is minimal. However, aquatic life is strongly influenced by the physical and chemical properties of the water system (Table 1.6). Temperature, transparency, and turbulence are the three main physical properties affecting aquatic life. Very low water temperatures result in very slow biological processes, whereas very high temperatures are fatal to most organisms. A difference of only a few degrees can produce large differences in the kinds of organisms present. For example, the thermal discharge of hot water from a power plants cooling water system can have serious adverse effects on temperature-sensitive fish while increasing the growth of algae and other microorganisms that are adapted to higher temperatures.

Table 1.6 Properties of water and the effects in water systems

Property	Effect
Color—transparent	Allows light required for photosynthesis to reach considerable depths
Density—maximum at 4 °C	Ice floats; vertical circulation restricted in stratified bodies of water
Dielectric constant—high	High solubility of ionic substances and their ionization in solution
Heat capacity—high	Stabilization of temperatures of organisms and geographical regions
Heat of evaporation—high	Determines transfer of heat and water molecules between the atmosphere and bodies of water
Latent heat of fusion—high	Temperature stabilized at the freezing point of water
Solvent power—high	Transport of nutrients and waste products, making biological processes possible in water
Surface tension—high	Controlling factor in physiology; governs drop and surface phenomena

The transparency of water is particularly important in determining the growth of algae. Turbid water may not be very productive of biomass, even though it has the nutrients and optimum temperature. Turbulence is an important factor in mixing and transport processes in water. Some small organisms (plankton) depend upon water currents for their own mobility. In fact, water turbulence is largely responsible for the transport of nutrients to living organisms and of waste products away from them. It plays a role in the transport of oxygen, carbon dioxide, and other gases through a body of water and in the exchange of these gases at the water-atmosphere interface. Moderate turbulence is generally beneficial to aquatic life.

Biochemical oxygen demand, another important water-quality parameter, refers to the amount of oxygen utilized when the organic matter in a given volume of water is degraded biologically. A body of water with a high biochemical oxygen demand, and no means of rapidly replenishing the oxygen, obviously cannot sustain organisms that require oxygen. The degree of oxygen consumption by microbial oxidation of contaminants in water is called the biochemical oxygen demand (or biological oxygen demand).

The levels of nutrients in water frequently determine its productivity. Aquatic plant life requires an adequate supply of carbon (carbon dioxide, CO_2), nitrogen (nitrate, NO_3^-), phosphorus (orthophosphate, HPO_4^-), and trace elements such as iron. In many cases, phosphorus is the limiting nutrient and is generally controlled in attempts to limit excess productivity. The salinity of water also determines the kinds of life forms present. Irrigation waters may pick up harmful levels of salt. Marine life obviously requires or tolerates salt water, whereas many freshwater organisms are intolerant of salt.

A majority of the important chemical reactions occurring in water, particularly those involving organic matter and oxidation-reduction processes, occur through bacterial intermediaries. Algae are the primary producers of biological organic matter (biomass) in water. Microorganisms are responsible for the formation of many sediment and mineral deposits; they also play the dominant role in secondary waste treatment.

Another example of waste control by oxidation involves oxidation of chemical waste materials in supercritical water. Supercritical water oxidation has been used to convert organic materials to carbon dioxide and is claimed to achieve destruction of 99.9+ of chemical such as polyols and amines that occur in liquid waste streams.

References

ASTM, 2019. Annual Book of Standards. ASTM International, West Conshocken, PA.

Benn, D.I., Evans, D.J.A., 2010. Glaciers and Glaciation, second ed. Taylor & Francis, CRC Press, Boca Raton, FL.

Boyd, C.E., 2015. Water Quality, second ed. Springer, New York.

Cartwright, F.F., Biddiss, M.D., 1972. Disease and History. Dorset Press, New York.

Cogley, G., 2012. The future of the world's glaciers. In: Henderson-Sellers, A., McGuffie, K. (Eds.), The Future of the World's Climate. Elsevier, New York, pp. 197–222.

Dodds, W.K., 2002. Freshwater Ecology: Concepts and Environmental Applications. Academic Press, New York.

Easterbrook, G., 1995. A Moment on the Earth: The Coming Age of Environmental Optimism. Viking Press, New York.

Eglinton, G. (Ed.), 1975. Environmental Chemistry. Volume 1. The Chemical Society, London, England. Specialist Periodical Reports.

Fitts, C.R., 2012. Groundwater Science, second ed. Academic Press, New York.

Freeze, R.A., Cherry, J.A., 1979. Groundwater. Prentice-Hall, Upper Saddle River, NJ.

Hiscock, K., 2005. Hydrogeology Principles and Practice. Blackwell Publishing, Jog Wiley & Sons Inc, Hoboken, NJ.

Hoekstra, A.Y., Mekonnen, M.M., 2012. The water footprint of humanity. Proceedings. National Academy of Sciences 109, 3232–3237.

Imam, T.S., 2009. The complexities in the classification of protozoa: a challenge to parasitologists. Bayero Journal of Pure and Applied Sciences 2 (2), 159–164.

James, P., Thorpe, N., 1994. Ancient Inventions. Ballantine Books, New York.

Jansson, P., Hock, R., Schneider, T., 2003. The concept of glacier storage: a review. J. Hydrol. 282, 116–129.

Jenne, E.D. (Ed.), 1979. Chemical Modeling in Aqueous Systems. American Chemical Society, Washington, DC. Symposium Series No. 93.

Lacy, W.J., 1983. In: Kent, J.A. (Ed.), Riegel's Handbook of Industrial Chemistry. Van Nostrand Reinhold, New York, p. 14.

Laws, E.A., 2017. Aquatic Pollution: An Introductory Text, fourth ed. John Wiley & Sons Inc, Hoboken, New Jersey.

Manahan, S.E., 2010. Environmental Chemistry, ninth ed. CRC Press, Taylor & Francis Group, Boca Raton, FL.

Melchior, D.C., Bassett, R.L. (Eds.), 1990. Chemical Modeling of Aqueous Systems II. American Chemical Society, Washington, DC. Symposium Series No. 416.

Mitsch, W.J., Gosselink, J.C., 2007. Wetlands, fourth ed. John Wiley & Sons Inc., Wiley, Hoboken, NJ.

Noyes, R. (Ed.), 1991. Handbook of Pollution Control Processes. Noyes Data Corp., Park Ridge, NJ.

Pickering, K.T., Owen, L.A., 1994. Global Environmental Issues. Routledge, Taylor & Francis, CRC Press, Boca Raton, FL.

Price, M., 2016. Introducing Groundwater, second ed. Routledge, Taylor & Francis Group, CRC Press, Boca Raton, FL.

Speight, J.G., 1996. Environmental Technology Handbook. Taylor & Francis, Washington, DC.

Speight, J.G., 2013a. The Chemistry and Technology of Coal, third ed. CRC Press, Taylor & Francis Groups, Boca Raton, FL.

Speight, J.G., 2013b. Coal-Fired Power Generation Handbook. In: Scrivener Publishing. Beverly, MA.

Speight, J.G., 2014. The Chemistry and Technology of Petroleum, fifth ed. CRC Press, Taylor & Francis Groups, Boca Raton, FL.

Speight, J.G., Lee, S., 2000. Environmental Technology Handbook, second ed. Taylor & Francis, New York.

Spellman, F.R., 2008. The Science of Water: Concepts and Applications, second ed. CRC Press, Taylor & Francis Group, Boca Raton, FL.

Weiner, R.F., Matthews, R.A., 2003. Environmental Engineering, fourth ed. Butterworth-Heinemann, Elsevier Science, Burlington, MA.

Weingärtner, H., Teermann, I., Borchers, U., Balsaa, P., Lutze, H.V., Schmidt, T.C., Franck, E.U., Wiegand, G., Dahmen, N., Schwedt, G., Frimmel, F.H., Gordalla, B.C., 2016. Water, 1. Properties, Analysis, and Hydrogeological Cycle. In: Ullmann's Encyclopedia of Industrial Chemistry. Wiley-VCH Verlag GmbH & Co. KGaA, Weinheim, Germany.

Further reading

Manahan, S.E., 2011. Water Chemistry: Green Science and Technology of Nature's Most Renewable Resource. CRC Press, Taylor & Francis Group, Boca Raton, FL.

The properties of water

2

Chapter outline

1 Introduction

The properties of water, which are to a large extent, due to the chemical and physical structure of the water molecule (Section 2, below). The study of water (hydrology) is divided into a number of sub categories—examples are: limnology, which is the branch of the water science that deals with the characteristics of fresh water, including biological properties as well as chemical and physical properties and (ii) oceanography,

Natural Water Remediation. https://doi.org/10.1016/B978-0-12-803810-9.00002-4

which is the branch of water science that deals with the ocean and its physical and chemical characteristics (Dodson, 2005; Graham and Farmer, 2007). Water is the solvent, the medium and the participant in most of the chemical reactions occurring in environment of the Earth.

In fact, a large part of the mass of most living organisms is water—in human tissues the percentage of water ranges from 20% *w*/w in bones to as much as 85% w/w in brain cells. Also, the water content is greater in embryonic and young cells and decreases as aging occurs. In total, approximately 70% of the total human body weight is water. Water is not only the major component of organisms but also one of the principal environmental factors affecting them. Many organisms live within the sea or in freshwater ponds, lakes, and rivers and, in fact, the physical and chemical properties of water have permitted living things to appear, to survive, and to evolve on the Earth.

The properties of water make it suitable for organisms to survive in different weather conditions. Ice expands as freezes becoming less dense per unit volume, which explains why ice is able to float on liquid water. During the winter when lakes begin to freeze, the surface of the water freezes and then moves down towards deeper water. Furthermore, the density of ice assures that the ice will from on top of the liquid water phase, if this was not the case, the lake would freeze from the bottom up killing all ecosystems living in the lake. In addition, the formation of the ice on top of the liquid water enables survival of the fish under the surface of the ice during the winter. The surface ice on top of lake water also shields lakes from the cold temperature outside and insulates the water beneath it, allowing the lake under the frozen ice to stay liquid and maintain a temperature adequate for the ecosystems living in the lake to survive.

However, before attempts are made to measure any of the properties of water, there is the need to insure and define the type of water: (i) pure or (ii) impure and, in the latter case, the nature and amount of the impurities must also be defined. Water can be contaminated by a number of gases, dissolved solids, and other chemical species and the presence contaminants can significantly alter the properties of water. Most (but not all) of the impurities in a sample of water can be removed by distillation—for example salts dissolved in the water will not be present in the water vapor but will concentrate in the solution left behind. On the other hand, volatile compounds will evaporate, but they may not condense as efficiently, and so their concentration in the final distilled solution will be reduced.

In this respect, the chemistry, physics, and biology of the ocean water systems are unique because of the high salt content, depth, and other factors. The environmental problems of the oceans have increased greatly in recent years because of ocean dumping of pollutants, oil spills, and increased utilization of natural resources from the oceans. In addition, many of the pollutants have considerable lifetimes in the ocean (Fig. 2.1).

But first, before dealing with the types of water (i.e. pure and impure), the next section deals with the physical structure of water to present to the reader an understanding of the relationship of the physical structure to the properties.

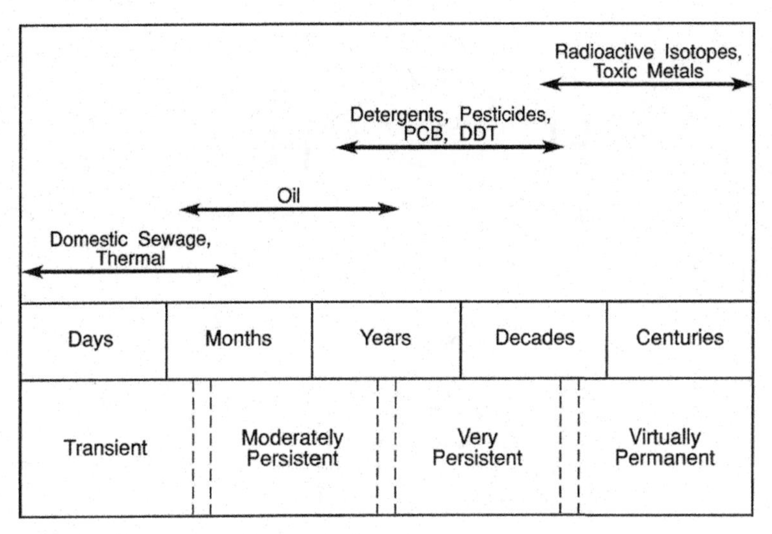

Fig. 2.1 Relative lifetimes of pollutants in the oceans.

2 Structure and bonding of water

An important feature of water is its polar nature because the structure has an angular molecular structure (angular molecular geometry) in terms of the relative positions of the two hydrogen atoms from the oxygen vertex. Furthermore, the oxygen atom also has two lone pairs of electrons. One effect usually ascribed to the lone pairs is that the hydrogen-oxygen-hydrogen (H-O-H) gas phase bend angle is 104.45°, which is smaller than the ideal tetrahedral angle of 109.47°. The lone pairs are closer to the oxygen atom than the electrons in the sigma bonds to the hydrogen atoms and, therefore, require more space. In addition, increased repulsion of the lone pairs of electrons forces the oxygen-hydrogen bonds closer to each other.

The water molecule consists of two oxygen bonds of length 0.096 nm (Fig. 2.2A) and the water molecule is approximately spherical as well as being electrically neutral but, because the electronegativity of oxygen is much greater than that of hydrogen, the electron distribution is concentrated more around the former, i.e. water is electrically polarized because of the higher electron density around the oxygen atom. A useful way to represent the polarity of a molecule is to assign a partial charge to each atom (Fig. 2.2B) in order to reproduce net charge on the molecule and the magnitude of the partial charge on an atom is a measure of the polarity of the atom. Thus, water is a very polar molecule with the ability to engage strong electrostatic interactions with itself, with other molecules, and with ions. Many of properties of liquid water including: (i) cohesiveness, (ii) the high heat of vaporization, (iii) the dielectric constant, and (iv) the surface tension can be explained with this simple molecular model.

Thus, the polarity of water is due to the occurrence of charge inequality on two ends of any molecule. Because of the unequal distribution of electrons in water molecule,

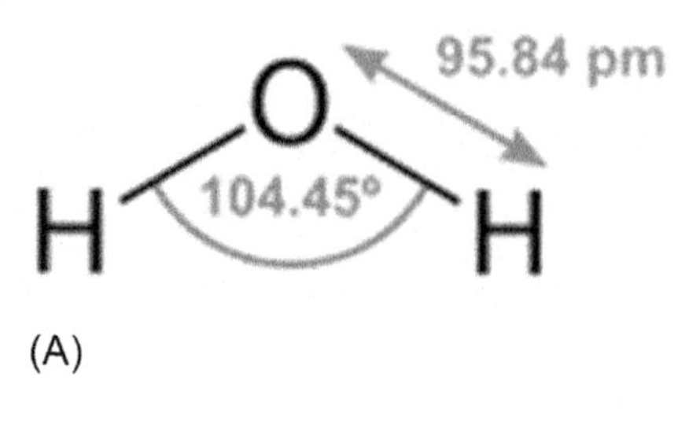

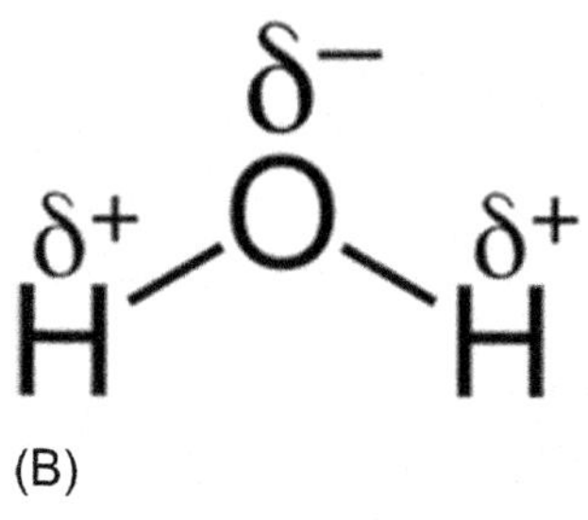

Fig. 2.2 (A) The structure of water also showing (B) the relative negativity of the hydrogen and oxygen atoms.

one part of the molecule is more negatively charged than the other part. This is caused by the property of *electronegativity*, which is the ability of an element to attract electrons. If the atoms that make up a molecule each have a different electronegativity which differ substantially, these atoms attract electrons unequally, thereby giving rise to a polar molecule.

Due to the difference in electronegativity of the atoms, a bond dipole moment exits as a result of the relative negativity of the hydrogen atoms and the oxygen atoms (Fig. 2.2) thereby rendering the oxygen partially negative and each hydrogen partially positive. The large molecular dipole exits because of the region between the two hydrogen atoms to the oxygen atom—the charge differences cause water molecules to aggregate (the relatively positive areas being attracted to the relatively negative areas). This attraction (hydrogen bonding) causes the existence of many of the unique properties of water, such as the solvent properties.

Although hydrogen bonding is a relatively weak attraction compared to the covalent bonds within the water molecule itself, it is responsible for a number of physical properties of water. These properties include the relatively high melting point and boiling point of water as a result of the hydrogen bonds between water molecules which requires more energy is required to break the hydrogen bonds between water molecules. In contrast, hydrogen sulfide (H_2S), has much weaker hydrogen bonding due to the lower electronegativity of sulfur. Also hydrogen sulfide is in the gaseous state at room temperature in spite of hydrogen sulfide having nearly twice the molar mass of water (Table 2.1). Also, the intermolecular bonding between water molecules also gives liquid water a relatively high specific heat capacity which makes water a good heat storage medium (coolant) and heat shield.

Table 2.1 Boiling points and freezing points of the hydrides related to water

Compound	Formula	Molar mass	Boiling point[a]		Freezing point[a]	
			°C	°F	°C	°F
Hydrogen telluride	H_2Te	129.6	−4	25	−49	−56
Hydrogen selenide	H_2Se	81	−42	−44	−64	−83
Hydrogen sulfide	H_2S	34	−62	−80	−84	−119
Water	H_2O	18	100	212	0	32

[a] Rounded to the nearest degree.

In terms of the properties of water, the structure conveys upon the molecule the following properties of state: (i) water can exist as a gas, i.e. steam, when water is heated to the boiling point, the water changes from the liquid phase to the gas phase or vapor phase, (ii) water can exist as a liquid, which is the familiar form of water, and (iii) water can exist as a solid, i.e. ice, which is less dense than water in the liquid state and, hence, ice will float on liquids water.

In the context of this book, the most important phase is the liquid phase, the structure of which is dominated by the hydrogen-bonding interaction.

2.1 The molecular symmetry of water

Water is a small molecule that is electrically neutral but polar, with the center of positive and negative charges located in different places. The molecule, the ten electrons pair up into five 'orbitals', one pair closely associated with the oxygen atom, two pairs associated with the oxygen atom as 'outer' electrons and two pairs forming each of the two identical oxygen-hydrogen covalent bonds. The eight outer electrons are often shown as the pairs where the pairs of electrons between the oxygen and hydrogen atoms represent the oxygen-hydrogen covalent bonds, and the other two pairs of electrons are often referred to as the *lone pairs* (Fig. 2.2A).

These electron pairs form electron clouds that are spread out approximately tetrahedrally around the oxygen nucleus as they repel each other and are the cause of the angular structure. The eight positive charges in the oxygen nucleus attract all these electrons strongly relative to the single positive charges on each of the hydrogen atoms. This leaves the hydrogen atoms partially denuded of electrons, and hence partially positively charged, and the oxygen atom partially negatively charged (Fig. 2.2B).

Due to the presence of these charges and the angular nature of the molecule, the center of the positive charge (half way between the two hydrogen atoms) does not coincide with the center of the negative charge (on the oxygen atom). As a result, in liquid water, this gives a molecular dipole moment from the center of negative charge to the center of positive charge, equivalent to a unit negative charge (that is, one electron) separated from a unit positive charge. The presence of this dipole moment in all water molecules causes its polar nature.

The opposite charges on the oxygen atom and the hydrogen atoms is the cause of the attraction between different water molecules. This attraction is particularly strong when the oxygen-hydrogen bond from one water molecule points directly at a nearby oxygen atom in another water molecule that is when the three atoms oxygen, hydrogen, and oxygen are in a straight line. This attraction between neighboring water molecules (hydrogen bonding), together with the high-density of molecules due to their small size, produces a great cohesive effect within liquid water that is responsible for the liquid state of water at ambient temperatures.

2.2 Hydrogen bonding

Water exists in three phases: (i) vapor, (ii) liquid, and (iii) solid, the last of which (ice) has at least eighteen crystalline phases (where the oxygen atoms are in fixed positions relative to each other, but the hydrogen atoms may or may not be disordered, and three amorphous (non-crystalline) phases (Bartels-Rausch et al., 2012) see [2145, 2349] for recent reviews of ice research). This large number is due to the open tetrahedrally arranged water molecular structure of hexagonal ice under normal atmospheric pressure and the large number of possible crystal structures that this ice can form as it is progressively crushed under high pressure.

All of the crystalline phases of ice involve hydrogen bonding of the water molecules to four neighboring water molecules (Zheligovskaya and Malenkov, 2006). In most cases, the two hydrogen atoms are equivalent with the hydrogen bonds of similar strength (Fig. 2.2). The water molecules retain their symmetry, for the most part, the ordering of the protons (in fixed positions with lower entropy) occurs at lower temperatures, whereas pressure reduces the distances between second shell neighbors (lower volume and greater van der Waals effects). The hydrogen-oxygen-hydrogen (H-O-H) angle in the ice phases is expected to be less (at approximately 107°) than the ideal tetrahedral angle (109.47°).

The hydrogen bond is a strong bond formed between a polar hydrogen and another heavy atom, usually carbon, nitrogen, oxygen or sulfur in biological molecules. In the gas phase the strength of a hydrogen bond is dependent on the molecular geometry and the surrounding molecules. In addition, the hydrogen bond is sometimes characterized as being intermediate between an ionic bond and a covalent bond, although the energy of the hydrogen bond is a function of the bond length and bond angle and can be accurately described by a Coulombic interaction (the electrostatic interactions between electric charges) between the partial atomic charges on the hydrogen, the heavy atom to which the hydrogen is covalently attached, and the oxygen atom, the nitrogen atom, the carbon atom, or the sulfur atom that are participating in the hydrogen bond.

Combined with a hydrogen-oxygen-hydrogen bond angle that is very close to the ideal tetrahedral angle of 109.58° and with the tendency for the four neighboring waters to repel each other electrostatically, the phenomenon is sufficient to explain the persistence of the tetrahedral pattern in liquid water. The open tetrahedral structure is also responsible for the anomalous temperature dependence of water density.

The hydrogen bond length of water varies with temperature and pressure and, since the covalent bond lengths vary much less with temperature and pressure, most of the densification of ice is due to reduced temperature or increased pressure must be due to a reduction in the hydrogen bond length. This hydrogen bond length variation can be shown from the changes in the volume of ice. As hydrogen bond strength depends almost linearly on its length (shorter length giving stronger hydrogen-bonding), the bond strength also depends almost linearly (outside extreme values) on the temperature and pressure.

Heavy water (deuterium oxide, D_2O) has similar, but not identical, properties to water. Heavy water (deuterium oxide, ^2H_2O, D_2O) is a form of water that contains a larger than normal amount of the hydrogen isotope deuterium (^2H or D, also known as heavy hydrogen), rather than the common hydrogen-1 isotope (^1H or H, also called protium) that makes up most of the hydrogen in normal water. The presence of deuterium gives the water different nuclear properties, and the increase of mass gives it slightly different physical and chemical properties when compared to normal water. Hydrogen bonds within heavy water (D_2O) are stronger than for water (H_2O).

Tritiated water is a radioactive form of water where the usual protium atoms are replaced with tritium (^3H). In the pure form it may be called tritium oxide (T_2O or ^3H_2O). Pure tritium is corrosive due to self-radiolysis and is used as a tracer for water transport studies in life-science research. The name super-heavy water helps distinguish the tritium oxide from heavy water which contains deuterium.

2.3 Structure of ice

As water cools, its molecular motion slows and the molecules move gradually closer to one another. The density of any liquid increases as its temperature decreases and for most liquids, this continues as the liquid freezes and the solid state is denser than the liquid state. However, water behaves differently and actually reaches the highest density at approximately 4 °C (39 °F). Between 4 °C (39 °F) and 0 °C (32 °F), the density of water shows a gradual decreases (Table 3.1) which is caused by the hydrogen bonds forming a network characterized by a hexagonal structure with open spaces in the middle of the hexagons.

Moreover, ice is less dense than liquid water and so the ice floats on the surface of the water. In water systems, such as lakes and ponds (i.e. non-flowing water systems), the water begins to freeze at the surface, closer to the cold air and where the water is not disturbed by continuous movement. As a result, a layer of ice forms, but does not sink as it would if water did not have this unique structure dictated by its shape, polarity, and hydrogen bonding. If the ice sank as it formed, the entire non-flowing water system would freeze solid. Since the ice does not sink, liquid water remains under the ice all winter long and is an important aspect of the non-flowing water system since it allows aquatic animals (fish and other organisms) to survive during the winter.

2.4 Self-ionization

The self-ionization of water (also called auto-ionization of water and the auto-dissociation of water) is an important aspect of water chemistry and is an ionization reaction that occurs in pure water in which a water molecule deprotonates (i.e. loses the nucleus of one of the hydrogen atoms) to form a hydroxide ion (OH^-). The hydrogen nucleus (H^+) immediately protonates another water molecule to forma hydronium ion (H_3O^+)—the reaction is reversible and is an example of autoprotolysis occurs, which applied not only to pure water but also to any aqueous solution, and exemplifies the amphoteric nature of water:

$$2H_2O \rightleftharpoons H_3O^+ + OH^-$$

If this equation is expressed on the basis of chemical activity rather than on the basis of concentration, the thermodynamic equilibrium constant (K_{eq}) for the ionization reaction is:

$$K_{eq} = \left(a_{H3O+} \times a_{OH-}\right) / a_{H2O}$$

This equation is sometimes written as:

$$K_{eq} = \left(a_{H+} \times a_{OH-}\right) / a_{H2O}$$

The assumption is that the sum of the chemical potentials of the hydrogen ion (H^+) and the hydroxonium ion (H_3O^+) are equal to twice the chemical potential of H_2O at the same temperature and pressure. However, because most acid-base solutions are typically very dilute, the activity of water is generally approximated as being equal to unity, which allows the ionic product of water to be expressed as:

$$K_{eq} = a_{H3O+} \times a_{OH-}$$

However, in dilute aqueous solutions, the activity of each solute particle is approximately equal to the concentrations of the particle. As a result, the *ionization constant*, the *dissociation constant*, the *self-ionization constant*, or the *ionic product* of water (K_w) is given by:

$$KW = \left[H_3O^+\right]\left[OH^-\right]$$

In this equation, $[H_3O^+]$ is the molality (i.e. the molar concentration of hydrogen or the hydronium ion and $[OH^-]$ is the concentration of the hydroxide ion.

2.5 Amphiprotic nature

Molecules or ions which can either donate or accept a proton, depending on their molecular circumstances amphiprotic and the most important amphiprotic species is water itself—the terms amphoteric and amphiprotic can be used interchangeably. When an acid donates a proton to water, the water molecule is a proton acceptor, and hence a base. On the other hand, when a base reacts with water, a water molecule donates a proton, and hence acts as an acid. However, a more subtle definition is that *amphiprotic substances* can react as either acids or bases in the Brønsted-Lowry sense while *amphoteric substances* are those substances that react in either manner as indicated by the broader Lewis definition.

The Brønsted-Lowry theory is s that when an acid and a base react with each other, the acid forms its conjugate base, and the base forms its conjugate acid by exchange of a proton (the hydrogen cation, or H+). On the other hand, a Lewis acid is a chemical species that contains an empty orbital that is capable of accepting an electron pair from a Lewis base to form a Lewis adduct. A Lewis base, then, is any species that has a filled orbital containing an electron pair which is not involved in bonding but may form a dative bond (a coordinate covalent bond) with a Lewis acid to form a Lewis adduct.

Water is a both a neutral molecule and an amphoteric molecule—the two are not mutually exclusive terms. An amphoteric molecule is simply one that can act as either an acid or a base, while a neutral molecule is one in which the total number of protons is equal to the number of electrons, such that there is zero net charge. Water can act as a Brønsted acid (i.e., an H+ donor), as per the following reaction:

$$H_2O + NH_2^- \rightarrow OH^- + NH3$$

However it can also act as a Brønsted base (i.e., an H+ acceptor):

$$H_2O + HCl \rightarrow H_3O^+ + Cl^-$$

When proton transfer is involved, as in the Brønsted definition of acids and bases, the more precise term is *amphiprotic*. However, water can also act as both a Lewis acid (an electron pair acceptor) and Lewis base (an electron pair donor). Notably, as a Lewis base, it is not limited merely to proton abstraction, but can act as a nucleophile with any number of electrophilic molecules.

In addition, a large number of protic substances exhibit this type of behavior to an appreciable degree, including concentrated solutions of various acids (e.g., sulfuric acid, hydrofluoric acid). Under the right conditions, a very wide array of molecules of diverse types can be amphoteric. The two necessary and sufficient criteria for a molecule to be amphoteric are: (i) the molecule possesses lone pairs, so that it can act as a base, and (ii) the molecule can accept electron pairs, so that it can act an acid.

3 General properties

Water is a polar inorganic compound that is, at ambient temperature, a tasteless, odorless, and colorless liquid that has been described as the universal solvent as well as solvent of life. It is the most abundant chemical compound on Earth and it is also the only common naturally-occurring chemical to exist as a gas, liquid, and solid on the surface of the Earth (Weingärtner et al., 2016). However, to put this in perspective, the water on the Earth, which seems so abundant at the surface, is actually <0.5% of the total mass of the Earth. In fact, the Earth, a so-called water planet, contains some 0.07% *w*/w water or 0.4% *v*/v which is located in (i) ice sheets and glaciers, (ii) lakes, (iii) rivers, (iv) oceans as well as in (v) groundwater systems (active and inactive), (vi) the atmosphere, (vii) in the soil, and (viii) in biological systems.

All components contained in natural waters give the water specific properties, such as: salinity, alkalinity, hardness, acidity, and corrosivity. Furthermore, knowledge of the chemical composition and properties of water is necessary for the resolution of scientific and applied problems concerned with water use for human domestic and industrial activities.

Thus, because of the potential uses of water, there are a number of interesting properties (Table 2.2) that make it suitable for living organisms and a solvent for many materials. Thus, it is the basic transport medium for nutrients and waste products in life processes. But, a point that must not be missed, there is also the very real potential for

Table 2.2 Common properties of water

IUPAC name	Water, oxidane
Other uncommon names	Hydrogen hydroxide, hydrogen oxide, dihydrogen monoxide, hydrogen monoxide, dihydrogen oxide, hydric acid, hydrohydroxic acid, hydroxic acid, hydrol
Molar mass	18.01528(33) g/mol
Appearance	White crystalline solid, colorless liquid
Odor	None
Density	Liquid: 0.9998396 g/mL at 0 °C, 0.9970474 g/mL at 25 °C, 0.961893 g/mL at 95 °C Solid: 0.9167 g/mL at 0 °C
Melting point	0 °C (32 °F; 273 K)
Boiling point	100 °C (212 °F; 373.13 K)
Solubility	Poorly soluble in haloalkanes, aliphatic and aromatic hydrocarbon derivatives, ethers. Miscible with methanol, ethanol, propanol, isopropanol, acetone, glycerol, 1,4-dioxane, tetrahydrofuran, sulfolane, acetaldehyde, dimethylformamide, dimethoxyethane, dimethyl sulfoxide, acetonitrile. Partially miscible with diethyl ether, methyl ethyl ketone, dichloromethane, ethyl acetate, bromine.
Vapor pressure	3.1690 kilopascals or 0.031276 atm
Conjugate acid	Hydronium
Conjugate base	Hydroxide
Thermal conductivity	0.6065 W/(m·K)
Refractive index (n_D)	1.3330 (20 °C)
Surface tension	Water-air: 72.86 ± 0.05 mN·m^{-1} at 20 °C
Viscosity	0.890 cP
Crystal structure	Hexagonal
Molecular shape	Bent
Dipole moment	1.8546 D
Heat capacity (C)	75.385 ± 0.05 J/(mol·K)
Std molar entropy (S^o_{298})	69.95 ± 0.03 J/(mol·K)
Std enthalpy of formation($\Delta_f H^o_{298}$)	−285.83 ± 0.04 kJ/mol
Gibbs free energy($\Delta_f G$)	−237.24 kJ/mol

water to transport toxins. The high dielectric constant of water affects its solvent properties, in that most ionic materials are dissociated in water. In addition, water has a high heat capacity; thus, a relatively large amount of her is required to change appreciably the temperature of a mass of water. Hence, the presence of a water system can have a stabilizing effect upon the temperature of nearby geographic regions—contributing to these proeprties are the various species dissolved in the water. These include: chloride ions, sulfate ions, bicarbonate ions, carbonate ions, sodium ions, potassium ions, calcium ions, and dissolved gases.

Chloride ions (Cl^-) have a large migratory ability in connection with the very high solubility of chloride salts of sodium, magnesium and calcium. The presence of these ions in water is associated with the processes of leaching from minerals, from rocks, and from saline deposits. Chloride ions are also present in atmospheric precipitation, and is particularly associated with industrial and municipal wastes.

Sulfate ions (SO_4^{2-}) are contained in all surface waters, and their content is limited by the presence of calcium ions together with which they form a slightly soluble calcium sulfate ($CaSO_4$). The main source of sulfate in water is various sedimentary rocks which include gypsum and anhydride. Water enrichment by sulfates takes place both by the process of oxidation of sulfide, which is abundant in the crust of the Earth, and oxidation of hydrogen sulfide (H_2S) which is created during volcanic eruption and is present in atmospheric precipitation. The processes of decompositions and oxidation of substances of vegetable and organic origin containing sulfur, and also human economic activity, have an effect on the sulfur content of water systems.

Bicarbonate ions and carbonate ions (HCO_3^- and $CO3^{2-}$) occur in natural waters in dynamic equilibrium with carbonic acid (H_2CO_3) in certain quantitative proportions and form a carbonate system of chemical equilibrium connected with the pH of water. When the pH of a water system is 7-8.5 (i.e. the water is mildly alkaline) the predominant ion is the bicarbonate ion but when pH is <5 (i.e. the water is acidic), the content of bicarbonate ions is close to zero. Carbonate ions dominate when the $pH > 8$. The sources of the bicarbonate ions and the carbonate ions are various carbonate rocks (limestones, dolomites, magnesites), from which dissolution takes place with the participation of carbon dioxide. Bicarbonate ions always dominate in water with low mineralization, and often in water system with moderate mineralization. Accumulation of the bicarbonate ions is limited by the presence of calcium ions, forming with the bicarbonate ion a poorly dissolved salt calcium bicarbonate which can dissociate to the difficulty-soluble calcium carbonate:

$$Ca^{2+} + 2HCO_3^- \rightarrow Ca(HCO_3)_2$$

$$Ca(HCO_3)_2 \rightarrow CaCO_3 + H_2O + CO_2$$

Sodium ions (Na^+) typically arise from sodium salts that have solubility in water. A high proportion of the sodium ions is balanced by the presence of chlorine ions, forming a stable mobile combination that migrates with high velocity in a solution. The sources of sodium ions in natural water systems are deposits of various salts (rock-salt), weathering products of limestone rocks, and its displacement from the absorbed complex of rocks and soils by calcium and magnesium.

Potassium ions (K^+), compared to sodium ions, occur in lower concentrations in surface waters because potassium has low migratory ability due to the active participation of potassium in biological processes, e.g. absorption by living plants and micro-organisms.

Calcium ions (Ca^{2+}) arise predominantly from The basic sources of calcium are carbonate rocks (limestone minerals and dolomite minerals) that are dissolved by carbonic acid (H_2CO_3) contained in water. When the availability of carbon dioxide (with which it in a balance), is low, however, the reaction begins to proceed in a reverse direction, accompanied by precipitation of calcium carbonate ($CaCO_3$). Another source of calcium ions in natural waters is gypsum ($CaSO_4.2H_2O$), which is common in many sedimentary rocks.

Magnesium ions (Mg^{2+}) are less abundant than calcium in the crust of the Earth and enters surface water systems as a result of the processes of chemical weathering and dissolution of dolomite minerals, marls (a carbonate-rich mudstone), and other rocks. Magnesium ions occur in all natural waters, but very seldom dominate. The weaker biological activity of magnesium, as compared with calcium, and also the higher solubility of magnesium sulfate and magnesium bicarbonate as compared to the equivalent compounds of calcium, favor increase of concentration of magnesium ions in water.

Dissolved gases occur in all but they differ in origin. The composition of gases connected with the exchange processes between water and atmosphere depends mainly on the content of the gases in the atmosphere. Processes that take place in water bodies, including biochemical ones, require the presence of oxygen (which is formed during photosynthesis), carbon dioxide, methane, and, to a lesser extent, hydrogen sulfide, ammonia, high molecular weight hydrocarbon derivatives, and nitrogen. Volcanic processes and degassing of the mantle of the Earth supply oxides and dioxides of carbon, methane, ammonia, hydrogen sulfide, hydrogen, hydrogen chloride, sulfurous gas and others into natural waters. Some other gases can appear and dissolve in water as a result of ultra-violet irradiation (ozone), thunderstorm discharges (nitric oxide), and anthropogenic pollution (such as sulfur-containing gas and ammonia).

Other substances (biogenous substances) dissolved in natural water systems include those from the various aquatic organisms that live in the water system. These include compounds of silicon, nitrogen, phosphorus and iron.

Silicon (Si) is a constant component in natural waters but the content of silicon relative to the total salt composition is low due to the low solubility of silicate minerals and their consumption by some organisms. Silicon occurs in waters in a fully dissolved state in the form of meta-silicic acid (H2SiO3) and ortho-silicic acid ($H4SiO_4$) and in a colloidal form of silicic acid (xSiO2.yH2O). Briefly, minerals and pollutants are suspended in water as very small particles and which are classified as *colloidal particles*. The properties and behavior of colloidal particles are strongly influenced by their physical and chemical characteristics and these particles play any important role in determining the properties and behavior of natural water and wastewater. Furthermore, contaminants bound to the surface of colloidal particles (*colloid-facilitated transport*) is an important method by which contaminants (that would otherwise be sorbed to sediments or, in the case of groundwater transport, to aquifer rocks) are transported in a water system.

Nitrogen (N) occurs in natural waters in the form of various inorganic ions (ammonium $NH4^{+}$ nitrite $NO2^{-}$ and nitrate $NO3^{-}$) and organic compounds (in the amino acids and proteins of organisms, and the products of their vital activity and decomposition). The transformation of complex organic forms into mineral ones happens in the process of biogenous element regeneration, the result of which is ammonia formation. Under oxidizing conditions with bacterial action, ammonia is oxidized into nitrite derivatives and nitrate derivatives.

Phosphorus (P) occurs in water in the form of inorganic and organic compounds in a dissolved state and in the form of suspended and colloidal substances. Phosphorus forms phosphorous acid (H3PO4) of neutral strength that dissociates into some derivative forms: H2PO4, $HPO4^{2-}$, $PO4^{3-}$, the relation between which is determined by the pH value of water. The main factor defining the concentration of phosphorus in the water system is the exchange between living organisms and inorganic forms. In natural water systems, phosphorus usually occurs in low concentrations due to the low solubility of its compounds and intensive consumption by the organisms that live in the water (hydrobionts).

In terms of thermal properties, water absorbs or releases more heat than many substances for each degree of temperature increase or decrease. Because of this property, water is widely used for cooling and for transferring heat in thermal and chemical processes. Differences in temperature between lakes and rivers and the surrounding air may have a variety of effects. For example, a local fog or a mist is likely to occur if a lake cools in the surrounding air enough to cause saturation in which small droplets of water are suspended in the air. Large bodies of water, such as the oceans or large lakes (an example is the Great Lakes of North America) have a significant influence on the climate of the Earth. These lakes are the large heat reservoirs and heat exchangers and the source of much of the moisture that falls as rain and snow over adjacent landmasses—for example, lake effect snow is an expected winter phenomenon in the adjacent eastern American states. When water is colder than the air, precipitation is curbed, winds are reduced, and fog banks are formed.

The high heat capacity of water also prevents sudden changes of temperature in large bodies of water and thereby protects aquatic organisms from the shock of abrupt temperature variations. The high heat of vaporization of water (585 cal/g at 20 °C, 68 °F) also stabilizes the temperature of bodies of water and influences the transfer of heat and water vapor between bodies of water and the atmosphere.

The physical condition of a body of water strongly influences the chemical and biological processes that occur in water. Surface water occurs primarily in streams, lakes, and reservoirs. Lakes may be classified as oligotrophic, eutrophic, or dystrophic, an order that often parallels the life of the lake. Oligotrophic lakes are deep, generally clear, deficient in nutrients, and without much biological activity. Eutrophic lakes have more nutrients, support more life, and are more turbid. Dystrophic lakes are shallow, clogged with plant life, and normally contain colored water with a low pH.

Wetlands (distinct ecosystems that is are inundated by (liquid) water, either permanently or seasonally, where oxygen-free processes prevail) in which the water is shallow enough to enable growth of bottom-rooted plants. Wet flatlands are areas where mesophytic vegetation (terrestrial vegetation that is neither adapted to particularly dry nor particularly wet environments) is more important than open water and which are commonly developed in filled lakes, glacial pits, and potholes, or in poorly drained coastal plains or flood plains. The term *swamp* is usually applied to a wetland where trees and shrubs are an important part of the vegetative association and the term *bog* generally implies a water-logged area that does not have a solid foundation. In fact, areas that have been designated as bogs consist of a thick zone of vegetation floating on water. Unique plant associations characterize wetlands in various climates and exhibit marked zonation characteristics around the edge in response to different thicknesses of the saturated zone above the firm base of soil material. Coastal marshes covered with vegetation adapted to saline water are common on all continents.

Some constructed reservoirs are very similar to lakes, while others differ a great deal from them. Reservoirs with a large volume relative to their inflow and outflow are called storage reservoirs. Reservoirs with a large rate of flow-through compared to their volume are called run-of-the-river reservoirs. The physical, chemical, and biological properties of water in the two types of reservoirs may vary appreciably. Water in storage reservoirs more closely resembles lake water, whereas water in run-of-the-river reservoirs is much like river water.

Impounding water in reservoirs may have some profound effects upon water quality which are the result of a variety of factors such as: (i) different velocity, (ii) changed detention time, and (iii) altered surface-to-volume ratios relative to the streams that were impounded. On the other hand, some resulting beneficial changes due to impoundment are a decrease in the level of organic matter, a reduction in turbidity, and a decrease in hardness (calcium and magnesium content).

Some detrimental changes are lower oxygen levels due to decreased re-aeration, decreased mixing, accumulation of pollutants, lack of a bottom scour produced by flowing water scrubbing a stream bottom, and increased growth of algae. Algal growth may be enhanced when suspended solids settle from impounded water, causing increased exposure of the algae to sunlight. Stagnant water in the bottom of a reservoir may be of low quality. Oxygen levels frequently go to almost zero near the bottom, and hydrogen sulfide is produced by the reduction of sulfur compounds in the low oxygen environment.

Insoluble iron (Fe) and manganese (Mn) species are reduced to soluble iron (Fe^{2+}) and manganese (Mn^{2+}) ions which must be removed prior to using the water. Estuaries constitute another type of body of water, consisting of arms of the ocean into which streams flow. The mixing of fresh and salt water gives estuaries unique chemical and biological properties. Estuaries are the breeding grounds of much marine life, which makes their preservation very important.

The unique temperature-density of water relationship results in the formation of distinct layers (stratification) within non-flowing bodies of water (Fig. 2.3). During the summer a surface layer (epilimnion) is heated by solar radiation and, because of its lower density, floats upon the bottom layer (hypolimnion) and is referred to as thermal stratification (Fig. 2.3). When an appreciable temperature difference exists between the layers, they do not mix but behave independently and have very different chemical and biological properties. The epilimnion, which is exposed to light, may have a heavy growth of algae. As a result of exposure to the atmosphere and (during daylight

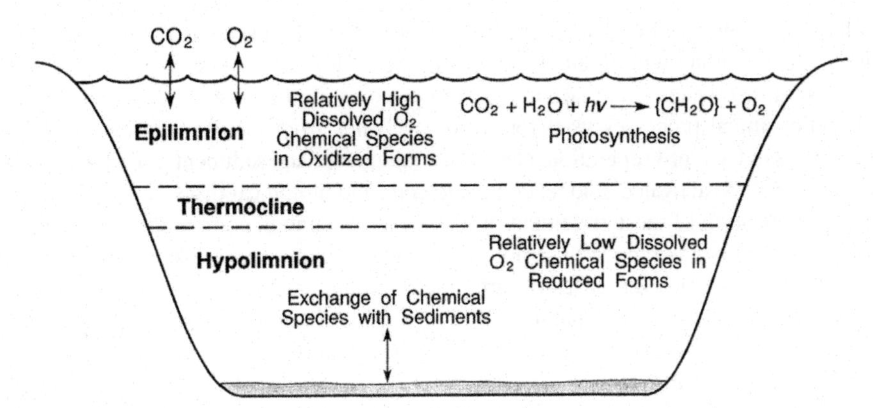

Fig. 2.3 Stratification of a lake caused by the unique temperature-density of water relationship.

hours) because of the photosynthetic activity of algae, the epilimnion contains relatively higher levels of dissolved oxygen and generally is aerobic. In the hypolimnion, bacterial action on biodegradable organic material may cause the water to become anaerobic. As a consequence, chemical species in a relatively reduced form tend to predominate in the hypolimnion.

The shear-plane, or layer between epilimnion and hypolimnion, is called the thermocline. During the autumn, when the epilimnion cools, a point is reached at which the temperatures of the epilimnion and hypolimnion are equal. This disappearance of thermal stratification causes the entire body of water to behave as a hydrological unit, and the resultant mixing is known as overturn, which generally occurs in the spring. During the overturn, the chemical and physical characteristics of the body of water become much more uniform, and a number of chemical, physical, and biological changes may result. Biological activity may increase from the mixing of nutrients. Changes in water composition during overturn may cause disruption in water treatment processes.

An increasing amount of attention has been given to thermal pollution, the raising of the temperature of a waterway by heat discharged from the cooling system or effluent wastes of an industrial installation. This rise in temperature may sufficiently upset the ecological balance of the waterway to pose a threat to the native life-forms. This problem has been especially noted in the vicinity of nuclear power plants. Thermal pollution may be combated by allowing wastewater to cool before emptying into the waterway. This is often done in large cooling towers.

In many cases, rivers, lakes, and oceans have become polluted form discharges of liquid wastes from residential, commercial, and industrial sources. Many of these bodies of water have been reclaimed because of the construction of new wastewater treatment facilities. Both physical-chemical and biological processes are used to remove organic matter from the liquid wastewater stream. It is this organic matter that causes a depletion of oxygen in rivers and lakes, with consequent anaerobic conditions leading to fish kills and noxious odors.

The various physical-chemical and biological processes remove an abundance of the organic matter, along with floatable scum and grease form the waste stream. Chemical disinfection inactivates bacteria, viruses, and protozoa in the waste stream. The physical and chemical processes used for removing solids from the waste stream include screening, sand and grit separation, chemical coagulation, and sedimentation. Biological processes include activated sludge, contact towers, and biological discs.

The application of these treatment processes to residential, commercial, and industrial waste waters has reduced the pollution load on many rivers, lakes, and harbors. There do remain, however, cases of large metropolitan areas that have not completed their cleanup campaigns, and in many other regions discharges of untreated mixtures of sanitary wastes and storm waters occur during heavy rains. These discharges emanate from large conduits that carry both sanitary and storm wastes. The impurities that run off the land during a storm mix with the sanitary wastes in the pipes, and the quantity of wastewater over taxes the carrying capacity of the conduits. A portion of this mixture flows into the nearest body of water before it can reach the wastewater treatment facility. Where wastewater must be discharged into lakes and dry streams, a higher level of treatment must be employed, including removal of nutrients and colloidal matter.

Metal ions in aqueous solution seek to reach a state of maximum stability through chemical reactions. Acid-base, precipitation, complex formation, and oxidation-reduction reactions all provide the means through which metal ions in water are transformed to more stable forms.

Hydrated metal ions, particularly those with a high positive charge tend to lose protons in aqueous solution, and fit the definition of Bronsted acids. Hydrated trivalent metal ions, such as iron (Fe^{3+}), generally are minus at least one hydrogen ion at neutral pH values or above. For tetravalent metal ions, the completely protonated forms are rare even at very low pH values. The tendency of hydrated metal ions to behave as acids may have a profound effect upon the aquatic environment. A good example is acidic mine water (the outflow of acidic water from metal mines or coal mines) which derives part of its acidic character from the character of hydrated iron.

Of the cations found in most freshwater systems, calcium (Ca^{2+})—a key element in many geochemical processes—generally has the highest concentration and minerals constitute the primary sources of calcium ions in waters. Among the primary contributing minerals are: gypsum ($CaSO_4.2H_2O$), anhydrite ($CaSO_4$), and dolomite ($CaCO_3.MgCO_3$), as well as calcite ($CaCO_3$) and aragonite ($CaCO_3$), which are different mineral forms (polymorphs) of calcium carbonate.

Water containing a high level of carbon dioxide readily dissolves calcium from its carbonate minerals. Calcium (Ca^{2+}) ions, along with magnesium (Mg^{2+}) ions and sometimes iron (Fe^{2+}) ions, account for water hardness. The most common manifestation of water hardness is the precipitate formed by soap in hard water. A number of other chemical species are present naturally, and are significant, in water (Table 2.3); some of these can also be pollutants. In particular, chelating agents are common potential water pollutants and occur in sewage effluent and industrial wastewater such as metal plating wastewater.

Chelating agents are complex-forming agents and have the ability to solubilize heavy metals. The formation of soluble complexes increases the leaching of heavy

Table 2.3 Chemical species commonly occurring in water

Chemical	Natural source	Other source
Sodium (Na^+)	Sea salt, soil dust	Biomass burning
Magnesium (Mg^+)	Sea salt, soil dust	Biomass burning
Potassium (K^+)	Sea salt, soil dust	Biomass burning, fertilizers
Calcium (Ca^+)	Sea salt, soil dust	Biomass burning, cement manufacture
Hydrogen (H^+)	Gas emissions	Fuel combustion
Chloride (Cl^-)	Sea salt	Industrial chemicals
Sulfate (SO_4^{2-})	Sea salt, soil dust, biological decay, volcanic activity	Biomass burning, fuel combustion
Nitrate (NO_3^{2-})	Lightning, biological decay	Biomass burning, fuel combustion, automobile emissions, fertilizers
Ammonium (NH_4+)	Biological decay	Fertilizers, biological decay
Phosphate (PO_4^{3-})	Soil dust, biological decay	Biomass burning, fertilizers
Bicarbonate (HCO_3^{2-})	Carbon dioxide in the air	Volcanic activity, biomass burning, automobile emissions, fuel combustion
SiO_2, Al, Fe	Soil dust	Fuel combustion

metals from waste disposal sites and reduces the efficiency with which heavy metals are removed with sludge in conventional biological waste treatment. Although chelating agents are never entirely specific for a particular metal ion, some complicated chelating agents of biological origin approach almost complete specificity for certain metal ions. One example of such a chelating agent is ferrichrome, synthesized by, and extracted from fungi, which forms extremely stable chelates with iron (Fe^3+). Ferrichrome is a cyclic hexa-peptide that forms a complex with iron atoms. It is composed of three glycine and three modified ornithine residues with hydroxamate groups [-N(OH)C(=O)C-]. The six oxygen atoms from the three hydroxamate groups bind the ferric iron (Fe^{3+}) in near perfect octahedral coordination.

Many organic compounds interact with suspended material and sediments in bodies of water. Settling of suspended material containing adsorbed organic matter carries organic compounds into the sediment of a stream or lake. This phenomenon is largely responsible for the presence of herbicides in sediments containing contaminated soil particles eroded from crop land. Some organic compounds are carried into sediments by the remains of organisms or by fecal pellets from the zooplankton that have accumulated organic contaminants. As might be anticipated, suspended particulate matter affects the mobility of organic compounds adsorbed on to particles. Furthermore, adsorbed organic matter undergoes chemical degradation and biodegradation at different rates and by different pathways compared to organic matter in solution. There is, of course, a vast variety of organic compounds that get into water by various (intentional and unintentional or circuitous) routes.

The most common types of sediments considered for their organic binding abilities are clays (Chapter 3), organic humic substances, and clay-humic complexes. Both

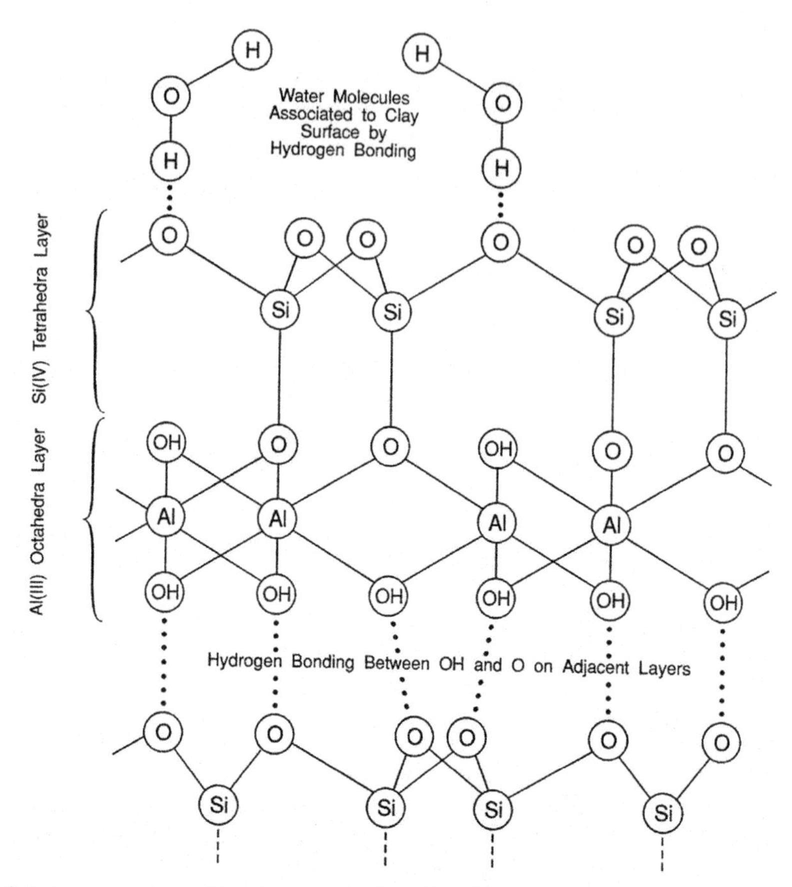

Fig. 2.4 Representation of the structure of clay minerals.

clays and humic substances act as cation exchangers. Therefore, these materials adsorb cationic organic compounds through ion exchange. This is a relatively strong sorption mechanism, greatly reducing the mobility and biological activity of the organic compound. When adsorbed by clays, cationic organic compounds are generally held between the layers of the clay mineral structure (Fig. 2.4) where the biological activity of the organic compounds is essentially zero.

4 Physical properties

Water is a tasteless, odorless liquid at ambient temperature and pressure and appears colorless. Ice also appears colorless, and water vapor is essentially invisible as a gas. Unlike other analogous hydrides of the oxygen family, water is primarily a liquid under standard conditions of temperature and pressure due to hydrogen bonding. The hydrogen bonds are continually breaking and reforming but these bonds are strong enough to create many of the peculiar properties of water. The strong cohesive interactions in water also result in: (i) a high viscosity, since for a liquid to flow interactions between neighboring molecules must constantly be broken, (ii) a high specific heat capacity—the ability to store a large amount of potential energy for a given increment in kinetic energy (temperature).

In part the high specific heat of water and high heat of vaporization of water relative to these properties shown by other liquids results from the small size of the water molecule. More intermolecular interactions are contained in a given volume of water than comparable liquids. When this is considered by expressing the specific heat and heat of vaporization on a molar basis, methanol and water are comparable. The surface tension of water, however, is still anomalously large after accounting for differences in size. Also, water has one of the highest dielectric constants of any nonmetallic liquid. It also has the remarkable properties of expanding when it is cooled to the freezing point, and again when it freezes. Both the expansion of water and its high dielectric constant reflect subtle structural features of liquid water at the molecular level.

The biological relevance of the physical properties of water, owing to its high boiling point, exists predominantly in its liquid form in the range of environments where life flourishes, although the other two phases, ice and vapor, play an essential role in shaping the environment. The high specific heat and heat of vaporization of water have important consequences for organisms at the cellular and physiological level, in particular for the efficiency of processes such as heat transfer, temperature regulation, and cooling. Viscosity is the major parameter of water that determines how fast molecules and ions can be transported and how rapidly they diffuse in aqueous solution. Thus, water provides a physical upper limit to the rates of many molecular level events, within which organisms must live and evolve—these include the rates of ion channel conductance, association of substrates with enzymes, binding rates, and rates of macromolecular assembly. It also sets an upper bound to the length scale over which biological processes can occur purely by diffusion. In many cases, for example in enzyme-substrate reactions, evolution has pushed the components of living systems to the limits set by the viscosity of water.

4.1 Boiling point, freezing point and melting point

The boiling point of a substance is the temperature at which the vapor pressure of the liquid equals the pressure surrounding the liquid and the liquids changes into a vapor. The boiling point depending upon the surrounding environmental pressure. For example, a liquid in a partial vacuum has a lower boiling point than when that liquid is at atmospheric pressure while that same liquid at high pressure has a higher boiling point than when that liquid is at atmospheric pressure. For example, water boils at 100 °C (212 °F) at sea level, but at 93.4 °C (200.1 °F) at 6250 ft above sea level. For a given pressure, different liquids will boil at different temperatures. The normal boiling point (also called the atmospheric boiling point or the atmospheric pressure boiling point) of a liquid is case in which the vapor pressure of the liquid is equal to the defined atmospheric pressure at sea level,

The normal boiling point of water is 100 °C (212 °F) because this is the temperature at which the vapor pressure of water is 760 mmHg, or 1 atm (14.7 psi). However, at 10,000 ft above sea level, the pressure of the atmosphere is only 526 mmHg and, at such as elevation, water boils when its vapor pressure is 526 mmHg, which occurs at a temperature of 90 °C (194 °F).

The freezing point is the temperature at which a liquid becomes a solid and an increase in pressure usually increases the freezing point. Although considered equal, the freezing point is lower than the melting point in the case of mixtures and for certain organic compounds. As a mixture freezes, the solid that forms first usually has a composition that is different from the composition of the liquid, and formation of the solid changes the composition of the remaining liquid, usually in a way that steadily lowers the freezing point. The heat of fusion is the heat that must be applied to melt a solid, must be removed from the liquid to freeze it. Some liquids can be supercooled (i.e. cooled below the freezing point) without solid crystals forming. Thus, because of the ability of some substances to supercool, the freezing point is not considered as a characteristic property of a substance. When the characteristic freezing point of a substance, such as water, is determined, in fact the actual methodology is almost always the principle of observing the disappearance rather than the formation of ice, that is, the melting point.

The melting point (sometime referred to as the liquefaction point) of a substance is the temperature at which the substance changes state from solid to liquid and at the melting point the solid phase and the liquid phase exist in equilibrium. The melting point of a substance depends on the surrounding pressure and is usually specified at a standard pressure, often atmospheric pressure. Thus, the melting point of water is the temperature at which it changes from solid ice into liquid water. The solid and liquid phase of water are in equilibrium at this temperature. The melting point depends slightly on pressure, so there is not a single temperature that can be considered to be the melting point of water. However, for practical purposes, the melting point of pure water ice at 1 atm of pressure is 0 °C (32 °F). The melting point and freezing point of water ideally are the same, especially if there are gas bubbles in water, but if the water is free of nucleating point, water can supercool to −42 °C (−43.6 °F) before freezing.

Because of the structure of water, there are often difference between water (the dihydride of oxygen) and other dihydrides of similar elements. For example, the boiling points of hydrides decrease as molecule size decreases (Table 2.1). So the hydride for tellurium: H_2Te (hydrogen telluride) has a boiling point of −4 °C (25 °F). Moving up, the next hydride would be H_2Se (hydrogen selenide) with a boiling point of −42 °C (108 °F). Also, hydrogen sulfide (H_2S) has a boiling point at −62 °C (−80 °F). So, despite its small molecular weight, water has an incredibly big boiling point. This is because water requires more energy to break its hydrogen bonds before it can then begin to boil. The same concept is applied to freezing point as well, as seen in the table below. The boiling and freezing points of water enable the molecules to be very slow to boil or freeze, this is important to the ecosystems living in water. If water was very easy to freeze or boil, drastic changes in the environment and so in oceans or lakes would cause all the organisms living in water to die.

The melting point of ice is 0 °C (32 °F) at standard pressure (760 mmHg, 1 atm., 14.7 psi) but, however, pure liquid water can be supercooled (the phenomenon when a liquid) below its freezing point without solidification or crystallization) well below that temperature without freezing if the liquid is not mechanically disturbed.

4.2 Compressibility

The compressibility (also known as the coefficient of compressibility or isothermal compressibility is a measure of the relative volume change of a fluid or solid as a response to the application of pressure and is expressed as:

$$\beta = (1/V)(\delta V/\delta p)$$

In this equation, where V is the volume and p is the pressure. The choice to define compressibility as the opposite of the fraction makes compressibility positive in the (usual) case that an increase in pressure induces a reduction in volume. It is also known as reciprocal of bulk modulus (k) of elasticity of a fluid. The compressibility of water is a function of pressure and temperature. The low compressibility of water means that even in the deep oceans at 2 miles depth, where pressures are on the order of 6000 psi, there is only a 1.8% decrease in volume. It is this low compressibility of water that leads to an incorrect assumption that water is incompressible.

4.3 Density of water and ice

The density of a chemical is the mass per unit value of the chemical and is characterized as the mass of the chemical divided by the volume of the chemical. Thus:

$$\rho = m/V$$

In this equation, ρ is the density, m is the mass, and V is the volume. For a pure substance the density has the same numerical value as the mass concentration. Different materials usually have different densities, and density may be relevant to purity.

Water has its maximum density at 4 °C (39 °F) a temperature above its freezing point (0 °C, 32 °F). The fortunate consequence of this fact is that ice floats, so that few large bodies of water ever freeze solid. Furthermore, the pattern of vertical circulation of water in lakes or the stratification of water in lakes (Fig. 2.3), a determining factor in the chemistry and biology of lake systems, is governed largely by the unique temperature-density relationship of water.

The density of water is about 1 g per cubic centimeter (62 pounds per cubic foot). The density varies with temperature, but not linearly: as the temperature increases, the density rises to a peak at 3.98 °C (39.16 °F) and then decreases. Regular, hexagonal ice is also less dense than liquid water—on freezing, the density of water decreases by about 9%. These effects are due to the reduction of thermal motion with cooling, which allows water molecules to form more hydrogen bonds that prevent the molecules from coming close to each other. However, at temperatures below 4 °C (39 °F) the breakage of hydrogen bonds due to heating allows water molecules to pack closer despite the increase in the thermal motion (which tends to expand a liquid), above 4 °C (39 °F) water expands as the temperature increases. Water near the boiling point is about 4% less dense than water at 4 °C (39 °F). Under increasing pressure, ice undergoes a number of transitions to other polymorphs—ppolymorphism is the ability of a solid material to exist in more than one form or crystal structure—and each of the polymorphs have a higher density than liquid water.

Under typical conditions, the lower density of ice than of water is vital to life—if water was most dense at the freezing point, in winter the very cold water at the surface of lakes and other water bodies would sink, the water system, i.e. the lake, would freeze from the bottom up, and all life in them would be killed. Furthermore, since water is a good thermal insulator (due to the high heat capacity), some frozen lakes might not completely thaw in summer. The layer of ice that floats on top insulates the water below—water at about 4 °C (39 °F) also sinks to the bottom, thus keeping the temperature of the water at the bottom constant (see diagram).

The density of salt water depends on the dissolved salt content as well as the temperature. Ice still floats in the oceans, otherwise they would freeze from the bottom up. However, the salt content of oceans lowers the freezing point by approximately 1.9 °C (3.4 °F) and lowers the temperature of the density maximum of water to the former freezing point at 0 °C (32 °F). Thus, in ocean water the downward convection of colder water is *not* blocked by an expansion of water as it becomes colder near the freezing point. The cold water of the ocean that is near the freezing point continues to sink.

As the surface of salt water begins to freeze (at approximately −2 °C, 28 °F) for normal salinity, i.e. 3.5% *w*/w salt) the ice that forms is essentially salt-free, with about the same density as freshwater ice. This ice floats on the surface, and the salt that is frozen out adds to the salinity and density of the sea water just below it (the brine rejection point). This denser salt water sinks by convection and the replacing seawater is subject to the same process. This produces essentially freshwater ice at approximately −2 °C (35 °F) on the surface. The increased density of the sea water beneath the forming ice causes it to sink towards the bottom. On a large scale, the process of brine rejection and sinking cold salty water results in ocean currents forming to transport such water away from the Polar Regions, leading to a global system of currents (the thermohaline circulation).

4.4 Miscibility and condensation

Miscibility occurs when two liquids with similar polarity (and, therefore, similar intermolecular interactions) are combined and the liquids mix to form a homogeneous solution.

Water is miscible with many liquids, including ethyl alcohol (ethanol, C_2H_5OH) in all proportions. On the other hand, water and most organic oils are immiscible and typically form layers according to increasing density from the top—for hydrocarbon oils, water is the bottom layer. More generally, on the basis of polarity, since water is a relatively polar compound, it will tend to be miscible with liquids of high polarity such as ethanol and acetone (CH_3COCH_3), whereas compounds with low polarity (such as hydrocarbon derivatives) will tend to be immiscible and poorly soluble in, or miscible with, water.

As a gas, water vapor is completely miscible with air. On the other hand, the maximum vapor pressure of water that is thermodynamically stable with the liquid (or solid) at a given temperature is relatively low compared with total atmospheric pressure. For example, if the partial pressure of the water vapor is 2% of atmospheric pressure and the air is cooled from 25 °C (77 °F), starting at approximately 22 °C (72 °F) water will start to condense, thereby defining the dew point (the temperature to which air must be cooled to become saturated with water vapor and, when further cooled, the airborne water vapor will condense to form liquid water), and creating fog or dew.

A saturated gas or one with 100% relative humidity occurs when the vapor pressure of water in the air is at equilibrium with vapor pressure due to (liquid) water; water (or ice, if cool enough) will fail to lose mass through evaporation when exposed to saturated air. Because the amount of water vapor in air is small, relative humidity, the ratio of the partial pressure due to the water vapor to the saturated partial vapor pressure, is much more useful. Vapor pressure above 100% relative humidity is called super-saturated and can occur if air is rapidly cooled, for example, by rising suddenly in an updraft.

4.5 Solid state, liquid state, and gaseous state

A state of matter is one of the distinct forms in which a substance can exist. Three states of matter are observable in everyday life: (i) solid, (ii) liquid, and (iii) gas. In some definitions, a fourth state—the plasma state—is added to the other three states (Table 2.4). Plasma, which is not considered by some observers to be a natural state of matter like the other three, can be artificially generated by heating or subjecting a neutral gas to a strong electromagnetic field to the point where an ionized gaseous substance becomes increasingly electrically conductive. Plasma and ionized gases have properties and display behavior unlike those of the other states, and the transition and is mostly a matter of nomenclature and subject to interpretation.

All substances, including water, become less dense when they are heated and denser when they are cooled and, thus, when water is cooled, it becomes denser and forms ice. Water is one of the few substances whose solid state can float on its liquid state because water continues to become denser until it reaches 4 °C (39 °F) and, after it reaches this temperature, water becomes less dense. When freezing, molecules within water begin to move around more slowly, making it easier for them to form hydrogen bonds and eventually arrange themselves into an open crystalline, hexagonal structure.

Table 2.4 Interrelationship of the various states of matter of water

Original state	Process	Product state
Gas	Condensation	liquid
Gas	Deposition	solid
Gas	Ionization	plasma
Liquid	Vaporization	gas
Liquid	Freezing	solid
Solid	Sublimation	gas
Solid	Melting	liquid
Plasma	Recombination	gas

Because of this open structure as the water molecules are being held further apart, the volume of water increases approximately 9%. Thus, the molecules are more tightly packed when water is in the liquid state than when water is in the solid state (ice).

It is very rare to find a compound that lacks carbon to be a liquid at standard temperatures and pressures and, this, it is unusual for water to be a liquid at room temperature! Water is liquid at room temperature and is more mobile quicker than ice (the solid state of water) thereby enabling the molecules to form fewer hydrogen bonds resulting in the molecules being packed more closely together. Each water molecule links to four others creating a tetrahedral arrangement, however they are able to move freely and slide past each other, while ice forms a solid, larger hexagonal structure.

As water boils, its hydrogen bonds are broken. On a molecular basis, steam particles move very far apart rapidly so any hydrogen bonds do not have the time to form. As a result, less and less hydrogen bonds are present as the particles reach the critical point above steam. The lack of hydrogen bonds explains why steam causes much worse burns that water. Steam contains all the energy used to break the hydrogen bonds in water. Then, in an exothermic reaction, steam is converted into liquid water and heat is released.

Water also forms a supercritical fluid and the critical temperature 374 °C (705 °F) and the critical pressure is 3190 psi. Briefly, the critical temperature is the temperature at and above which vapor of the substance cannot be liquefied, no matter how much pressure is applied. The critical pressure of a substance is the pressure required to liquefy a gas at its critical temperature.

In nature, critical conditions occurs only rarely, and in extremely hostile conditions. A likely example of naturally occurring supercritical water is in the hottest parts of deep water hydrothermal vents, in which water is heated to the critical temperature by volcanic plumes and the critical pressure is caused by the weight of the ocean at the extreme depths where the vents are located. This pressure is reached at a depth on the order of 7000 ft.

Water has a very high specific heat capacity—the second highest among all the heteroatomic species (after ammonia, NH_3), as well as a high heat of vaporization, both of which are a result of the extensive hydrogen bonding between the individual molecules. These two unusual properties allow water to moderate the climate of the Earth by buffering large fluctuations in temperature.

4.6 Specific heat

The specific heat of a substance is the amount of energy required to raise the temperature of water by one degree Celsius and water has a relatively large specific heat. The calorie is defined as the amount of heat energy required to raise the temperature of 1 g of water at 4 °C by 1 °C.

The specific heat of water is five times greater than the specific heat of sand which explains why beach sand may quickly warm to the point that it is too hot to stand on while ocean water warms only a little. Because so much heat loss or heat input is required to lower or raise the temperature of water, the oceans and other large bodies of water have relatively constant temperatures. Thus, many organisms living in the oceans are provided with a relatively constant environmental temperature. The high water content of plants and animals living on land helps them to maintain a relatively constant internal temperature. During the evening the temperature of sand will decrease while the temperature of the ocean remains relatively constant.

4.7 Surface tension

The surface tension of a fluid is the elastic tendency of the fluid surface which makes the fluid acquire the least surface area possible. Surface tension is an important factor in the phenomenon of capillarity. At liquid-air interfaces, surface tension results from the greater attraction of liquid molecules to each other (due to cohesion) than to the molecules in the air (due to adhesion). The overall effect is an inward force at the surface of the fluid that causes the liquid to behave as if its surface was covered by a stretched elastic membrane.

Water has an unusually high surface tension because of the relatively high attraction of water molecules to each other through a web of hydrogen bonds, water has a higher surface tension than most other liquids. Besides mercury, water has the highest surface tension for all liquids, which is due to the hydrogen bonding in water molecules. The surface tension of water causes water molecules at the surface of the liquid (in contact with air) to hold closely together, forming an invisible film. Surface tension is essential for the transfer of energy from wind to water to create waves. Waves are necessary for rapid oxygen diffusion in lakes and seas. Next to mercury, water has the highest surface tension of all commonly occurring liquids.

The high surface tension of water is relevant at two levels. First, below a length scale of about 1 mm surface tension forces dominate gravitational and viscous forces, and the air-water interface becomes an effectively impenetrable barrier. This becomes a major factor in the environment and life style of small insects, bacteria and other microorganisms. Second, at the molecular (0.1 to 100 nm; a nanometer is one billionth of a meter) scale the surface tension plays a key role in the solvent properties of water. The high dielectric constant of water also plays an important role in its action as a solvent. The biological significance of the expansion of water upon cooling and upon freezing, though crucial, is largely indirect through geophysical aspects such as ocean and lake freezing, the formation of the polar ice cap, and in weathering by freeze-thaw cycles.

4.8 Thermal conductivity

The thermal conductivity (k) of a substance is the rate at which heat passes through a specified material, expressed as the amount of heat that flows per unit time through a unit area with a temperature gradient of one degree per unit distance. The unit for k is watts (W) per meter (m) per degree kelvin (K). Typically metals have a high conductivity whereas other materials have a much lower thermal conductivity. As an illustration, water has a thermal conductivity of 0.6089 at 26.85 °C whereas copper has a thermal conductivity of 384.1 at 18.05 °C. Furthermore, when a material undergoes a phase change (e.g. from solid to liquid), the thermal conductivity may change abruptly. For example, when ice melts to form liquid water at 0 °C (32 °F), the thermal conductivity changes from 2.18 W/(m·K) to 0.56 W/(m·K).

If the material is a poor conductor of heat, this property is referred to as thermal resistance, or *R*-value, which describes the rate at which heat is transmitted through the material. The *R*-value is given in units of square feet times degrees Fahrenheit times hours per British thermal unit (ft^2 °F h/Btu) for a 1-in.-thick slab.

4.9 Triple point

The triple point of a substance is the temperature and pressure at which the three phases (gas, liquid, and solid) of that substance coexist in thermodynamic equilbrium. It is also the temperature and pressure at which the sublimation curve, the fusion curve, and the vaporization curve meet. The triple point of water was used to define the kelvin, which is base unit of thermodynamic temperature.

The single combination of pressure and temperature at which liquid water, solid ice, and water vapor can coexist in a stable equilibrium occurs at exactly 273.1600 K (0.0100 °C, 32.0180 °F) and a partial vapor pressure of 0.00603659 atm). At that point, it is possible to change all of the substance to ice, water, or vapor by making arbitrarily small changes in pressure and temperature. Even if the total pressure of a system is well above the triple point of water. Thus, the gas-liquid-solid triple point of water corresponds to the minimum pressure at which liquid water can exist. At pressures below the triple point (as in outer space), solid ice when heated at constant pressure is converted directly into water vapor without the intervention of the liquid phase (sublimation). Above the triple point, solid ice when heated at constant pressure first melts to form liquid water, and then evaporates or boils to form vapor at a higher temperature.

Thus, in the case of water, the temperature and pressure at which ordinary gaseous, liquid, and solid water coexist in equilibrium. Due to the existence of many polymorphs (forms) of ice, water has other triple points (Table 2.5), which have either three polymorphs of ice or two polymorphs of ice and liquid in equilibrium.

Table 2.5 The various triple points of water

Phases in stable equilibrium	Pressure	Temperature
Liquid water, ice, and water vapor	611.657 Pa[47]	273.16 K (0.01 °C)
Liquid water, ice, and ice III	209.9 MPa	251 K (−22 °C)
Liquid water, ice III, and ice V	350.1 MPa	−17.0 °C
Liquid water, ice V, and ice VI	632.4 MPa	0.16 °C
Ice, ice II, and ice III	213 MPa	−35 °C
Ice II, ice III, and ice V	344 MPa	−24 °C
Ice II, ice V, and ice VI	626 MPa	−70 °C

Ice II, Ice III, Ice V, and Ice VI are the various polymorphs (crystal forms) of ice.

4.10 Vapor pressure and heat of vaporization

Vapor pressure (or equilibrium vapor pressure) is the pressure exerted by a vapor in thermodynamic equilibrium with the condensed phases (solid or liquid) at a given temperature in a closed system. The equilibrium vapor pressure is an indication of the evaporation rate of the liquid and relates to the tendency of particles to escape from the liquid (or a solid). A substance with a high vapor pressure at normal temperatures is often referred to as volatile. As the temperature of a liquid increases, the kinetic energy of its molecules also increases and as the kinetic energy of the molecules increases, the number of molecules transitioning into a vapor also increases, thereby increasing the vapor pressure. When a sample of a liquid is introduced into a container, the liquid will tend to evaporate. Molecules will escape from the relative confinement of the liquid state into the gaseous state. If the container is closed, this conversion will appear to stop when equilibrium is achieved. Under equilibrium conditions, the rate of evaporation equals the rate of condensation.

A measure of the extent of vaporization is the vapor pressure which is the partial pressure exerted by the gas phase in equilibrium with the liquid phase. Other gases, such as air, can be present, but what matters here is the pressure of the substance involved in the gas-liquid equilibrium. Vapor pressure can be expressed in any convenient unit such as: mm mercury, atmospheres, bars, psi, Pascals, and Kilopascals. The higher the vapor pressure of a substance, the greater the concentration of the compound in the gaseous phase and the greater the extent of vaporization. Liquids vary considerably in their vapor pressures. If two substances are compared at the same temperature, the more volatile one will have the higher vapor pressure.

The vapor pressure of water is the pressure at which water vapor is in thermodynamic equilibrium with the condensed state. At higher pressures water would condense. The water vapor pressure is the partial pressure of water vapor in any gas mixture in equilibrium with solid or liquid water. Like all liquids, water boils when the vapor pressure is equal to the surrounding pressure. In nature, the atmospheric pressure is lower at higher elevations and water boils at a lower temperature—for example at 7000 ft above sea level, water boils at 92 °C (198 °F) instead of 100 °C (212 °F).

The heat of vaporization is the amount of heat needed to turn one gram of a liquid into a vapor, without a rise in the temperature of the liquid. The heat of vaporization is a latent heat which is the additional heat required to change the state of a substance from solid to liquid at its melting point, or from liquid to gas at its boiling point, after the temperature of the substance has reached either of these points. Note that a latent heat is associated with no change in temperature, but a change of state. Because of the high heat of vaporization, evaporation of water has a pronounced cooling effect and condensation has a warming effect.

The heat of vaporization of water is the highest known—as the molecules evaporate, the surface they evaporate from gets cooler (evaporative cooling) because the molecules with the highest kinetic energy are lost to evaporation. The process of evaporation in a closed container will proceed until there are as many molecules returning to the liquid as there are escaping. At this point the vapor is said to be saturated, and the pressure of that vapor (usually expressed in mm mercury) is the saturated vapor pressure.

4.11 Viscosity

The viscosity of a fluid is the property of fluid that is a measure of the resistance to flow by the fluid. Liquids with stronger intermolecular interactions are usually more viscous than liquids with weak intermolecular interactions. Related to viscosity, cohesion is intermolecular forces between like molecules; this is why water molecules are able to hold themselves together in a drop. Water molecules are very cohesive because of the polarity of the molecule. Thus, water molecules stay in close proximity to each other (cohesion), due to the collective action of hydrogen bonds between water molecules. These hydrogen bonds are constantly breaking and new bonds being formed with different water molecules; but at any given time in a sample of liquid water, a large portion of the molecules are held together by such bonds.

Water also has high adhesion properties because of its polar nature. On extremely clean/smooth glass, the water may form a thin film because the molecular forces between glass and water molecules (adhesive forces) are stronger than the cohesive forces. In order to dehydrate hydrophilic surfaces (i.e. to remove the strongly held layers of water of hydration) requires doing substantial work against these forces, called hydration forces. These forces are very large but decrease rapidly over a nanometer or less. They are important in biology, particularly when cells are dehydrated by exposure to dry atmospheres or to extracellular freezing.

Most plants have adapted to take advantage of the adhesion properties of water that helps move water from the roots to the leaves (often referred to as capillary action). As an example, in the redwood tree (one of the tallest plants on Earth) water moves from the roots of the tree to the leaves, which can be a distance of two hundred feet or more. As a plant loses water through pores in the leaves, more water moves up from roots and stems to replace the lost water—the process of water loss by leaves is known as *transpiration.*

5 Chemical properties

Water—a polar molecule—tends to be slightly positive on the hydrogen side and slight negative on the oxygen side (Fig. 2.2). This polarity of the water molecule allows it to form an electrostatic bond (a hydrogen bond) between the positive hydrogen side of the water molecule and other negative ions or polar molecules. Molecules and ions with which water forms hydrogen bonds (such as sodium chloride) are hydrophilic while, on the other hand, ions and molecules that do not form hydrogen bonds with water are hydrophobic. When ionic compounds such as sodium chloride are added to water, hydrogen bonding will tend to pull those ionic compounds apart and dissolve as ions (e.g. Na+and Cl^-). Once ionic compounds dissolve, the anions and cations circulate through the water allowing further reactions to occur. Thus, at standard conditions, water is a polar liquid that slightly dissociates or self ionizes into a hydronium ion (H_3O^+) and a hydroxide ion (OH^-):

$$2H_2O \rightleftharpoons H_3O^+ + OH^-$$

It is the behavior of water in chemical systems that gives it the name of a *universal solvent* since it can solve more substances than any other liquid. This is why the waters of lakes, rivers, seas, and oceans that contain in fact a huge number of solved elements and minerals released by rocks or by the atmosphere. Wherever it flows, water dissolves and carries an extremely high amount of substances and thus performs an important task: that of carrying, sometimes to long distances, the substances it encounters along its way.

5.1 Geochemistry

The action of water on rock over long periods of time typically leads to weathering and erosion, which are physical processes that convert solid rocks and minerals into soil and sediment. However, under some conditions chemical reactions with water occur as well, resulting in metasomatism (chemical alteration of a rock by hydrothermal and other fluids) or mineral hydration which is a type of chemical alteration of a rock which produces clay minerals. Thus, water systems are essentially solutions of soluble mineral salts and the chemical character of the water, like that of solutions in general, depends on the nature and proportion of the mineral they contain. The interpretation of the chemical character of water from any water system may be, subject to the analytical test methods used, uncertain and unsatisfactory if it is based merely on the amounts of the soluble species determined to be in the water.

All minerals and, consequently, all rocks are soluble to some extent in water, but the solubility of a mineral (i.e. the quantity of dissolved minerals in water) depends on physical variables (such as temperature and pressure) but dissolution of a mineral is not an instantaneous process. In fact, the mineral content of a water system results from three factors: (i) the type of rock, for example, the mineralogy and grain size in contact with the flowing water, (ii) the climate and environmental conditions which determining flow rate, the temperature, and the pressure, and (iii) the flow conditions, which determine the time of contact between water and rock. The first factor imparts the main geochemical characteristics of water. For example, the total dissolved solids (TDS) is strongly related to the aquifer rock geochemistry—salt rocks such as halite and gypsum will give rise to a high values of the total dissolved solids while water flowing through granitic rocks or basalts or through quartz sands or sandstones will have lower values for the total dissolved solids. The second and third factors are particularly responsible for the solution rate of rock, depending on flow rate, and for ion content, depending on residence time and chemical content. Waters with a very long residence time, such as mineral and thermal groundwaters, show higher ion concentrations than surface waters or shallow groundwaters, when in contact with the same rocks (Bakalowicz, 1994).

Furthermore, in groundwater systems in which residence time (i.e. the water-rock contact time) is short, changes of flow conditions are typically responsible for changes in chemical content. On the other hand, waters of deep or thermal aquifers do not generally show any seasonal variation in their chemical content, except when they mix with shallow waters. Also, a part of the dissolved solids may originate outside of the aquifer, having been supplied by rainwater and air dust. In industrial areas, air pollution will enrich rainwater in various acidic species which are responsible for acid rain (Fig. 2.5) and in other chemicals such as heavy metals (Pb) and organic compounds (hydrocarbon

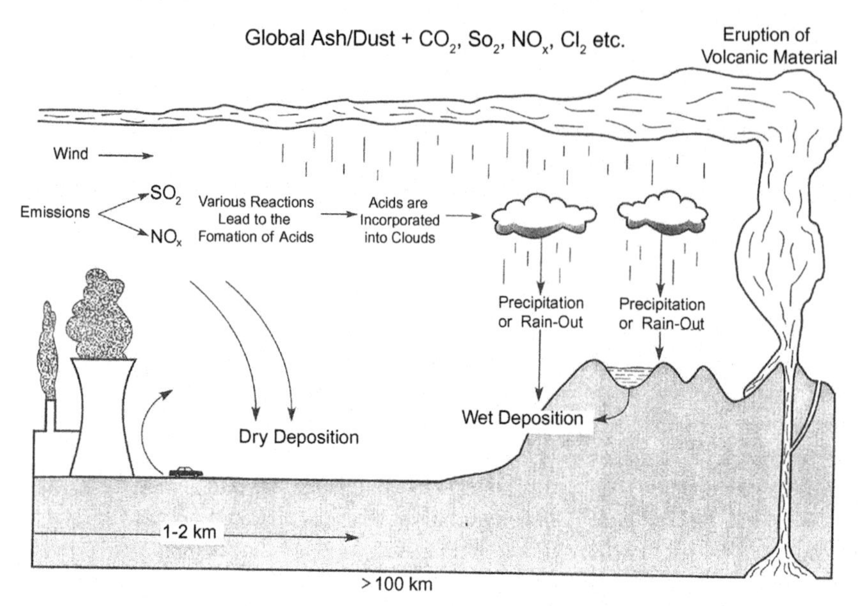

Fig. 2.5 General illustration of acid rain formation and deposition.

derivatives). Human land use activities may introduce additional pollutants into water systems. For example, domestic wastewaters may concentrate sodium chloride, potassium chloride, and borate (from washing powder) and also ammonia, nitrate (from fertilizers). These salts are biologically active and, depending on the oxygen content of groundwater, may react in accordance with redox gradients (Table 2.6).

Table 2.6 Redox potential of various aqueous solutions—anoxic waters may have negative redox potentials

Aqueous solution	Redox potential, mV	Aqueous solution	Redox potential, mV
Electrolytic catholyte (H_2, alkaline)	−600 ~ −650	Deep well water	0
Water associated with oil deposits	−500	Degassed pure water	+200
Organic-rich saline	−400	Distilled water	+250
Euxinic water[a] (H_2S)	−250	Groundwater	+250
Anaerobic yeast fermentation	−180	Mineral water	+200 to ca. +400
Anaerobic water-logged soil	−100	Tap water	+220 to ca. +380
Deep seawater (ca. 7000 ft depth)	ca. +430	Surface seawater	ca. +400
Rainwater	+600	Electrolytic anolyte (O_2, acidic)	+600

[a] Water is both anoxic and sulfidic with a raised level of hydrogen sulfide.

Another aspect of water geochemistry is that ice can participate in the formation of hydrocarbon hydrates (often referred to as clathrate compounds or clathrate hydrates) with a variety of low molecular weight hydrocarbon derivatives, such as methane. In these compounds, the hydrocarbon is embedded in a spacious ice crystal lattice. The most notable of these is methane clathrate ($4CH_4.23H2O$) that naturally found in large quantities on the ocean floor (Speight, 2019).

5.2 Ligand chemistry

In coordination chemistry, a ligand is an ion or molecule that binds to a central metal atom to form a coordination complex. The bonding with the metal generally involves formal donation of one or more of the electron pair(s) of the ligand. The nature of metal-ligand bonding can range from covalent to ionic and the metal-ligand bond order can range from one to three. In general, ligands are viewed as electron donors and the metals as electron acceptors and while ligands are considered to be Lewis bases, cases are known to involve Lewis acids as ligands.

The Lewis base character of water makes it a common ligand in transition metal complexes. Water is typically a monodentate ligand, insofar as it forms only one bond with the central atom. Examples of these complexes of which include: $Fe(H_2O)^{2+}{}_6$ and perrhenic acid $[Re_2O_7(OH_2)_2]$ which contains two water molecules coordinated to a rhenium center. In solid hydrate derivatives, water can be either a ligand or simply lodged in the framework, or both.

Metal aquo complexes are coordination compounds containing metal ions with only water as a ligand (Ogden and Beer, 2006). These complexes are the predominant species in aqueous solutions of many metal salts, such as metal nitrates, sulfates, and perchlorates. They have the general stoichiometry $[M(H_2O)_n]^{z+}$. The Most common complexes are the octahedral complexes with the formula $[M(H_2O)_6]^{2+}$ and $[M(H_2O)_6]^{3+}$:

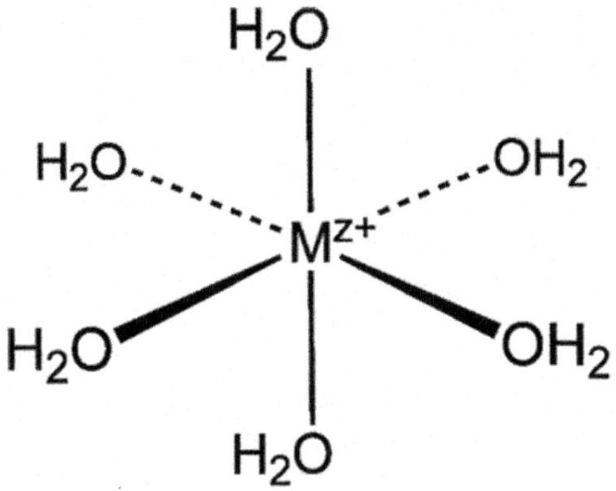

H_2O

H_2O OH_2

M^{z+}

H_2O OH_2

H_2O

The behavior of the metal aquo compounds is an important aspect of environmental, biological, and industrial chemistry. For example, ligand exchange involve replacement of a water ligand (coordinated water) with water in solution (bulk water), which is shown (for clarification) in the equation below as (*H_2O*):

$$\left[M(H_2O)_n\right]^{z+} + (H_2O) \rightarrow \left[M(H_2O)_{n-1}(H_2O)\right]^{z+} + H_2O$$

The rates of such reactions vary considerably and the main factor affecting the reaction rate is affected by two factors: (i) highly charged metal aquo cations exchange their water more slowly than singly charged species and (ii) electron configuration. Water exchange usually follows a dissociative substitution pathway so the rate constants indicate first order reactions.

The electron exchange reaction usually applies to the interconversion of di- and trivalent metal ions, which involves the exchange of only one electron. In the process (*self-exchange*) the ion *appears* to exchange electrons with itself.

$$\left[M(H_2O)_6\right]^{2+} + \left[M(H_2O)_6\right]^{3+} \rightleftharpoons \left[M(H_2O)_6\right]^{3+} + \left[M(H_2O)_6\right]^{2+}$$

The rates of electron exchange reactions vary widely, the variations being attributable to differing reorganization energies: when the divalent (2+) and trivalent (3+) ions differ widely in structure, the rates tend to be slow.

Solutions of metal aquo complexes are acidic owing to the ionization of protons from the water ligands:

$$\left[Cr(H_2O)_6\right]^{3+} \rightleftharpoons \left[Cr(H_2O)_5(OH)\right]^{2+} + H^+$$

The aquo ion is a weak acid, of comparable strength to acetic acid. In more concentrated solutions, some metal hydroxo complexes undergo condensation reactions to form polymeric species. The hydrolyzed species often exhibit very different properties from the precursor hexa-aquo complex.

Relevant to homogeneous water oxidation catalysis, some metal complexes catalyze the oxidation of water:

$$2H_2O \rightarrow O_2 + 4H^+ + 4e^-$$

5.3 Organic chemistry

Water has emerged as a versatile solvent for organic chemistry and is not only environmentally benign, but also gives completely new reactivity (li and Chen, 2006). The types of organic reactions in water include (i) reactions of carbanions, (ii) reactions of carbocations, and (iii) reactions of radicals. In fact, aqueous organic reactions have broad applications such as synthesis of biological compounds from carbohydrate derivatives and the chemical modification of biomolecules. Also, *on water reactions* are a group of organic reactions that take place as an emulsion in water and that exhibit an unusual acceleration of the reaction rate compared to the same reaction in an organic solvent or compared to the corresponding reaction in dry media. A dry media reaction (also called a solid-state reaction or a solvent-less reaction) is a chemical reaction system in the absence of a solvent. Thus, water can often be a viable and green solvent in organic chemical reactions.

As a base, water reacts readily with organic carbocations; for example in a hydration reaction, a hydroxyl group (OH^-) and an acidic proton (H^+) are added to the two carbon atoms bonded together in the carbon-carbon double bond, resulting in an alcohol. When addition of water to an organic molecule cleaves the molecule in two, hydrolysis occurs.

Organic reactions that occur at the water interface for water-insoluble compounds, and reactions in water solution for water soluble compounds, has added an option for organic synthesis under more beneficial economic and environmental conditions. Many organic molecules are partially soluble in water and reactions that appear as heterogeneous mixtures and suspensions may involve on-water and in-water reaction modes occurring simultaneously. The on-water catalytic effect, relative to neat reactions or organic solvents, depends on the properties of reactant compounds. In some cases when on-water reactions produce quantitative yields of water-insoluble products they can reach ideal synthetic aspirations.

6 Electrical properties

The behavior and movement of the electrons in water, a V-shaped molecule composed of two hydrogen atoms and an oxygen atom (Fig. 2.2), are vital to understanding the behavior of water. Chemically, the simple approach is to assign a positive charge to a hydrogen atom and a negative charge to an oxygen atom (Fig. 2.2b) as an aid to understanding the behavior.

Pure water, which does not contain ions from any external source, is an excellent insulator. However, as a consequence of the structure of water, which is a polar molecule (Fig. 2.2), there is a difference in electronegativity and a bond dipole moment points from each hydrogen atom to the oxygen atom rendering the oxygen partially negative and each hydrogen partially positive. The charge differences cause water molecules to aggregate (the relatively positive areas being attracted to the relatively negative areas).

Not surprisingly, water undergoes auto-ionization in the liquid state, when two water molecules form one hydroxide anion (OH^-) and one hydronium cation (H_3O^+).

$$2H_2O \rightleftharpoons H_3O^+ + OH^-$$

However, because water is such a good solvent, it almost always has some solute dissolved in it, often a salt which produced positively-charged and negatively-charged ions which can carry charges back and forth, thereby allowing the water to conduct electricity far more readily.

6.1 *Electrolysis*

Electrolysis is a technique that uses a direct electric current (DC) to drive an otherwise non-spontaneous chemical reaction. The electrolysis of water is the decomposition of water into oxygen and hydrogen by the passage of an electric current. In the process, a DC electrical power source is connected to two electrodes, or two plates (typically made from some inert metal such as platinum, stainless steel, or iridium) which are placed in the water.

Water can be split into its constituent elements, hydrogen and oxygen, by passing an electric current through it. In pure water at the negatively charged cathode, a reduction reaction takes place, with electrons (e^-) from the cathode being donated to hydrogen cations to form hydrogen gas. Conversely, at the positively charged anode, an oxidation, reaction which generates oxygen gas and giving electrons to the anode to complete the equation (energy) balance:

$$2H^+(aq) + 2e^- \rightarrow H_2(g)(\text{reduction at the cathode})$$

$$2H_2O(1) \rightarrow O_2(g) + 4H^+(aq) + 4e^-(\text{oxidation at the anode})$$

$$\text{Overall reaction}: 2H_2O(l) \rightarrow 2H_2(g) + O_2(g)(\text{overall reaction})$$

The acid-balanced reactions (above) predominate in acidic (low pH) solutions, while the base-balanced reactions predominate in basic (high pH) solutions. Thus:

$$2H_2O(1) + 2e^- \rightarrow H_2(g) + 2OH^-(aq)(\text{cathode reduction})$$

$$2OH^-(aq) \rightarrow 1/2O_2(g) + H_2O(1) + 2e^-(\text{anode oxidation})$$

The number of hydrogen molecules produced is thus twice the number of oxygen molecules and, assuming equal temperature and pressure for both gases, the volume of the hydrogen gas produced is twice volume of the oxygen gas produced. The gases bubble to the surface, where they can be collected.

6.2 Electrical conductivity

In the current context, the electrical conductivity of a substance is the ability or power of the substance to conduct electricity. More specifically, the electrical conductivity is defined as the ratio between the current density (J) and the electric field intensity (e) and it is the opposite (or reciprocal_ of the resistivity (r):

$$s = J/e = 1/r$$

The commonly used units for measuring electrical conductivity of water are: μS/cm (micro-Siemens/cm), or dS/m (deci-Siemens/m), where 1000 μs/cm = 1 dS/m.

An electrical current results from the motion of electrically charged particles in response to forces that act on them from an applied electric field. Within most solid materials a current arise from the flow of electrons (electronic conduction) and, in all conductors, semiconductors, and many insulated materials only electronic conduction exists, and the electrical conductivity is strongly dependant on the number of electrons available to participate to the conduction process.

In the current context, the electrical conductivity is a measure of the capability of water to pass electrical flow, which is directly related to the concentration of ions in the water. These conductive ions come from dissolved salts and inorganic materials such as alkalis, chlorides, sulfides and carbonate compounds. Distilled or deionized water can act as an insulator due to its very low (if not negligible) conductivity value. Sea water, on the other hand, has a very high conductivity. Pure water is not a good conductor of electricity and, because the electrical current is transported by the ions in solution, the conductivity increases as the concentration of ions increases. As examples:

Water type	**Conductivity**
Ultra-pure water	5.5×10^{-6} Siemens per meter
Drinking water	0.005–0.05 Siemens per meter
Sea water	5 Siemens per meter

The total dissolved solids (TDS) is a measure of the total ions in solution and, as the solution becomes more concentrated (TDS >1000 mg/L), the proximity of the solution ions to each other depresses their activity and consequently their ability to transmit current, although the physical amount of dissolved solids is not affected. With the reverse osmosis process, water is forced through a semi-impermeable membrane leaving the impurities behind. This process is capable of removing 95–99% *w*/w of the total dissolved solids thereby providing pure water.

The specific conductance is a conductivity measurement made at or corrected to 25 °C (77 °F) and is the standardized method of reporting conductivity. Since the temperature of water will affect conductivity readings, reporting conductivity at 25 °C (77 °F) allows direct comparison of the data (usually reported in micro-Siemens/cm at 25 °C, 77 °F). If a conductivity measurement is made at 25 °C (77 °F), it can be reported as the specific conductance. If a measurement is made at a different temperature and corrected to 25 °C (77 °F), the temperature coefficient must be considered. The specific conductance temperature coefficient can range depending on the measured temperature and ionic composition of the water.

Resistivity is a measurement of the opposition of water to the flow of a current over distance. The resistivity decreases as the ionic concentration in water increases. As already stated, resistivity and conductivity are reciprocals:

$$\text{Resistivity} = 1/\text{conductivity}$$

$$\text{Conductivity} = 1/\text{resistivity}$$

6.3 Dielectric constant

The dielectric constant (symbol ε, also known as the relative permittivity) is a measure of the ease with which a material is polarized by an electric field relative to vacuum. It is defined by the magnitude of the dielectric polarization (dipole moment per unit volume) induced by a unit field. Dielectric constants have no units - they are coefficients that multiply the capacitance with free space (or vacuum) as dielectric.

Water is an order of magnitude more polarizable than most organic solvents. The dielectric constant of a polar liquid such as water depends on four major factors: (i) the permanent dipole moment of the molecule, (ii) the density of dipoles, (iii) the ease with which the dipoles can reorient in response to a field, and (iv) the efficiency of the reorientation. The abnormally high dielectric constant of water is often explained as being due to a large liquid phase dipole moment.

Water has a high dipole moment, it is small so there are a large number of dipoles per unit volume, and in the liquid state they are easily and rapidly (within 10 ps) reoriented. In addition, because water is extensively H bonded, the polarization response is cooperative: water molecules cannot simply reorient independently of their neighbors. They effectively reorient in groups of about three. Finally, there is a small contribution to the dielectric constant (c. 2–3) from the polarizability and flexibility of water. All these factors explain the very high dielectric constant of water. Decreasing the temperature increases the dielectric constant since it reduces the randomizing thermal fluctuations that oppose dipole alignment by an electrostatic field.

6.4 Ionization

Ionization is the process by which an atom or a molecule acquires a negative or positive charge by gaining or losing electrons to form ions (cations and anions), often in conjunction with other chemical changes.

Cations are ions with a net positive charge—examples are silver: Ag^+, hydronium: H_3O^+, and ammonium: NH_4^+—while anions are ions with a net negative charge—examples are: hydroxide anion: OH^-, oxide anion: O^{2-}, and sulfate anion: SO_4^{2-}. Because they have opposite electrical charges, cations and anions are attracted to each other. Cations repel other cations, while anions repel other anions.

Usually in a molecule if atoms are bonded with high electronegative difference the bonded pair of electrons will be unequally shared between two atoms which are bonded together. In that the atom which is comparatively more electronegative will attract the bonded pair of electrons towards itself and the atoms which is less electronegative attracts less bonded pairs towards itself. There will be a separation of charges between the atoms in which bonding takes place.

Because the hydrogen-oxygen bond of water is strongly polarized, the electron density around the hydrogen atom is very low and the hydrogen-oxygen bond is rather weak compared with most covalent bonds. Thermal fluctuations in the liquid often result in sufficient further polarization of the hydrogen-oxygen bond that the hydrogen nucleus can dissociate as a proton, or H^+ ion.

Remarkably, the mobility of a proton in ice is higher still, clearly demonstrating that proton transport occurs not so much by movement of a single proton, but by a hopping mechanism between hydrogen-bonded water molecules, whereby a water molecule accepts a proton on one side, and releases a proton on the other side.

The unique ability of water to ionize easily and to solvate the ions allows it to partake in proton exchange reactions with many polar solutes. Furthermore, acid-base and proton exchange reactions are pervasive in the environment due to the facile ionization of water and the high proton mobility.

References

Bakalowicz, M., 1994. Water geochemistry: water quality and dynamics. In: Groundwater Ecology. Academic Press, New York.

Bartels-Rausch, T., Bergeron, V., Cartwright, J.H.E., Escribano, R., Finney, J.L., Grothe, H., Gutirrez, P.J., Haapala, J., Kuhs, W.F., Pettersson, J.B.C., Price, S.D., Sainz-Daz, C.I., Stokes, D.J., Strazzulla, G., Thomson, E.S., Trinks, H., Uras-Aytemiz, N., 2012. Ice structures, patterns, and processes: a view across the icefields. Review Mod. Phys. 84, 885–944.

Dodson, S.I., 2005. Introduction to Limnology. McGraw-Hill, New York.

Graham, M.C., Farmer, J.G., 2007. Chemistry of freshwaters. In: Harrison, R.M. (Ed.), Principles of Environmental Chemistry. Royal Society of Chemistry, Cambridge, United Kingdom, pp. 80–169.

Ogden, M.I., Beer, P.D., 2006. Water & O-donor ligands. In: Encyclopedia of Inorganic Chemistry. Wiley-VCH, Weinheim, Germany.

Speight, J.G., 2019. Natural Gas: A Basic Handbook, second ed. Gulf Publishing Company, Elsevier, Cambridge, Massachusetts.

Weingärtner, H., Teermann, I., Borchers, U., Balsaa, P., Lutze, H.V., Schmidt, T.C., Franck, E.U., Wiegand, G., Dahmen, N., Schwedt, G., Frimmel, F.H., Gordalla, B.C., 2016. Water, 1. Properties, analysis, and hydrogeological cycle. In: Ullmann's Encyclopedia of Industrial Chemistry. Wiley-VCH Verlag GmbH & Co. KGaA, Weinheim, Germany.

Zheligovskaya, E.A., Malenkov, G.G., 2006. Crystalline water ices. Russ. Chem. Rev. 75, 57–76.

Further reading

Boyd, C.E., 2000. pH, Carbon Dioxide, and Alkalinity. Water Quality. Springer. Boston, Massachusetts.

Li, C.-J., Chen, L., 2006. Organic chemistry in water. Chem. Soc. Rev. 35, 68–82.

Water chemistry

Chapter outline

1 Introduction

Water is a chemical compound of hydrogen and oxygen and, although the same formula (H_2O) also represents the compositions of steam, liquid water, and ice, the molecules in these forms are associated structurally, more correct to consider the various condensed phases (gas, liquid, solid) of water in terms of these associations rather than as simple aggregates of molecules. Moreover, since there are three isotopes of hydrogen (^1H, ^2H, and ^3H) that exist in nature with ^1H being the most abundant (99.98% natural abundance) and three isotopes of oxygen (^{16}O, ^{17}O and ^{18}O) that exist in nature with ^{16}O being the most abundant (99.76% natural abundance) giving rise to eighteen possible isotopic varieties of the water molecule are possible. The most common, of course, is $^1H_2^{16}O$.

Isotopic abundances aside, the physical properties of water (Chapter 2) are unique in a number of respects and are of great importance because of the departures from what might be considered typical properties normal for such a compound (Table 3.1), with respect both to the development and continued existence of life forms and to the

Natural Water Remediation. https://doi.org/10.1016/B978-0-12-803810-9.00003-6

Table 3.1 Boiling points of the hydrides related to water

Compound	Formula	Molar mass	Boiling point[a]		Freezing point[a]	
			°C	°F	°C	°F
Hydrogen telluride	H_2Te	129.6	−4	25	−49	−56
Hydrogen selenide	H_2Se	81	−42	−44	−64	−83
Hydrogen sulfide	H_2S	34	−62	−80	−84	−119
Water	H_2O	18	100	212	0	32

[a] Rounded to the nearest degree.

Table 3.2 Density of water and ice

Temperature (°C)	Density (g/cm^3)
100 (liquid)	0.9584
50	0.9881
25	0.9971
10	0.9997
4	1.000
0 (liquid)	0.9998
0 (solid)	0.9168

shape and composition of the of the Earth (Stumm and Morgan, 1996). The boiling and freezing points of water are higher than would be expected for a compound having such a low molecular weight (Chapter 2), and the surface tension and dielectric constant of liquid water are also much greater than might be expected. When water freezes, its density decreases; in fact, the maximum density of water at 1 atm pressure occurs near 4 °C (39 °F) (Table 3.2). Although this type of behavior is not unique in liquid-solid transitions, it is an attribute of water that is most fortunate for all life forms. The physical properties of liquid water are best understood by considering the structure of the water molecule. The two chemical bonds formed between the oxygen atom and the hydrogen atoms are at an angle of approximately 105° to each other (Fig. 3.1). As a result, the hydrogen atoms ions are on the same side of the molecule, giving it a dipolar character.

Besides the simple electrostatic effect, attributable to the dipolar property, the attached hydrogen ions retain a capacity for specific interaction with electronegative ions and between water molecules. This effect, known as hydrogen bonding, is present in both liquid and solid forms of water and results in the well-defined crystal structure of ice (Rapaport, 1983). In liquid water there is much disorder, but the attractive forces between molecules are strongly evident. The energy required to separate the molecules is indicated by the high heat of vaporization of water, and in another way by its high surface tension. Liquid water has some of the properties of a polymer. The presence of dissolved ions in water changes some of its physical properties, notably its ability to conduct electricity. The dipolar nature of the water molecule, however, is an important factor in the behavior of the solute ions as well as the solvent.

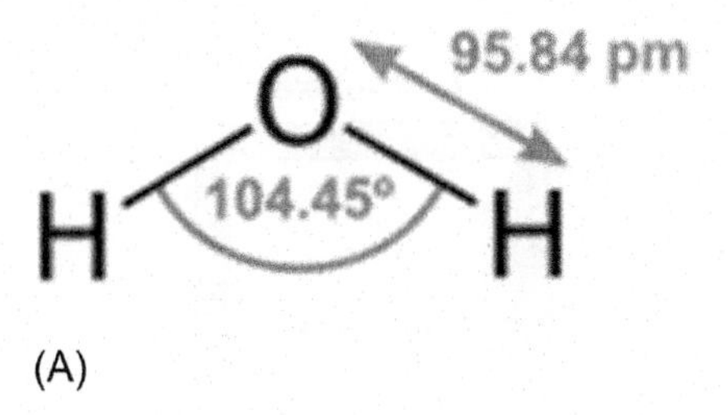

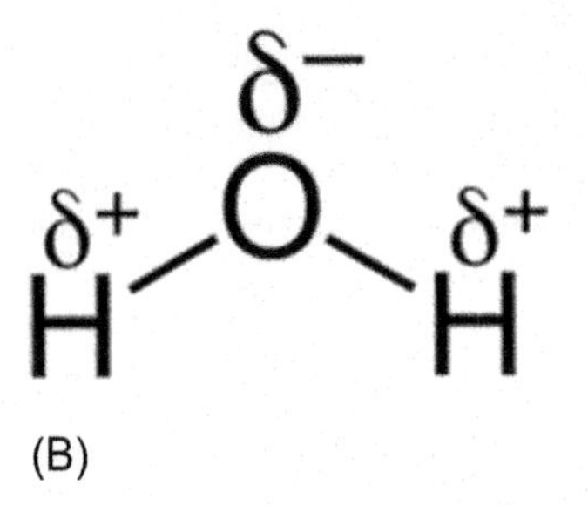

Fig. 3.1 (A) The structure of water also showing (B) the relative negativity of the hydrogen and oxygen atoms.

The dipolar water molecules are strongly attracted to most mineral surfaces, form sheaths arranged in an orderly pattern around many forms of dissolved ions, and insulate the electrical charges on the ions from other charged species. The effectiveness of water as a solvent is related to such activities. Its effectiveness in weathering rocks is also increased by the ability of this cohesive liquid to wet mineral surfaces and penetrate into small openings.

2 The hydrosphere

The hydrosphere (often referred to as the aquasphere) is generally defined by geochemists as the vapor, liquid, and solid water present at and near the land surface, and its dissolved constituents. Water vapor and condensed water of the atmosphere are usually included, but water that is immobilized by incorporation into mineral structures in rocks is usually not thought of as part of the hydrosphere. In fact, in the processes of the hydrological cycle on the Earth, connecting hydrosphere with atmosphere, lithosphere and biosphere (Fig. 3.2) the chemical composition of water is formed. Interacting with all the components of the natural landscape and being influenced by natural and man-made factors, water, a universal solvent, is enriched by a wide variety of different substances in gaseous, solid and liquid states that create an enormous variability of natural water types from the perspective of their chemical composition.

The hydrosphere contains all the solid, liquid, and gaseous water of the Earth and ranges in thickness from (approximately) 6 to 12 miles. The hydrosphere extends

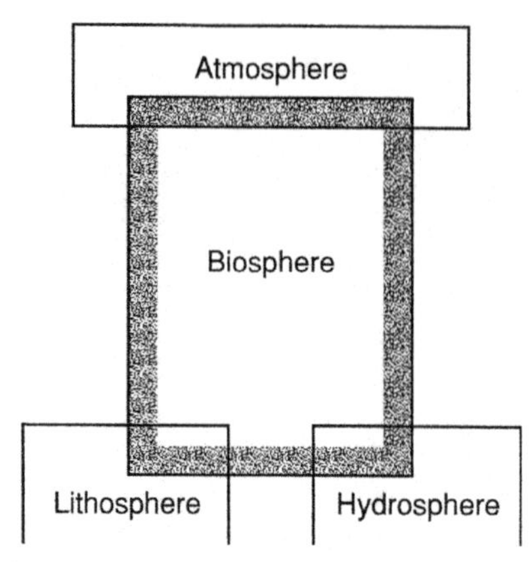

Fig. 3.2 Representation of the Interrelationship of the biosphere with the atmosphere, the lithosphere, and the hydrosphere.

from the surface of the Earth downward several miles into the lithosphere and upward approximately 7 miles into the atmosphere. A small portion of the water in the hydrosphere is fresh water (non-salty water, non-saline water). This water flows as precipitation from the atmosphere down to the surface of the Earth, as rivers and streams along the surface of the Earth and also as groundwater beneath the surface of the Earth. Most of fresh water of the Earth, however, is frozen in the form of ice sheets and glaciers (Chapter 1).

The oceans constitute approximately 98% *v*/v of the hydrosphere, and thus the average composition of the river water. Sphere is, for all practical purposes, that of seawater. The Obviously, the chemical composition of surface runoff water of the ocean basins is generally fairly well mixed waters of the Earth is highly variable through both time with regard to major constituents, although concentrations and space, and this book discusses the variations and of most minor elements are not uniform with depth or reasons for them at some length. The average concentrations of the major dissolved global average has little significance except, perhaps, as a elements or ions, and of some of the minor ones, are baseline for comparison. On the basis of stability of each of the complex species, the predominant forms in which the dissolved constituents occur.

Substantial differences in concentration between water near the surface and water at depth, as well as on an area basis, are characteristic of solutes that are used as nutrients by marine life. Some of the minor elements have distributions that resemble those of the nutrients. In addition, the chemical composition of surface runoff water of the ocean basins is generally fairly well mixed waters of the Earth is highly variable through both time with regard to major constituents, although concentrations and space, and this book discusses the variations and of most minor elements are not uniform with depth or reasons for them at some length.

The average concentrations of the major dissolved global average has little significance except, perhaps, as a elements or ions, and of some of the minor ones, are baseline for comparison.

Table 3.3 Illustration of the contact angle as it relates to wettability

Contact angle	Degree of wetting	Interaction strength	
		Solid-liquid	Liquid-liquid
$\theta=0$	Perfect wetting	Strong	Weak
$0<\theta<90°$	High wettability	Strong	Strong
		Weak	Weak
$90° \leq \theta < 180°$	Low wettability	Weak	Strong
$\theta=180°$	Non-wetting	Weak	Strong

Finally, the property of water known as *wettability* (Table 3.3) is an important aspect of the properties of water. Briefly, the wettability of a solid surface is the ability of a solid surface to reduce the surface tension of the liquid in contact with the surface such that the liquid spreads over the surface and wets it. Thus, wettability refers to the interaction between fluid and solid phases. in a reservoir rock the liquid phase can be water or oil (gas is also included within the "oil" term), and the solid phase is the rock mineral assemblage.

When water, or any other liquid for that matter, is in contact with a solid surface (such as a mineral stratum), the behavior of the liquid depends on the relative magnitudes of the surface tension forces and the attractive forces between the molecules of the liquid and of those comprising the surface. If a water molecule is more strongly attracted to its own kind (intramolecular forces), the surface tension forces will dominate the interaction by increasing the curvature of the interface.

On the other hand, many minerals have hydrophilic groups at the surface which readily attach to water molecules through hydrogen bonding (intermolecular forces). Thus causes the water to spread evenly over the surface of the mineral—the mineral surface is described as being to be *wet*. The extent to which the water wets the surface (the degree of wetting or wettability) is determined by a force balance between adhesive and cohesive forces. A liquid will wet a surface if the angle at which the liquid contacts the surface (the *contact angle*) is >90° (Fig. 3.3). Typically, the value of the contact angle can be predicted from the properties of the liquid and solid separately. By reducing the surface tension with surfactants, a nonwetting material can be made to become partially or completely wetting.

For water, a wettable surface may also be termed hydrophilic and a non-wettable surface may be termed hydrophobic. Superhydrophobic surfaces have contact angles >150°, showing almost no contact between the liquid drop and the surface (the Lotus effect). For non-aqueous liquids (i.e. organic liquids, such as hydrocarbon liquids),

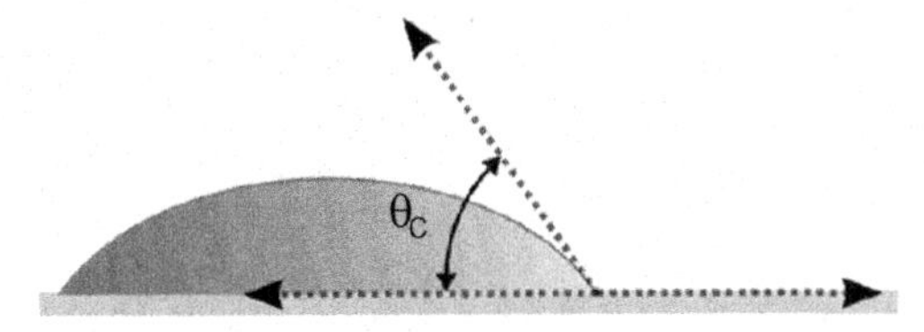

Fig. 3.3 Illustration of the contact angle between a liquid and a solid surface.

the term *lyophilic* is used for low contact angle conditions and *lyophobic* is used when higher contact angles result. Similarly, the terms *omniphobic* and *omniphilic* apply to both polar and non-polar liquids.

3 Composition of water

The composition of natural water is derived from many different sources of solutes, including gases from the atmosphere, weathering and erosion of rocks and soil, solution or precipitation reactions occurring below the land surface, and cultural effects resulting from human activities (Table 3.4). Broad interrelationships among these processes and their effects can be discerned by application of principles of chemical thermodynamics (Chapter 4). Some of the processes of solution or precipitation of minerals can be closely evaluated by means of principles of chemical equilibrium, including the law of mass action. Other processes are irreversible and require consideration of reaction mechanisms and rates. The chemical composition of the crustal rocks of the Earth and the composition of the ocean and the atmosphere are significant in evaluating sources of solutes in natural freshwater.

The ways in which solutes are taken up or precipitated and the amounts present in solution are influenced by many environmental factors, especially climate, structure and position of rock strata, and biochemical effects associated with life cycles of plants and animals, both microscopic and macroscopic. Taken together and in application with the further influence of the general circulation of all water in the hydrologic cycle, the chemical principles and environmental factors form a basis for the developing science of natural-water chemistry.

Slightly acidic rainwater reacts with land-derived dust particles in the atmosphere which results in the rainwater gaining dissolved calcium (Ca^{2+}), magnesium (Mg^{2+}),

Table 3.4 Examples of chemical species in water and the sources

Substance	Source
Aluminum	Aluminum-containing minerals
Chloride	Minerals, pollution
Fluoride	Minerals, water additive
Iron	Minerals, mine water
Magnesium	Minerals, such as dolomite
Manganese	Minerals, decay of nitrogenous organic matter, pollution
Potassium	Mineral forest fire runoff
Phosphorus	Minerals, fertilizer runoff, domestic wastes (from detergents)
Silicon	Minerals, such as sodium feldspar pollutants
Sulfur	Minerals, pollutants, acid mine water, acid rain
Sodium	Minerals, pollution

sodium (Na^+), potassium (K^+), and other elements. Although carbonic acid (H_2CO_3) is a weak acid, it is very effective over geologic time. Carbonic acid is largely responsible for the breakdown of rocks to soil during chemical weathering and the formation of limestone caverns and sink holes. Also, sea spray, carried aloft by winds blowing across the ocean, contributes to dissolved constituents in rainwater. Although dissolved minerals from spray are more abundant in coastal areas, they occur throughout the atmosphere. Sea spray is the primary source of chloride (Cl^-) in rainwater and a significant amount of sodium (Na^+).

Natural water is a dynamic chemical system containing in its composition a complex group of gases, mineral and organic substances in the form of true solutions as well as suspended matter and colloidal matter. The variety and complexity of natural water composition is defined not only by the occurrence of a large number of chemical elements in it, but also by the difference of forms and the values and presence of each. In fact, almost all known chemical elements occur in natural water.

The composition of stream and lake water varies from one place to another, and within a single watershed varies both seasonally and along the stream's path. The major source of dissolved minerals in streams and lakes is the rocks the water moves over and through along its path from where it falls as precipitation to where it exits the watershed or enters the lake. As the slightly acidic water encounters rocks, the minerals begin to dissolve and contribute their elements to the water. The type of rocks in the watershed influence stream-water composition. A stream flowing over sedimentary rocks will have a different composition than a stream flowing over igneous rocks.

Also contributing to stream-water and lake-water composition are reactions between the water and the biomass, particularly in forests. Leaves and branches help neutralize the pH of the precipitation and contribute dissolved elements. Biologic activity in the stream or lake (such as photosynthesis) can change the pH and the dissolved oxygen content. In addition, temperature influences the amount of dissolved gases (such as oxygen).

Stream-water composition changes from the headwaters to the outlet because the water is in contact with the rocks and sediments of the streambed for cumulatively longer times. Also, tributaries draining different geologic areas may enter the stream, and groundwater may seep into the stream. In pools or other slow-moving stream segments, oxidizing and reducing reactions may occur where organic matter accumulates. Seasonal variations in stream-water composition may reflect differing precipitation amounts, as well as the portion of the stream's flow that is contributed by groundwater. In the drier times of the year the proportion of groundwater contribution is greater than in the wet season.

Lake-water composition is influenced by evaporation, among many other factors. As water evaporates, the dissolved minerals are left behind. The greater the extent of the evaporation, the higher the concentration of dissolved minerals (salts) in the water. If evaporation continues far enough, minerals such as calcite ($CaCO_3$) or gypsum ($CaSO_4.2H_2O$) may precipitate from the solution.

Many of the factors that influence the surface water composition also influence groundwater composition. Groundwater is always in contact with rocks and minerals and moves more slowly than surface water—on the order of inches per day instead of miles per hour. As a result, groundwater often contains more dissolved minerals than surface water. When water seeps below the surface, it passes through the soil where microbial respiration processes release carbon dioxide. As water encounters the carbon dioxide, the pH is lowered, and the water is more acidic and can dissolve more minerals. Also at higher temperatures, minerals dissolve more readily, thus deep groundwater tends to be warmer (e.g., the source of water from hot springs) and, as a result, has higher content of dissolved minerals. Ultimately, the composition controls the composition of groundwater is (i) the geologic materials groundwater is moving through, (ii) the type of reactions taking place, and iii) the contact time, or length of time groundwater has been in contact with the rocks. The contact time may vary from several days to >10,000 years.

Typically, groundwater has a total dissolved solids (TDS) content of <250 mg/L (mg/L). In some areas, however, groundwater with >100,000 mg/L of total dissolved solids is found. For example, sea water has a content of total dissolved solids on the order of approximately 35,000 mg/L. Saline groundwater has been found in a variety of geologic environments, commonly in marine sedimentary rocks, but also in ancient metamorphic and igneous rocks. Saline groundwater can form in at least three ways: (i) from trapped sea water, (ii) from dissolving highly soluble minerals, and (iii) as a result of a long contact time with rocks, and thus chemical reaction time with surrounding rocks (Drever, 1997).

Trapped sea water (connate water) occurs when marine sediments are deposited and some of the sea water commonly remains trapped between the mineral grains. Connate water later may migrate through the rocks as groundwater. *Highly soluble* minerals occur in groundwater after the relatively mineral-free when groundwater encounters easily-dissolved minerals such as gypsum ($CaSO_4.2H_2O$) or halite (NaCl), will become saltier. High *contact times* occur when groundwater that follows deep paths below the ground may be in contact and able to react with rocks for thousands or tens of thousands of years. This allows the groundwater to acquire an increasingly higher content of total dissolved solids with time.

Thus, water is a necessary, but complex, chemical entity (Chapter 1, Chapter 2) that belies the simple formula (H_2O) insofar as water plays a role in all aspects of life on Earth. The chemical reactions in all plants and animals that support life take place in a water medium. Water not only provides the medium to make these life sustaining reactions possible, but water itself is often an important reactant or product of these reactions. In short, the chemistry of life is water chemistry.

The components of water chemical composition are the main characteristics of water quality that define its fitness for particular kinds of water use. Water quality assessment is performed according to certain parameters of water properties and composition, including concentrations of polluting harmful and toxic substances, which are categorized according to a harmfulness index. The standards of pollutant concentrations adopted in various countries are variously known as quality criteria and include but are not limited to (i) concentration standards, (ii) maximum allowable concentrations, and (iii) maximum allowable levels. Other measures of chemical substances dangerous for biota include classifications based on their type of effect—toxicity, carcinogenicity, and mutagenicity. Different requirements specified for water quality are regulated by state normative documents, regulations on surface water protection from pollution, and standing standards. For this reason and also because of natural differences in chemical composition of water bodies in different regions, establishment of common, strict norms for water quality can be problematic.

Water is also one of the most important resources in the chemical process industries (CPI)—presenting both opportunities as well as environmental compliance obligations. It is involved in a broad range of operations, including separations, product recovery, wastewater treatment, and disinfection, to name a few. Because of its pervasiveness, water requires a new technological definition through which environmental scientists and environmental engineers view their study and work.

Rocks are composed primarily of minerals, naturally crystallized materials having a periodic structure. The long-range structure of crystals, expressed internally as the periodic lattice, determines their fundamental physical and chemical properties. Water, in addition to supplying the basis for life on Earth, is also its critical solvent. It is the dominant medium through which rocks and minerals communicate during chemical precipitation and dissolution reactions. However, at room temperature the mobility of ions via diffusion to and from sites in the solid bulk crystal is extremely limited. Dissolution and precipitation reactions thus usually occur at the *mineral-water interface*) and this interface is the locus of exchange and interaction between the surface atoms of the solid and the overlying aqueous phase.

In addition to water molecules, the fluid contains dissolved components, such as inorganic salts, hydrogen and hydroxyl ions, gases such as carbon dioxide, oxygen, and organic molecules. These components interact with each other as well as with the mineral surface, yielding a complex distribution of species and functional groups (*moieties*) that characterize even compositionally simple solutions. This fluid-solid interaction alters both the surface layers of the crystal and the *boundary layer* of the fluid. As used here, the term boundary layer applies to that fluid in direct contact with the mineral surface. Although the *bulk* fluid may be in turbulent motion, intermolecular attractive forces between the mineral surface and the fluid bring the fluid velocity to zero (sometimes referred to as the *no-slip condition*). This constraint can reduce advection and turbulent mixing within the boundary layer, whose thickness is a function of the flow characteristics prevailing in the overlying bulk fluid. The demands of reactive fluxes from precipitation or dissolution of the underlying mineral surface must be satisfied by the diffusive flux of components through the boundary layer.

This section presents the properties of water as they relate to composition and chemical reactions, particular water-rock interactions.

3.1 Chemical composition

Historically, the nature and composition of water has been a perennially-asked question since at least Biblical times and may even go back even further to more ancient cultures (Wisniak, 2004). The answer has been debated intensely with some light being shed upon the true nature and composition of water.

Chemically pure water is composed of only two substances, hydrogen and oxygen. If the composition of water is referred back to the gaseous state of its two components, two volumes of hydrogen combine with one volume of oxygen to form water. However, the chemical composition of water in the Earth-bound water systems is determined by various existence factors (such as seasonal factors) and processes in the area in which the water exists (Spencer et al., 2008; Dragon and Marciniak, 2010; Guseva, 2016). The chemical compositions of rivers, lakes and groundwaters (waters of active layer) form in natural conditions of landscape.

Thus, although water has the simple formula H_2O, it is a complex chemical solution—in the natural environment pure water is, essentially, non-existent in. Natural water, whether in the atmosphere, on the ground surface, or under the ground, always contains dissolved minerals and gases as a result of its interaction with the atmosphere, minerals in rocks, organic matter, and living organisms.

The acidity of water is gauged by the pH, which is a measure of the concentration of the hydrogen ion (H^+) in the solution according to the relationship $pH = -\log(H^+)$. The higher the concentration of the hydrogen ion in the water, the lower the pH (pH <7, neutral pH is 7), and the greater the acidity with the most acid waters on the order of at pH = 1. Basic (alkaline) water has pH >7.

Natural rainwater is slightly acidic because it interacts with carbon dioxide (CO_2) in the atmosphere to form carbonic acid (H_2CO_3). Some of the carbonic acid in the rainwater then breaks down (dissociates), producing more hydrogen ion and bicarbonate ion, both of which are dissolved in the rainwater. Thus:

$$H_2O + CO_2 \rightarrow H_2CO_3$$

$$H_2CO_3 \rightarrow HCO_3^- + H^+$$

The hydrogen ion produced by the second reaction lowers the pH of rainwater the final pH of the water depends on the amount of carbonic acid in the water which, in turn, depends on the amount of carbon dioxide in the atmosphere.

Although carbonic acid is a weak acid, it is very effective over geologic time (measured in hundreds of years and even in millennia). Carbonic acid is largely responsible for the breakdown of rocks to soil during chemical weathering and the formation of limestone caverns and sink holes. Slightly acidic rainwater reacts with land-derived dust particles (typically mineral dust particles) in the atmosphere which result in the rainwater gaining dissolved calcium (Ca^{2+}), magnesium (Mg^{2+}), sodium (Na^+), potassium (K^+), and other elements. Also, sea spray, carried aloft by winds blowing across the ocean, contributes to dissolved constituents in rainwater. Although dissolved minerals from sea spray are more abundant in coastal areas, they occur throughout the atmosphere. In fact, sea spray is the primary source of chloride (Cl^-) in rainwater and a significant amount of sodium (Na^+).

The salinity of water (the example used here is rover water) can be defined as the concentration of all cations, significantly sodium (Na^+), potassium (K^+), magnesium (Mg^{2+}) and calcium (Ca^{2+}) and of the anions carbonate (CO_3^{2-}) and SO_4^{2-}) and the halides (simply represented as X^-). In soft waters the calcium cations and the carbonate anions will be significantly reduced while in acidic water the sulfate anion may be dominant. Finally, a word about natural mineral water.

The term *natural mineral water* refers to water that is *microbiologically wholesome* but ensures the absence of the main contamination indicators (parasites and pathogenic microorganisms, *Escherichia coli* and fecal streptococci, sporulated sulfite-reducing anaerobes, *Pseudomonas aeruginosa*) both at source and at the point of use. The characteristics of a natural mineral water have to be proved from different points of view: (i) geological and hydrological, (ii) physical, chemical and physicochemical, (iii) microbiological, and (iv) pharmacological, physiological and clinical effects.

In the case of geology and hydrology, there must be detailed description of the catchment site, considering the nature of the terrain, the stratigraphy of the hydrogeological layer and a description of the catchment operations. For the physical, chemical and physicochemical implications, there must be a report about the main physical and chemical analysis to describe the final characteristics of the mineral water (i.e. rate of flow of the spring, temperature at source, dry residues at 180°, pH, anions and cations, trace elements, toxicity of certain constituent elements). In terms of so the possibility of any microbiological effects, test should be conducted in order to determine the possibility of any physiological effects and benefits on human health.

3.2 Water-rock interactions

The term rock-water interaction suggests many possible processes in nature, such as river water smoothing down rocks over the ages of time, or ocean waves crashing into a rocky shoreline. Within the Geological Sciences, however, the term has a more specific meaning: it refers to the dominantly chemical and thermal exchanges (reactions) that occur between groundwater and rocks. On a chemical basis, the term water-rock interactions is indicative of many possible processes in nature, such as river water smoothing down rocks over the ages of time, or ocean waves crashing into a rocky shoreline. On a geological basis, the term has a more specific meaning: it refers to the dominantly chemical and thermal exchanges (reactions) that occur between groundwater and rocks (or minerals). Whatever the basis of the term, the water-rock interaction is an important effect that influences the composition of water and, in fact, the chemical composition of water is predominantly dominated by rock weathering processes (Reynolds Jr. and Johnson, 1972; Drever and Zobrist, 1992; Millot et al., 2002; Stober and Bucher, 2002; Hinman, 2013).

While tepid rainwater or chilly water in a cave can cause significant changes with (especially geological) time, the chemical aggressiveness of such waters is trivial compared to the saline, gassy, pressurized, searing-hot waters that circulate deep below the surface of the Earth. An impressive example of a water-rock reaction is the subterranean dissolution of limestone ($CaCO_3$) by carbon-dioxide-enriched groundwaters to create caves. The dissolved limestone is often redeposited downstream along the flow path of the groundwater, wherever the groundwater contacts air and is able to degas. The resulting accumulations (spectacular stalactites or terraces of travertine) are conceptually the equivalents of the stain that flowing rainwater leaves as it runs over rusting metal.

When aluminosilicate rocks are mostly predominant, carbonate rocks tend to be less common. Thus, the main process of water chemical composition formation is hydrolysis of aluminosilicate minerals and less the dissolution of carbonates (Morel and Hering, 1993). These processes lead to the release of chemical elements in solution and secondary mineral precipitation. These processes lead to the release of chemical elements in solution and secondary mineral precipitation. The aluminosilicates hydrolyses processes can be described by following reactions:

$$NaAlSi_3O_8 + H^+ + 7H_2O \rightarrow 3H_4SiO_4 + Al(OH)_3 + Na_+$$

$$CaAl_2Si_2O_8 + 2H^+ + 6H_2O \rightarrow Ca^{2+} + 2H_4SiO_4 + 2Al(OH)_3$$

$$KAlSi_3O_8 + H_+ + 7H_2O \rightarrow 3H_4SiO_4 + Al(OH)_3 + K_+$$

In the first stage of water-rock interaction, the primary aluminosilicates of various compositions are dissolved by water which leads to the release of a great quantity of chemical elements in solution.

Briefly, muscovite (also known as common mica, isinglass, or potash mica) is a hydrated phyllosilicate mineral of aluminum and potassium [$KAl_2(AlSi_3O_{10}(FOH)_2$], or $(KF)_2(Al_2O_3)(SiO_2)_6(H_2O)$] which has a highly perfect basal cleabage that yields thin laminae (sheets) which are often highly elastic. Therefore, waters are likely to promote dissolution of primary aluminosilicate minerals and constantly dissolve them. Weathering of aluminosilicate minerals is a slow process, releasing chemical elements in solution from primary rocks and also producing secondary minerals; removing chemical elements from solution. The processes of water-rock interaction have a stage character, one of the main factors in which is water-rock interaction time, determined by water exchange intensity.

Rivers, lakes and waters of stone pits and exploration trenches are characterized by different time of interaction with rocks and therefore they reach different water-rock interaction stages—primary minerals are minerals in which dissolution proceeds relatively readily and secondary minerals are minerals for which dissolution which processes at a slower rate that the primary minerals. During the processes, the accumulation of aluminum (Al) and silicon (Si) and other chemical elements will occur, providing there are no interruptions in the progress of the dissolution, until saturation with respect to the least soluble secondary mineral. When the water reaches the state of saturation with respect to kaolinite, the second stage of water-rock interaction, the following equations are representative of the dissolution process:

$$2Al(OH)_3 + 2H_4SiO_4 \rightarrow Al_2Si_2O_5(OH)_4 + 5H_2O$$

or

$$CaAl_2Si_2O_8 + 2H^+ + H_2O \rightarrow Al_2Si_2O_5(OH)_4 + Ca^{2+}$$

$$2NaAlSi_3O_8 + 4Al(OH)_3 + H_2O \rightarrow 3Al_2Si_2O_5(OH)_4 + 2Na^+ + 2OH^-$$

The main source of sulfate ion in many waters is the oxidation processes. It is the result of oxidation processes of sulfide ores, which are in contact with the atmosphere:

$$MS + 2O_2 + H_2O + Me(OH)^+ + SO_4^{2-} + H_+$$

Reaction rates are controlled partly by solution chemistry which, in turn partly controls the formation of excited states and stability of intermediate states (Brantley et al., 2008; Birkle and Torres Alvarado, 2010; Ibrahim et al., 2019). Biological processes circumvent these controls by routing reactions through lower excitation thresh-holds and different intermediates. Enzymes control the sequence of steps, leading to more limited reaction products and minimizing the effects of solution chemistry on mineralogy. For example, iron oxides, oxyhydroxides, and sulfates form during weathering of igneous rocks, among other types. Iron oxidation under acidic conditions is extremely slow (Morgan and Lahav, 2009). Also, pH-dependent hydrolysis products react more quickly than the aqueous ion, and reduced iron can persist for long periods in low humidity regardless of pH. Biological processes mediate oxidation under acidic conditions, so ferric iron oxides in low-temperature acidic environments could be an indicator of the presence of life and water, although such evidence would be insufficient to demonstrate the presence of either life or water (Hinman, 2013).

Furthermore, many low-temperature reactions have solution-precipitation mechanisms and water provides a medium for isotopic exchange of carbon, sulfur, oxygen, hydrogen, and nitrogen as a reaction progresses. The amount of water relative to the amount of rock contributes to the fractionation of isotopes in geological systems. This water-rock ratio determines the isotopic reservoir from which life removes, selectively or not, elements for energy, structure, and metabolic purposes.

Morphological signatures, e.g., stromatolites (calcareous mounds built up of layers of lime-secreting cyanobacteria and trapped sediment, found in Precambrian rocks), most often form from mineral precipitation from aqueous solutions. The solutions are either inherently saturated or reach saturation through biological processes. Examples of the former include silica-saturated hot spring solutions that precipitate sinter or geyserite, entombing microbes in the process, or iron oxide accumulation in drainages made oversaturated and acidic by microbial oxidation of pyrite. Examples of the latter generally involve precipitation of carbonate minerals, made oversaturated by removal of carbon dioxide during synthesis of organic matter. In either case, morphological signatures preserve microbial cells, filaments, and extracellular polymeric substances during mineralization process.

Chemical signatures in water often take the form of organic compounds, although many types of organic compounds, some of which are large and complex, are found in meteorites and are formed abiotically (Botta and Bada, 2002). The maturation of natural organic matter is believed to proceed by (i) condensation reactions and (ii) loss of functional groups. Immiscibility between water and non-polar organic compounds serves to concentrate the latter as the former forces it to migrate according to density, leaving more recalcitrant organic matter in place.

Minerals, with or without structural defects, are also catalysts for reactions that modify organic matter or change the chemical. Minerals also can store biological information in the form of biomolecules incorporated into or onto the minerals, making these minerals excellent targets for exploration for life. Precipitation from water provides a mechanism by which life can influence the properties of minerals (Hinman, 2013).

Whilst part of this discussion has included the effect of dissolved carbon dioxide on water-rock interactions, collectively water and gas, as the two most common fluids and primary geologic forces, are crucial components in various geological processes. Gas-water-rock interactions play indispensable roles in the evolution of geo-environmental issues. For example, the accurate prediction of groundwater flow and contaminant transport requires a profound understanding of physicochemical processes that occur among liquid, solid, and gas phases.

At present, gas-water-rock interactions related to the transport and retention of toxic contaminants such as heavy metals and organic contaminants in aquifers and vadose zones should be paid more attention, due to the release of toxic contaminants from the intensive human activities. In addition, gas-water-rock reaction might change stress field, groundwater seepage field, and properties of rocks and soils, which subsequently leads to instability of slope and landslide hazards. A rapid sliding rock mass is likely to trigger waves or stir atmosphere generating air blasts and facilitating its transport, both aggravating the hazards.

In summary, the chemical interaction of water and rock is one of the most fascinating and multifaceted process in water geochemistry. The composition of surface water and groundwater is largely controlled by the reaction of water with rocks and minerals. At elevated temperature, hydrothermal features, hydrothermal ore deposits and geothermal fields are associated with chemical effects of water-rock interaction. Surface outcrops of rocks from deeper levels in the crust, including exposures of lower crustal and mantle rocks, often display structures that formed by interaction of the rocks with a supercritical aqueous fluid at very high temperature.

Understanding water-rock interaction is also of great importance to applied geology and geochemistry, particularly in areas such as water chemistry and hydrogeology which can lead to even further understanding of the behavior of water.

4 Acidity and alkalinity

The acidity of water is its base-neutralizing capacity while the alkalinity of water is its acid-neutralizing capacity. Both parameters are related to the buffering capacity of water (the ability to resist changes in pH when an acid or base is added). Water with high alkalinity can neutralize a large quantity of acid without large changes in pH while, on the other hand, water with high acidity can neutralize a large quantity of base without large changes in pH. Both properties are a consequence of the structure and boning of water (Chapter 2).

Acidity is determined by measuring the amount of standard base that must be added to the water sample raise the pH to a specified value. Acidity is a net effect of the presence of several constituents, including dissolved carbon dioxide, dissolved multivalent metal ions, strong mineral acids such as sulfuric, nitric, and hydrochloric acids, and weak organic acids such as acetic acid. Dissolved carbon dioxide (CO_2) is the main source of acidity in unpolluted waters. Acidity from sources other than dissolved carbon dioxide is not commonly encountered in unpolluted natural waters and can be used as an indicator of pollution.

On the other hand, alkalinity is determined by measuring how much standard acid must be added to a given amount of water in order to lower the pH to a specified value. As is the case for acidity, alkalinity is a net effect of the presence of several constituents, but the most important are the bicarbonate (HCO_3^-), carbonate (CO_3^{2-}), and hydroxyl (OH^-) anions. Alkalinity can be sed as an indicator for the concentration of these constituents, although there are contributors (usually minor contributors) to alkalinity, such as ammonia, and other basic substances.

Water molecules tend to ionize and dissociate into ions (charged particles) hydrogen ions (H^+) and hydroxide ions (OH^-). In pure water a very small number of water molecules form ions and the tendency of water to dissociate is balanced by the tendency of hydrogen ions and hydroxide ions to reunite to form water. A neutral solution contains an equal number of hydroxide ions and hydrogen ions. A solution with a greater concentration of hydrogen ions (H^+) is acidic whereas a solution with a greater concentration of hydroxide (OH^-) ions is alkaline (basic). Thus, the term *acidity* as applied to natural water and wastewater is the capacity of the water to neutralize hydroxyl ions (OH^-).

In natural waters that are not highly polluted, alkalinity is more commonly found than acidity and is often a good indicator of the total dissolved inorganic carbon (bicarbonate and carbonate anions) present. Since all natural waters contain dissolved carbon dioxide, they all will have some degree of alkalinity contributed by carbonate species—unless acidic pollutants would have consumed the alkalinity. Alkalinity in environmental waters is beneficial because it minimizes pH changes, reduces the toxicity of many metals by forming complexes with them, and provides nutrient carbon for aquatic plants.

Alkalinity is important to fish and other aquatic life because it buffers both natural and human-induced changes in the pH of the water. The chemical species that cause alkalinity, such as carbonate (CO_3^{2-}), bicarbonate (HCO_3^-), hydroxyl (OH^-), and phosphate ions PO_4^{3-}), can form chemical complexes with many toxic heavy metal ions, often reducing the toxicity of the heavy metals in a water system.

Thus, although virtually all water has some alkalinity, acidic water is not frequently encountered, except in cases of severe pollution. Acidity generally results from the presence of weak acids such as phosphoric acid carbon dioxide, hydrogen sulfide, proteins, fatty acids, and acidic metal ions, particularly Fe^{3+}. Acidity is more difficult to determine than is alkalinity One reason for the difficulty in determining acidity is that two of the major contributors are carbon dioxide and hydrogen sulfide, both volatile solutes which are readily lost from solution.

Pure water, like distilled water, has a pH of 7 (neuter). Seawater is essentially alkaline, having a pH of around 8. Most fresh water has a pH between 6 and 8, apart from acid rain, of course, that has a pH that is below 7. Rain water is generally mildly acidic, with a pH between 5.2 and 5.8 if not having any acid stronger than carbon dioxide. If high amounts of nitrogen oxides and sulfur oxides are present in the air, they too will dissolve into the cloud and rain drops, producing acid rain. Thus:

$$SO_2 + H_2O \rightarrow H_2SO_3 \left(\text{sulfurous acid}\right)$$

$$SO_3 + H_2O \rightarrow H_2SO_4 \text{ (sulfuric acid)}$$

$$NO + H_2O \rightarrow HNO_2 \text{ (nitrous acid)}$$

$$3NO_2 + 2H_2O \rightarrow HNO_3 \text{ (nitric acid)}$$

From the pollution standpoint, acids are the most important contributors to acidity. The term free mineral acid is applied to strong acids such as sulfuric acid (H_2SO_4) and hydrochloric acid (hydrogen chloride, HCl) in water. Acidic water from mines is a common water pollutant that contains an appreciable concentration of free mineral acid. Excessively high pH and low pH can be detrimental for the use of water. High pH causes a bitter taste, water pipes and water-using appliances become encrusted with deposits, and it depresses the effectiveness of the disinfection of chlorine, thereby causing the need for additional chlorine when pH is high. Low-pH water will corrode or dissolve metals and other substances.

Pollution can change a the pH of the water which, in turn, can harm animals and plants living in the water. For instance, water coming out of an abandoned coal mine can have a pH of 2, which is very acidic and would definitely affect any wildlife that got into the water. Also, the pH of water determines the solubility (amount that can be dissolved in the water) and biological availability (amount that can be utilized by aquatic life) of chemical constituents such as nutrients (phosphorus, nitrogen, and carbon) and heavy metals (such as lead, copper, and cadmium). For example, in addition to affecting how much and what form of phosphorus is most abundant in the water, pH also determines whether aquatic life can use it. In the case of heavy metals, the degree to which they are soluble determines their toxicity and the effects (Table 3.5). The danger of heavy metal pollutants in water lies in two aspects of their impact: (i) heavy metals have the ability to persist in natural ecosystems for an extended period and (ii) heavy metals have the ability to accumulate in successive e levels of the biological chain thereby causing acute and chronic diseases. For example, cadmium and zinc can lead to acute gastrointestinal and respiratory damages to the brain, the heart, and the kidneys (Akpor and Muchie, 2010).

The capacity of water to accept hydrogen ions (H^+) is called alkalinity which is an important aspect of water treatment and in the chemistry and biology of natural waters. Alkalinity is the capacity of water to resist changes in pH that would make the water more acidic and should not be confused with the term basicity which is an absolute measurement on the pH scale. Alkalinity is measured by titrating the solution with a monoprotic source such as hydrochloric acid (HCl aq) until the pH changes abruptly, or it reaches a known endpoint where that happens. Alkalinity is expressed in units of meq/L (milliequivalents per liter), which corresponds to the amount of monoprotic acid added as a titrant in millimoles per liter.

Frequently, the alkalinity of water must be known to calculate the quantities of chemicals to be added in treating the water. Highly alkaline water often has a high pH and generally contains elevated levels of dissolved solids. These characteristics may be detrimental for water to be used in boilers, food processing, and municipal water systems. Also, alkalinity serves as a pH buffer and reservoir for inorganic carbon, thus

Table 3.5 Effects of heavy metals in water

Metal	Effects
Cadmium (Cd)	Decreases seed germination, lipid content, and plant growth Induces phytochelatins production
Chromium (Cr)	Decreases enzyme activity and plant growth Produces membrane damage, chlorosis and root damage
Copper (Cu)	Inhibits photosynthesis, plant growth and reproductive process Decreases thylakoid surface area
Mercury (Hg)	Decreases photosynthetic activity and water uptake Decreases antioxidant enzymes Accumulates phenol and proline
Nickel (Ni)	Reduces seed germination Reduces dry mass accumulation Reduces protein production Increases free amino acids
Lead (Pb)	Reduces chlorophyll production and plant growth Increases superoxide dismutase
Zinc (Zn)	Reduces Ni toxicity and seed germination Increases plant growth and ATP/chlorophyll ratio

helping to determine the ability of a water to support algal growth and other aquatic life. It is used by biologists as a measure of water fertility. Generally, the basic species responsible for alkalinity in water are bicarbonate ions, carbonate ions, and hydroxide ions. Other, usually minor, contributors to alkalinity are ammonia and the conjugate bases of phosphoric, silicic, boric, and organic acids.

Water is amphoteric insofar as it has the ability to act as either an acid or a base in chemical reactions. According to the Brønsted-Lowry definition, an acid is a proton (H^+) donor and a base is a proton acceptor. When reacting with a stronger acid, water acts as a base but when reacting with a stronger base, water acts as an acid. For examples, instance, water receives a hydrogen ion (H^+) from hydrogen chloride (HCl gas) when hydrochloric acid (aqueous) is formed:

$$HCl(\text{acid}) + H_2O(\text{base}) \rightleftharpoons H_3O^+ + Cl^-$$

On the other hand, in the reaction with ammonia (NH_3,) water donates a hydrogen ion (H^+) ion, and is thus acting as an acid:

$$NH_3(\text{base}) + H_2O(\text{acid}) \rightleftharpoons NH_4^+ + OH^-$$

In addition, and because the oxygen atom in water has two lone pairs of electrons, water often acts as a Lewis base (an electron pair donor) in reactions with Lewis acids. However, water can also react with Lewis bases, forming hydrogen bonds between the electron pair donors and the hydrogen atoms of water. Also, water can be described as both a weak hard acid and a weak hard base (HSAB concept, HSAB theory), meaning that it reacts preferentially with other hard species:

$$H^+ \left(\text{Lewis acid}\right) + H_2O\left(\text{Lewis base}\right) \rightarrow H_3O^+$$

$$Fe^{3+} \left(\text{Lewis acid}\right) + H_2O\left(\text{Lewis base}\right) \rightarrow Fe\left(H_2O\right)^{3+}_6$$

$$Cl^- \left(\text{Lewis base}\right) + H_2O\left(\text{Lewis acid}\right) \rightarrow Cl\left(H_2O\right)^-_6$$

Briefly, the HSAB concept is an acronym for *hard and soft (Lewis) acids and bases* and is also known as the Pearson acid-base concept, the HSAB concept is widely used in chemistry to explain the stability of compounds, reaction mechanisms, and reaction pathways. The theory/concept assigns the terms *hard* or *soft*, and *acid* or *base* to chemical species. *Hard* applies to species which are small, have high charge states (the charge criterion applies mainly to acids, to a lesser extent to bases), and are weakly polarizable while the term *soft* applies to species which are big, have low charge states and are strongly polarizable.

When a salt of a weak acid or of a weak base is dissolved in water, water can partially hydrolyze the salt, producing the corresponding base or acid, which renders aqueous solutions of sodium carbonate a basic solution (pH >7.0):

$$Na_2CO_3 + H_2O \rightleftharpoons NaOH + NaHCO_3$$

In typical groundwater or seawater, the measured alkalinity (A) is derived from the equations:

$$A_T = \left[HCO_3^-\right]_T + 2\left[CO_3^{2-}\right]_T + \left[B(OH)_4^-\right]_T + \left[OH^-\right]_T + 2\left[PO_4^{3-}\right]_T + \left[HPO_4^{2-}\right]_T + \left[SiO(OH)_3^-\right]_T - \left[H^+\right]_{sws} - \left[HSO_4^-\right]$$

In this equation, the subscript T indicates the total concentration of the ionic species in the solution as measured. This is opposed to the free concentration, which considers the significant amount of ion pair interactions that occur in seawater).

Alkalinity can be measured by titrating a sample with a strong acid until all the buffering capacity of the aforementioned ions above the pH of bicarbonate or carbonate is consumed. This point is functionally set to pH 4.5 and, at this point, all the bases of interest have been protonated to the zero level species, hence they no longer cause alkalinity. In the carbonate system the bicarbonate ions [HCO_3^-] and the carbonate ions [CO_3^{2-}] have become converted to carbonic acid [H_2CO_3] at this pH. This pH (also called the carbon dioxide equivalence point) is the point where the major component in water is dissolved carbon dioxide which is converted to an aqueous solution carbonic acid (H_2CO_3 aq) in an aqueous solution. Since there are no strong acids or bases at this point, the alkalinity is modeled and quantified with respect to the carbon dioxide equivalence point equivalence point. Because the alkalinity is measured with respect to the carbon dioxide equivalence point equivalence point, the dissolution of carbon dioxide, although it adds acid and dissolved inorganic carbon, does not change the alkalinity.

In natural conditions, the dissolution of basic rocks and addition of ammonia [NH_3] or organic amines leads to the addition of base to natural waters at the carbon dioxide equivalence point equivalence point. The dissolved base in water increases the pH and titrates an equivalent amount of carbon dioxide to bicarbonate ion and carbonate ion. At equilibrium, the water contains a certain amount of alkalinity contributed by the concentration of weak acid anions. Conversely, the addition of acid converts weak acid anions to carbon dioxide and continuous addition of strong acids can cause the alkalinity to become less than zero. For example, the following reactions take place during the addition of acid to a typical seawater solution:

$$B(OH)_4^- + H^+ \rightarrow B(OH)_3 + H_2O$$

$$OH^- + H^+ \rightarrow H_2O$$

$$PO_4^{-3} + 2H^+ \rightarrow H_2PO_4^-$$

$$HPO_4^{-2} + H^+ \rightarrow H_2PO_4^-$$

$$\left[SiO(OH)_3^-\right] + H^+ \rightarrow \left[Si(OH)_4^0\right]$$

From the protonation reactions (above), that most bases consume one proton (H^+) to become a neutral species, thus increasing alkalinity by one per equivalent. However, the carbonate ion (CO_3^{-2}) however, will consume two protons before becoming a zero level species (carbon dioxide), thus it increases alkalinity by two per mole of the carbonate ion. The proton (H^+) and the bisulfate ion (HSO_4^-) decrease the alkalinity because they act as sources of protons and both of these ions are often represented collectively as $[H^+]_T$.

Alkalinity is typically reported as mg/L as $CaCO_3$ which can be converted into milliequivalents per Liter (meq/L) by dividing by 50 which is the approximate molecular weight of calcium carbonate divided by two (i.e. [(MW $CaCO_3$)]/2).

Finally, a word on the issues of pH. The pH is an indication for the acidity of a substance. It is determined by the number of free hydrogen ions (H+) in the water and serves as an indicator that compares some of the most water-soluble ions. The outcome of a pH-measurement is determined by a consideration between the number of hydrogen (H^+) ions and the number of hydroxide (OH^-) ions. When the number of hydrogen ions is equal to the number of hydroxyl ions, the water is neutral and will have a pH of 7. The further the pH lies above or below 7, the more basic or acid (respectively) the solution is.

5 Reactivity of water

The reactivity of water is a combination of the chemical make-up of water and a water-reactive substance, particularly a substance that spontaneously undergoes a chemical reaction with water (Table 3.6) (Stumm and Morgan, 1996). Notable examples include metals (such as lithium, sodium potassium, rubidian, cesium, and

Table 3.6 General summary of the reactions of water with various chemicals

Reactant	Reactant	Products
Group 1 metal in period 3 or higher	Cold water	Metal hydroxide & molecular hydrogen
Group 2 metal in period 3 or higher	Cold water	Metal hydroxide & molecular hydrogen
Nonmetal element (excluding halogens)	Cold water	No reaction
Fluorine (F_2)	Water	Hydrogen fluoride (HF) and molecular oxygen (O_2)
Halogen	Water	Hydrohalic acid or hypohalous acid
Nonmetal Halide	Water	Nonmetal oxide and hydrogen halide
Metal oxide	Water	Metal hydroxide
Nonmetal oxide	Water	Oxoacid

Table 3.7 The periodic table of the elements showing the groups and periods including the lanthanide elements and the actinide elements

↓Period	Group→1	2		3	4	5	6	7	8	9	10	11	12	13	14	15	16	17	18
1	1 H																		2 He
2	3 Li	4 Be												5 B	6 C	7 N	8 O	9 F	10 Ne
3	11 Na	12 Mg												13 Al	14 Si	15 P	16 S	17 Cl	18 Ar
4	19 K	20 Ca		21 Sc	22 Ti	23 V	24 Cr	25 Mn	26 Fe	27 Co	28 Ni	29 Cu	30 Zn	31 Ga	32 Ge	33 As	34 Se	35 Br	36 Kr
5	37 Rb	38 Sr		39 Y	40 Zr	41 Nb	42 Mo	43 Tc	44 Ru	45 Rh	46 Pd	47 Ag	48 Cd	49 In	50 Sn	51 Sb	52 Te	53 I	54 Xe
6	55 Cs	56 Ba	*	71 Lu	72 Hf	73 Ta	74 W	75 Re	76 Os	77 Ir	78 Pt	79 Au	80 Hg	81 Tl	82 Pb	83 Bi	84 Po	85 At	86 Rn
7	87 Fr	88 Ra	**	103 Lr	104 Rf	105 Db	106 Sg	107 Bh	108 Hs	109 Mt	110 Ds	111 Rg	112 Cn	113 Nh	114 Fl	115 Mc	116 Lv	117 Ts	118 Og
			*	57 La	58 Ce	59 Pr	60 Nd	61 Pm	62 Sm	63 Eu	64 Gd	65 Tb	66 Dy	67 Ho	68 Er	69 Tm	70 Yb		
			**	89 Ac	90 Th	91 Pa	92 U	93 Np	94 Pu	95 Am	96 Cm	97 Bk	98 Cf	99 Es	100 Fm	101 Md	102 No		

*Lanthanide elements.
**Actinide elements.

francium) and the alkali earth metals (magnesium though barium) (Table 3.7). Some water-reactive substances are also pyrophoric (a tendency to ignites spontaneously in air at or below 55 °C (130 °F) and should be kept away from moisture.

5.1 Alkali metals

In the pure form, an alkali metal (lithium, sodium, potassium, rubidium, and cesium) is a soft, shiny metal with a low melting point. The alkali metals (lithium, sodium potassium., cesium, and francium) are the most reactive metals in the Periodic Table

(Table 3.7) insofar as they react vigorously or even explosively with cold water, resulting in the displacmet of hydrogen, itself a flammable gas that can (will) contribute to the conflagration. Hydrogen.

In the reaction, the Group 1 metal (represented as M) is oxidized to the metal ion, and water is reduced to hydrogen gas (H_2) and hydroxide ion (OH^-). Thus:

$$2M_{(s)} + 2H_2O_{(l)} \rightarrow 2M^+_{(aq)} + 2OH^-_{(aq)} + H_{2(g)}$$

The Group 1 metals or alkali metals become more active in the higher periods of the periodic table.

Thus, the alkali metals react with air to form caustic metal oxides. The heavier alkali metals (rubidium and cesium) will spontaneously ignite upon exposure to air at room temperature. In the reaction, heat, heat, hydrogen gas, and the corresponding metal hydroxide are produced. The heat produced by this reaction may ignite the hydrogen or the metal itself, resulting in a fire or an explosion. The heavier alkali metals will react more violently with water (Urben, 2007).

5.2 Alkaline earth metals

The alkaline earth metals (beryllium, magnesium, calcium, strontium, barium, and radium) are the second most reactive metals in the periodic table (Table 3.7), and, like the Group 1 metals, have increasing reactivity in the higher periods. Beryllium is the only alkaline earth metal that does not react with water or steam, even when the metal is heated to red heat (approximately 700–800 °C, 1290–1470 °F). in addition, beryllium forms an outer oxide layer (BeO) protects the metal and lowers the reactivity of the metal.

Magnesium exhibits an insignificant reaction with water, but burns reacts with steam to produce white magnesium oxide and hydrogen gas:

$$Mg_{(s)} + 2H_2O_{(l)} \rightarrow Mg(OH)_{2(s)} + H_{2(g)}$$

Typically, the reaction between a metal and cold water will produce the corresponding metal hydroxide. However, if a metal reacts with steam, like magnesium, the metal oxide is produced as a result of the metal hydroxide decompositon when heated. Thus:

$$M(OH)_2 \rightarrow MO + H2O$$

The hydroxide derivatives of calcium, strontium, and barium are only slightly water-soluble but produce sufficient hydroxide ions to make the environment basic and the general equation is:

$$M_{(s)} + 2H_2O_{(l)} \rightarrow M(OH)_{2(aq)} + H_{2(g)}{}^{[9]}$$

If the water is hard, the fact that there are two types of hard water: (i) include temporary hard water and (ii) permanent hard water. Temporary hard water contains the bicarbonate ion (HCO_3^-) which forms a carbonate ion (CO_3^{-2}) when heated:

$$2HCO_3^- \rightarrow CO_3^{-2}(aq) + CO_2(g), + H_2O$$

The bicarbonate ions react with alkaline earth cations and precipitate out of solution, causing boiler scale and problems in water heaters and plumbing.

Common cations in hard water include magnesium (Mg^{+2}) and calcium (Ca^{+2}). In order to produce soft water, treatment includes the addition of an alkaline earth metal hydroxide, such as *slaked lime* [$Ca(OH)_2$] which dissolves in the water to produce a metal ion (M^{2+}) and hydroxide ions (OH^-). The hydroxide ions combine with the bicarbonate ions in the water to produce water and a carbonate ion.

$$HCO_3^- + OH- \rightarrow CO_3^{2-} + H_2O$$

The carbonate ion then precipitates out with the metal ion to form the metal carbonate (MCO_3) which forms a precipitate.

The other type of hard water—permanent hard water—contains bicarbonate ions (HCO_3^-) as well as other anions, such as sulfate ions (SO_4^{-2}). To soften permanent water, sodium carbonate (Na_2CO_3) is added which results in the precipitation of the magnesium ions (Mg^{+2}) and the calcium ions (Ca^{+2}) as the respective metal carbonates and introduces sodium (Na^+) ions into the solution.

5.3 Halogens

The halogen group in the Periodic Table consists of five chemically-related elements: fluorine, chlorine, bromine, iodine, and astatine (Table 3.8) and the halogen group is the only group that contains elements in three of the main states of matter (gas, liquid, solid) and standard temperature, and pressure (STP, 0°C, 32°F, 273.15 K and an absolute pressure of exactly 14.7 psi, 10^5 Pa, 100 kPa, 1 bar). Most halogens are typically produced from minerals and are dangerous (to the point of exhibiting lethal toxicity) and all of the halogens form acids when bonded to hydrogen. Generally halogens react with water to give their halides and hypohalite (–OX) derivatives. The halogen gases vary in their reactions with water due to their different electronegativities.

Fluorine is one of the most reactive elements, attacking otherwise-inert materials such as glass, and fit is a corrosive and highly toxic gas. The reactivity of fluorine is such that, if used or stored in laboratory glassware, it can react with glass in

Table 3.8 Physical data for the halogens

Halogen	Atomic weight	Melting point, °C (°F)	Boiling point, °C (°F)	Density (g/cm^3 at 25°C (77°F)
Fluorine	18.998	−219.62	−188.12	0.0017
Chlorine	35.457	−101.5	−34.04	0.0032
Bromine	79.904	−7.3	58.8	3.1028
Iodine	126.904	113.7	184.3	4.933
Astatine	210	302	337	6.2–6.5

the presence of small amounts of water to form silicon tetrafluoride (SiF_4). The high reactivity of fluorine allows participation in some of the strongest bonds possible, especially to carbon.

Fluorine reacts vigorously with water to produce oxygen (O_2) and hydrogen fluoride (HF):

$$2F_2(g) + 2H_2O(l) \rightarrow O_2(g) + 4HF(aq)$$

Because fluorine (F_2) is highly electronegative, it can displace oxygen gas from water. The products of this reaction include oxygen gas and hydrogen fluoride. The hydrogen halides react with water to form *hydrohalic* acids (HX). With the exception of hydrofluoric acid (HFaq), the hydrohalic acids are strong acids in water.

Chlorine dissolved in water at ambient temperature (21 °C, 72 °F) reacts to form hydrochloric acid (HCl) and hypochlorous acid (HOCl), a solution that can be used as a disinfectant or bleach:

$$Cl_2(g) + H_2O(l) \rightarrow HCl(aq) + HOCl(aq)$$

Hypochlorous (HOCl) acid is a strong bleaching agent and is not very stable in solution and readily decomposes, especially when exposed to sunlight, yielding oxygen.

$$2HOCl \rightarrow 2HCl + O_2$$

Bromine reacts slowly to form a yellow-brown solution of hydrobromic acid (HBr) and hydrobromous acid (HOBr):

$$Br_2(g) + H_2O(l) \rightarrow HBr(aq) + HOBr(aq)$$

Hypobromous (HOBr) acid is a weak bleaching agent:

However, iodine (which has minimum solubility in water does not react but iodine will form an aqueous solution in the presence of iodide ion, such as by addition of potassium iodide(KI). The hypoiodous acid is a very weak bleaching agent.

$$2I(g) + 2H_2O(l) \rightarrow HI(aq) + HOI(aq)$$

5.4 *Hydrides*

A hydride is the anion of hydrogen, H^-, or, more commonly, it is a compounds in which one or more hydrogen centers have nucleophilic, reducing, or basic properties. In compounds that are regarded as hydrides, the hydrogen atom is bonded to a more electropositive element or group. However, the basic metal-hydrogen bond polarity causes hydrides to react vigorously with water, often in an irreversible manner.

There are three basic types of hydrides (i) saline hydride or ionic hydride, (ii) metallic hydride, and (iii) covalent hydride which may be distinguished on the basis of type of chemical bond involved. A fourth type of hydride, the dimeric hydride (of which borane, BH_3, is an example), may also be identified on the basis of structure. Aqueous solutions of borane are extremely unstable:

$$BH_3 + 3H2O \rightarrow B(OH)_3 + 3H_2$$

Saline, or ionic, hydrides are defined by the presence of hydrogen as a negatively charged ion (i.e., H^-). The saline hydrides are generally considered to be the hydrides of the alkali metals and the alkaline earth metals (*i*th the possible exception of beryllium hydride, BeH_2, and magnesium hydride, MgH_2). These metals enter into a direct reaction with hydrogen at elevated temperatures (30–700 °C [570–1300 °F]) to produce hydrides of the general formulas MH and MH_2. Such compounds are white crystalline solids when pure but are usually gray, owing to trace impurities of the metal.

The alkaline-earth metals beryllium and magnesium also form stoichiometric MH_2 hydrides, which are less ionic and more covalent than the other alkaline earth metal hydrides. The transition metals and inner transition metals form a large variety of compounds with hydrogen, ranging from stoichiometric compounds to extremely complicated non-stoichiometric systems. (Stoichiometric compounds have a definite composition whereas non-stoichiometric compounds have a variable composition.

Metallic hydrides are formed by heating hydrogen gas with the metals or their alloys. The most thoroughly studied compounds are those of the most electropositive transition metals (the scandium, titanium, and vanadium families). For example, in the titanium family, titanium (Ti), zirconium (Zr), and hafnium (Hf) form nonstoichiometric hydrides when they absorb hydrogen and release heat. These hydrides have a chemical reactivity similar to the finely divided metal itself, being stable in air at ambient temperature but reactive when heated in air or with acidic compounds. They also have the appearance of the metal, being grayish black solids. The metal appears to be in a+3 oxidation, and the bonding is predominantly ionic.

Covalent hydrides are primarily compounds of hydrogen and nonmetals, in which the bonds are evidently electron pairs shared by atoms of comparable electronegativities. For example, most nonmetal hydrides are volatile compounds, held together in the condensed state by relatively weak van der Waals intermolecular interactions. Covalent hydrides are liquids or gases that have a low melting point and allow boiling point, except in those cases (such as water) where their properties are modified by hydrogen bonding. Covalent hydrides can be formed from boron (B), aluminum (Al), and gallium (Ga) of group 13 in the Periodic Table (Table 3.7). Ionic hydrogen species of both boron (BH_4^-) and aluminum (AlH_4^-) are extensively used as hydride sources.

Each of the halogen forms a binary compound with hydrogen, HX—at ambient temperature and pressure, these compounds are gases, with hydrogen fluoride having the highest boiling point because of intermolecular hydrogen bonding. The hydrogen halides are proton donors in aqueous solution. However, these compounds are, as a class, much stronger acids with the acid strength of the HX compounds increasing down the group—hydrofluoric acid (HF_{aq}) is a very weak acid and hydriodic acid (HI_{aq}) is the strongest proton donor. With the exception of hydrogen fluoride, all of the hydrogen halides dissolve in water to form strong acids. The difference in the proton-donating ability of hydrogen fluoride and the other hydro-halogen compounds is due to a variety of factors, among them being the strong bond that forms between hydrogen and fluorine.

5.5 Methane

The reaction of methane gas with steam is a common process (methane reforming) in the refining industry (Speight, 2014, 2017). The steam reforming of methane (typically, purified natuaal gas) is the most common method of producing hydrogen. At high temperature (700–1100°C, 1290–2010°F) and in the presence of a metal-based catalyst (e.g. nickel) steam reacts with methane to yield carbon monoxide and hydrogen:

$$CH_4 + H_2O \rightarrow CO + 3H_2$$

5.6 Oxides

Sodium oxide is a simple strongly basic oxide—it contains the oxide ion (O^{2-}) which is a very strong base with a high tendency to combine with hydrogen ions. Sodium oxide reacts exothermically with cold water to produce sodium hydroxide solution. A concentrated solution of sodium oxide in water will have pH 14.

$$Na_2O + H_2O \rightarrow 2NaOH$$

As a strong base, sodium oxide also reacts with acids. For example, it reacts with dilute hydrochloric acid to produce sodium chloride solution.

$$Na_2O + 2HCl \rightarrow 2NaCl + H_2O$$

Magnesium oxide is another simple basic oxide, which also contains oxide ions but it is not as strongly basic as sodium oxide because the oxide ions are not as weakly-bound as in the sodium oxide.

Initially, magnesium oxide powder does not appear to react with water but a quick test will demonstrate that the pH of the resulting solution is approximately 9, indicating that hydroxide ions have been produced. In fact, some magnesium hydroxide is formed in the reaction, but as the species is almost insoluble, few hydroxide ions actually dissolve. Thus:

$$MgO + H_2O \rightarrow Mg(OH)_2$$

Magnesium oxide reacts with acids as predicted for a simple metal oxide. For example, it reacts with warm dilute hydrochloric acid to give magnesium chloride solution:

$$MgO + 2HCl \rightarrow MgCl_2 + H_2O$$

Aluminum oxide can be confusing because it exists in a number of different forms and one of those forms is very unreactive (known chemically as alpha-Al_2O_3) and is produced at high temperatures. However, aluminum oxide is amphoteric—it react both as a base and as an acid.

Reaction with water: Aluminum oxide is insoluble in water and does not react like sodium oxide and magnesium oxide. The oxide ions are held too strongly in the solid lattice to react with the water.

Aluminum oxide contains oxide ions, and thus reacts with acids in the same way sodium oxide or magnesium oxide. Aluminum oxide reacts with hot dilute hydrochloric acid to give aluminum chloride solution:

$$Al_2O_3 + 6HCl \rightarrow 2AlCl_3 + 3H_2O$$

Aluminum oxide also displays acidic properties, as exhibited in reactions with bases such as sodium hydroxide. Various aluminates (compounds in which the aluminum is a component in a negative ion) exist, which is possible because aluminum can form covalent bonds with oxygen. This is possible because the electronegativity difference between aluminum and oxygen is small, unlike the difference between sodium and oxygen, for example (electronegativity increases across a period of the Periodic Table).

Aluminum oxide reacts with hot, concentrated sodium hydroxide solution to produce a colorless solution of sodium tetrahydroxoaluminate:

$$Al_2O_3 + 2NaOH + 3H_2O \rightarrow 2NaAl(OH)_4$$

Silicon is too similar in electronegativity to oxygen to form ionic bonds. Therefore, because silicon dioxide does not contain oxide ions, it has no basic properties. In fact, it is very weakly acidic, reacting with strong bases.

Silicon dioxide does not react with water, due to the thermodynamic difficulty of breaking up its network covalent structure. Silicon dioxide reacts with hot, concentrated sodium hydroxide solution, forming a colorless solution of sodium silicate:

$$SiO_2 + 2NaOH \rightarrow Na_2SiO_3 + H_2O$$

Phosphorus(III) oxide (often represented as either P_2O_3 or P_4O_6) reacts with cold water to produce a solution of the weak acid, H_3PO_3 (phosphorous acid, orthophosphorous acid or phosphonic acid):

$$P_4O_6 + 6H_2O \rightarrow 4H_3PO_3$$

The fully-protonated acid structure is shown below:

O
||
HO — P — OH
|
H

Phosphorus(III) oxide is unlikely to be reacted directly with a base. In phosphorous acid, the two hydrogen atoms in the –OH groups are acidic, but the third hydrogen atom is not. Therefore, there are two possible reactions with a base like sodium hydroxide, depending on the amount of base added:

$$NaOH + H_3PO_3 \rightarrow NaH_2PO_3 + H_2O$$

$$2NaOH + H_3PO_3 \rightarrow Na_2HPO_3 + 2H_2O$$

In the first reaction, only one of the protons reacts with the hydroxide ions from the base. In the second case (using twice as much sodium hydroxide), both protons react.

On the other hand, if phosphorus(III) oxide is reacted directly with sodium hydroxide solution, the same salts are possible:

$$4NaOH + P_4O_6 + 2H_2O \rightarrow 4NaH_2PO_3$$

$$9NaOH + P_4O_6 \rightarrow 4Na_2HPO_3 + 2H_2O$$

Phosphorus(V) oxide (phosphorus pentoxide, P_2O_5 or P_4O_{10}) reacts violently with water to give a solution containing a mixture of acids, the nature of which depends on the reaction conditions. Only one acid is commonly considered, phosphoric(V) acid, H_3PO_4 (also known as phosphoric acid or as orthophosphoric acid).

$$P_4O_{10} + 6H_2O \rightarrow 4H_3PO_4$$

This time the fully protonated acid has the following structure:

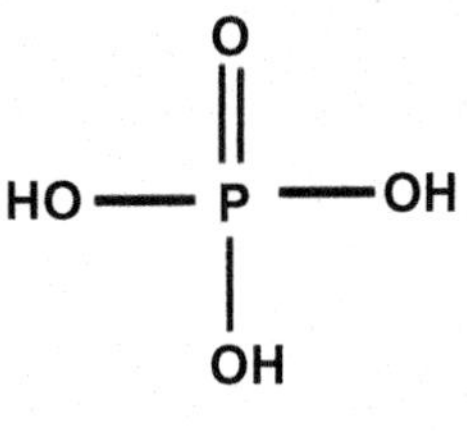

O
||
HO — P — OH
|
OH

Phosphoric (V) oxide is also unlikely to be reacted directly with a base, but the hypothetical reactions are considered. In the acid form, molecule has three acidic –OH groups, which can cause a three-stage reaction with sodium hydroxide:

$$NaOH + H_3PO_4 \rightarrow NaH_2PO_4 + H_2O$$

$$2NaOH + H_3PO_4 \rightarrow Na_2HPO_4 + 2H_2O$$

$$3NaOH + H_3PO_4 \rightarrow Na_3PO_4 + 3H_2O$$

Similar to phosphorus (III) oxide, if phosphorus(V) oxide is reacted directly with sodium hydroxide solution, the same possible salt as in the third step (and only this salt) is formed:

$$12NaOH + P_4O_{10} \rightarrow 4Na_3PO_4 + 6H_2O$$

Sulfur dioxide (SO_2) is fairly soluble in water, reacting to give a solution of sulfurous acid, H_2SO_3, which only exists in solution, and any attempt to isolate it gives off sulfur dioxide:

$$SO_2 + H_2O \rightarrow H_2SO_3$$

The protonated acid has the following structure:

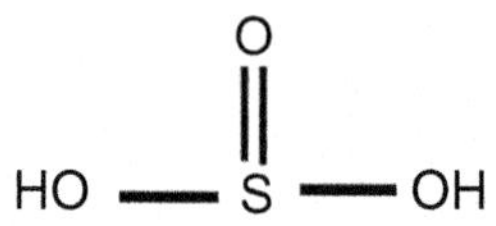

Sulfurous acid is also a relatively weak acid, with a pK_a of around 1.8, but slightly stronger than the two phosphorus-containing acids above.

Sulfur dioxide also reacts directly with bases such as sodium hydroxide solution. Bubbling sulfur dioxide through sodium hydroxide solution first forms sodium sulfite solution, followed by sodium hydrogen sulfite solution if the sulfur dioxide is in excess:

$$SO_2 + 2NaOH \rightarrow Na_2SO_3 + H_2O$$

$$Na_2SO_3 + H_2O \rightarrow 2NaHSO_3$$

Another important reaction of sulfur dioxide is with the base calcium oxide to form calcium sulfite. This is one of the important methods of removing sulfur dioxide from flue gases in power stations:

$$CaO + SO_2 \rightarrow CaSO_3$$

Sulfur trioxide (SO_3) reacts violently with water to produce concentrated sulfuric acid:

$$SO_3 + H_2O \rightarrow H_2SO_4$$

Sulfuric acid is a strong acid, and solutions will typically have a pH of approximately 0. The acid reacts with water to give a hydronium ion (a hydrogen ion in solution) and a hydrogen sulfate ion:

$$H_2SO_4(aq) + H_2O \rightarrow H_3O^+ + HSO_4^-(aq)$$

The second proton is more difficult to remove. In fact, the hydrogen sulfate ion is a relatively weak acid, similar in strength to the acids discussed above. This reaction is more appropriately described as an equilibrium:

$$HSO_4^-(aq) + H_2O \rightleftharpoons H_3O^+(aq) + SO4^{2-}(aq)$$

Sulfuric acid displays all the reactions characteristic of a strong acid. For example, a reaction with sodium hydroxide forms sodium sulfate; in this reaction, both of the acidic protons react with hydroxide ions as shown:

$$2NaOH + H_2SO_4 \rightarrow Na_2SO_4 + 2H_2O$$

In principle, sodium hydrogen sulfate can be formed by using half as much sodium hydroxide; in this case, only one of the acidic hydrogen atoms is removed.

Sulfur trioxide itself also reacts directly with bases such as calcium oxide, forming calcium sulfate:

$$CaO + SO_3 \rightarrow CaSO_4$$

This reaction is similar to the reaction with sulfur dioxide presented.

Chlorine forms several oxides, but only two (chlorine(VII) oxide, Cl_2O_7, and chlorine(I)oxide, Cl_2O) are considered here. Chlorine(VII) oxide is also known as dichlorine heptoxide, and chlorine(I) oxide as dichlorine monoxide. Chlorine(VII) oxide is the highest oxide of chlorine—the chlorine atom is in its maximum oxidation state of +7.

Chlorine(VII) oxide reacts with water to give the very strong acid, chloric(VII) acid, also known as perchloric acid:

$$Cl_2O_7 + H_2O \rightarrow 2HClO_4$$

As in sulfuric acid, the pH of typical solutions of perchloric acid are approximately 0. Neutral chloric(VII) acid has the following structure:

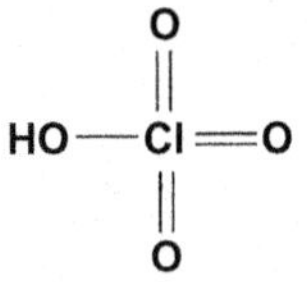

```
        O
        ||
HO —— Cl ══ O
        ||
        O
```

When the chlorate (VII) ion (perchlorate ion) forms by loss of a proton (in a reaction with water, for example), the charge is delocalized over every oxygen atom in the ion. That makes the ion very stable, making chloric(VII) acid very strong.

Chloric(VII) acid reacts with sodium hydroxide solution to form a solution of sodium chlorate(VII):

$$NaOH + HClO_4 \rightarrow NaClO_4 + H_2O$$

Chlorine(VII) oxide itself also reacts directly with sodium hydroxide solution to give the same product:

$$2NaOH + Cl_2O_7 \rightarrow 2NaClO_4 + H_2O$$

Chlorine(I) oxide is far less acidic than chlorine(VII) oxide. It reacts with water to some extent to give chloric(I) acid, HOCl (also known as hypochlorous acid).

$$Cl_2O + H_2O \rightleftharpoons 2HOCl$$

The structure of chloric(I) acid is exactly as shown by its formula, HOCl. It has no doubly-bonded oxygens, and no way of delocalizing the charge over the negative ion formed by loss of the hydrogen. Therefore, the negative ion formed not very stable, and readily reclaims its proton to revert to the acid. Chloric(I) acid is very weak ($pK_a = 7.43$) and reacts with sodium hydroxide solution to give a solution of sodium chlorate(I) (sodium hypochlorite):

$$NaOH + HOCl \rightarrow NaOCl + H_2O$$

Chlorine(I) oxide also reacts directly with sodium hydroxide to give the same product:

$$2NaOH + Cl_2O \rightarrow 2NaOCl + H_2O$$

5.7 Oxygen ions

Oxygen dissolves naturally when water comes in contact with air. Oxygen is highly reactive, and can therefore be applied to break down hazardous substances. It may also be applied as a bleach. Oxygen in ozone compounds is applied for drinking water disinfection. Waters are not contaminated by oxygen when it is applied industrially.

Gaseous oxygen does not react with water. It is water soluble and functions as an oxidizer:

$$O_2 + 2H_2O + 4e^- \rightarrow 4OH^-$$

Oxygen may oxidize organic matter. This is principally a biological process as might occur in a variety of environmental situations. Each individual compound has a reaction mechanism that can be described by means of an electron balance. Examples are given below (water is excluded from these equations):

$$Fe^{2+} + 0.25O_2 \rightarrow Fe(OH)_3 + 2.5H^+$$

$$Mn^{2+} + O_2 \rightarrow MnO_2 + 2H^+$$

$$NH_4^+ + 2O_2 \rightarrow NO_3^- + 6H^+$$

$$CH_4 + 2O_2 \rightarrow CO_2 + 4H^+$$

These mechanisms show that ammonium and methane apply large amounts of oxygen, and the resulting oxidation reactions form higher or lower amounts of acid. Under normal conditions acid in water reacts with HCO_3^-, forming CO_2.

The oxygen atom is very reactive and forms oxides with virtually all other elements, with the exception of helium, neon, argon, and krypton. There are also a large amounts of compounds that react with water.

Dissolved oxygen is an important determinant for stability of waters and survival of water organisms. Microorganisms may decompose organic substances in water by means of oxygen. Oxygen application per unit of time is indicated by biochemical oxygen demand (BOD). Organic pollutants may negatively influence water organisms, because they decrease the biochemical oxygen demand. Thermal pollution causes the same problem, because oxygen solubility is lower in warmer water. This may be a consequence of cooling water discharge on surface waters.

In eutrophic lakes and relatively enclosed sea areas, oxygen concentrations decrease strongly with depth. In some cases conditions may even be anaerobic. Natural examples of influences of temperature on oxygen concentrations in water and environmental impact are seasonal temperature changes in lakes. In winter the water has the same temperature and oxygen concentration everywhere. In summer water in surface layers in warmer than deeper water, resulting in lower oxygen solubility. Algae and plants in the surface layers work oppositely. They produce a high amount of oxygen at high temperatures, causing the water to become oxygen saturated. These plants die off pretty quickly, and are decomposed by microorganisms applying oxygen, which is now abundant in surface layers of the water source. However, organic matter often settles and remains on the bottom of a water body as sediment. This may cause oxygen deficits from decomposition. When an ecological equilibrium is established in lakes, these problems may be solved. However, when effects such as wastewater or contaminant discharge and/or the over fertilization of the soil add nutrient that must be decomposed and increase algal blooms, oxygen concentration may decrease to a level where no organism survives (eutrophication; eutrophic = nutrient rich, oligotrophic = nutrient poor).

Oxygen atoms can be found in a number of toxic organic and inorganic compounds. Toxic compounds are for example hyperoxides and peroxides. Some substances are toxic under low oxygen conditions in water, because breathing of organisms increases and consequently substances are absorbed more rapidly. For obligatory anaerobic organisms, high oxygen concentrations are toxic.

Ozone is an environmental pollutant when it is present in the troposphere. In the stratosphere it functions as a protective layer that reflects solar UV-radiation. Without this ozone layer, life on earth would be impossible. A number of plant species are susceptible to high ozone concentrations in air. This does not show as visible stress symptoms, but rather as growth limitations.

5.8 Redox chemistry

An oxidation-reduction (redox) reaction is a type of chemical reaction that involves a transfer of electrons between two species. An oxidation-reduction reaction is any chemical reaction in which the oxidation number of a molecule, atom, or ion changes by gaining or losing an electron. Redox reactions are common and vital to some of the basic functions of life, including photosynthesis, combustion, and corrosion or rusting.

Water contains hydrogen in the oxidation state +1 and oxygen in the oxidation state −2. Water oxidizes chemicals such as hydrides, alkali metals, and some alkaline earth metals. One example of an alkali metal reacting with water is:

$$2Na + 2H_2O \rightarrow H_2 + 2Na^+ + 2OH^-$$

Some other reactive metals, such as aluminum and beryllium are also oxidized by water but the oxides of these metals adhere to the metal and form a passive protective layer.

Water to emit oxygen gas, but very few oxidants react with water even if the reduction potential of theses oxidants is greater than the potential of O_2/H_2O. Almost all such reactions require a catalyst. An example of the oxidation of water is:

$$4AgF_2 + 2H_2O \rightarrow 4AgF + 4HF + O_2$$

The redox potential of liquid water varies over a range according to the solute(s) in the water (Table 2.5). This potential can be determined using an oxidation-reduction potential (ORP) electrode and is a measure of the collective redox potential of all of the chemical species in the everything in the water, including dissolved gasses such as oxygen. The potential of the solution is determined relative to the standard potential generated by the reference electrode and then corrected for that potential.

6 Water as a solvent

Because of the unique properties (Table 3.9), water has the power to dissolve more substances than any other liquid and, because of this power, water is often referred to as the *universal solvent*. Some substances, such as common table salt (sodium chloride, NaCl) dissolve in water very easily. When placed in water, sodium chloride molecules dissociate and the positively charged sodium ion (Na+) binds to the oxygen atom of the water while the negatively charged chloride ion (Cl-) binds to a hydrogen atom of water. This property of water allows for the transport of nutrients vital to life in animals and plants—when rainwater falls through the atmosphere to the surface of the Earth, the water dissolves atmospheric gases which, in turn, can affect the quality of the land, lakes and rivers, depending the gases dissolved in the rainwater. If the gases are sulfur oxides (sulfur dioxide, SO_2, and sulfur trioxide, SO_3) and/or nitrogen oxides (nitric oxide, NO, and nitrogen dioxide, NO_2), the rainwater is termed *acid rain*

Table 3.9 Summary of the unique properties of water

Property	Effects and significance
Excellent solvent	Transport of nutrients and waste products, making biological processes possible
Highest dielectric constant	High solubility of ionic substances and their ionization in solution
Higher surface tension than any other liquid	Controlling factor in physiology; governs drop and any surface phenomena
Transparent to visible and longer-wavelength fraction of UV light	Colorless, allowing light required for photosynthesis to reach considerable depths
Maximum density as a liquid at 4°C	Ice floats; vertical circulation restricted in stratified bodies of water
Higher heat of evaporation any other material	Determines transfer of heat
Higher latent heat of fusion than any other liquid except ammonia	Temperature stabilized at the freezing point of water
Higher heat capacity than any other liquid except ammonia	Stabilization of temperatures of organisms and any geographical regions

$$SO_2 + H_2O \rightarrow H_2SO_3\ (\text{sulfurous acid})$$

$$SO_3 + H_2O \rightarrow H_2SO_4\ (\text{sulfuric acid})$$

$$NO + H_2O \rightarrow HNO_2\ (\text{nitrous acid})$$

$$3NO_2 + 2H_2O \rightarrow HNO_3\ (\text{nitric acid})$$

Because of polarity of water, it is able to dissolve or dissociate many particles. Oxygen has a slightly negative charge, while the two hydrogens have a slightly positive charge. The slightly negative particles of a compound will be attracted to theh ydrgoenatom s of the water molecule, while the slightly positive particles will be attracted to oxygen of the water molecule; this causes the compound to dissociate.

Water is an excellent solvent due to the high dielectric constant. Substances that mix well and dissolve in water are hydrophilic (water-loving) substances, while those that do not mix well with water are hydrophobic (water-fearing) substances. The ability of a substance to dissolve in water is determined by whether or not the substance can match or exceed the strong attractive that water molecules generate between other water molecules. If a substance has properties that do not allow it to overcome these strong intermolecular forces, the molecules are precipitated from the water. Contrary to the common misconception, water and hydrophobic substances do not repel, and the hydration of a hydrophobic surface is energetically, but not entropically, favorable.

When an ionic or polar compound enters water, it is surrounded by water molecules (hydration). The relatively small size of water molecules (approximately 3 Å, 3 Å) allows many water molecules to surround one molecule of solute. The partially negative dipole ends of the water are attracted to positively charged components of the solute, and vice versa for the positive dipole ends.

In general, ionic and polar substances such as acid derivatives, alcohol derivatives, and salts are relatively soluble in water, and non-polar substances such as fats and oils are not. Non-polar molecules stay together in water because it is energetically more favorable for the water molecules to hydrogen bond to each other than to engage in van der Waals with non-polar molecules. Van der Waals forces are the relatively weak attractive forces that act on neutral atoms and molecules and that arise because of the electric polarization induced in each of the particles by the presence of other particles.

Dissolved gases (oxygen for fish and carbon dioxide for photosynthetic algae) are crucial to the welfare of living species in water. Carbon dioxide is produced by respiratory processes in waters and sediments and can also enter water from the atmosphere. Carbon dioxide is required for the photosynthetic production of biomass by algae and in some cases is a limiting factor. High levels of carbon dioxide produced by the degradation of organic matter in water can cause excessive algal growth and productivity. Dissolved oxygen frequently is the key substance in determining the extent and kinds of life in a body of water. Oxygen deficiency is fatal to many aquatic animals such as fish. The presence of oxygen can be equally fatal to many kinds of anaerobic bacteria. Without an appreciable level of dissolved oxygen, many kinds of aquatic organisms cannot exist in water. Dissolved oxygen is consumed by the degradation of organic matter in water. Many fish kills are caused not from the direct toxicity of pollutants but a deficiency of oxygen because of its consumption in the biodegradation of pollutants.

The solubility of oxygen in water depends upon water temperature, the partial pressure of oxygen in the atmosphere and the salt content of the water. Water in equilibrium with air cannot contain a high level of dissolved oxygen compared to many other solute species. If oxygen-consuming processes are occurring in the water, the dissolved oxygen level may rapidly approach zero unless some efficient mechanism for the re-aeration of water is operative, such as turbulent flow in a shallow stream or air pumped into the aeration tank of an activated sludge secondary waste treatment facility. The problem becomes largely one of kinetics, in which there is a limit to the rate at which oxygen is transferred across the air-water interface. This rate depends upon turbulence, air bubble size, temperature, and other factors.

The temperature effect on the solubility of gases in water is especially important in the case of oxygen. At higher temperatures, the decreased solubility of oxygen, combined with the increased respiration rate of aquatic organisms, frequently causes a condition in which a higher demand for oxygen accompanied by lower solubility of the gas in water results in severe oxygen depletion.

In addition to a wide array of other chemicals, carbon dioxide, bicarbonate ion, and carbonate ion have an extremely important influence upon the chemistry of water and also of wastewater. Many minerals are deposited as salts of the carbonate ion. Algae in water utilize dissolved carbon dioxide in the synthesis of biomass. Carbon dioxide is a weak acid in water. Because of the presence of carbon dioxide in air and its production from microbial decay of organic matter, dissolved carbon dioxide is present in virtually all natural waters and waste waters. Carbon dioxide is a weak acid so that rainfall even from an unpolluted atmosphere is slightly acidic due to the presence of dissolved carbon dioxide.

A large share of the carbon dioxide found in water is a product of the breakdown of organic matter by bacteria. Even algae, which utilize carbon dioxide in photosynthesis produce it through their metabolic processes in the absence of light. As water seeps through layers of decaying organic matter while infiltrating the ground, it may dissolve a great deal of carbon dioxide produced by the respiration of organisms in the soil. Later, as water goes through limestone formations, it dissolves calcium carbonate because of the presence of the dissolved carbon dioxide. This process is the one by which limestone caves are formed.

6.1 Effect of solutes

A solution is a homogenous (uniform throughout) mixture, on a molecular level, of two or more substances—typically the dissolved phase (the solute) and the dissolving phase (in the current context—water). The solvent typically determines the physical state of the solvent while the solutes (solids, liquids, or gasses) in a solution have an effect on the properties. Unlike pure substances, solutions do not have a definite composition and the composition is dependent on the amount of solute dissolved in the solvent. Concentrated solutions have relatively high amounts of solute dissolved in the solvent while dilute solutions have relatively low amounts. The concentration of a solution is typically expressed in terms of the weight of the solute (such as grams of solute) per volume of solvent (i.e. per liter of solvent).

Solutes perturb the structure of water, primarily in the first hydration shell (the layer of water in contact with the solvent)of the solute, with a lesser effect on more distant waters. In contrast, solutes have a large effect on the angular structure of water (Fig. 3.1). Non-polar solutes and groups shift the bimodal distribution of water-water Hydrogen bond angles towards the more ice-like, linear form, effectively increasing the ordering of water by decreasing the less ordered population of hydrogen bonds. These solutes lack the ability to make strong electrostatic interactions with water, and they interact primarily through the van der Waals potential. Their effect is essentially geometric: they tend to displace the more weakly hydrogen bonded facial water in the coordination shell, thus reducing the population of more bent hydrogen bonds. Ions and polar solutes and groups have the opposite effect. They shift the distribution of water-water hydrogen bond angles towards the more bent form.

This is a consequence of the strong electrostatic interactions they can make with water. Water dipoles tend to align towards or away from the atoms with large atomic partial charges, consequently distorting the water-water hydrogen bond.

6.2 Hydrophobic and hydrophilic interactions

The word *hydrophobic* describes the fact that nonpolar substances don't combine with water molecules. Water is a polar molecule because of the partial charge between the atoms (Fig. 3.1) and, therefore, any materials with a charge (negative or positive) will be able to interact with water molecules to dissolve. Thus, hydrophobic molecules (molecules that do not have a charge and are non-polar) do not have the possibility of charge-to-charge interactions that will allow them to interact with water or if a surface is hydrophobic, it tends to be not to be wetted by water. Hydrophobic materials often do not dissolve in water or in any solution that is predominantly aqueous. On the other hand, a hydrophilic molecule has an affinity for water and tends to dissolve in or mix with water or if a surface is hydrophilic tends to be easily wetted by water.

Water can dissolve a remarkable variety of important molecules, ranging from simple salts through small molecules such as sugars and metabolites to very large molecules such as proteins and nucleic acids. In fact water is sometimes called the universal solvent. Practically all the molecular processes essential to life—chemical reactions, association and binding of molecules, diffusion-driven encounters, ion conduction—will only take place at significant rates in solution, hence the importance of solvent properties of water. Equally important as the ability of water to act as a good solvent is the differential effect as a solvent—the fact that water dissolves some molecular species much better than others.

At the low solubility end are aliphatic amino acids such as leucine, the aromatic amino acids such as phenylalanine, and the hydrocarbon 'tails' of lipids. These solutes are hydrophobic. Other solutes such as nucleic acid bases and the amino acid tryptophan have intermediate solubility, and cannot be simply classified as hydrophobic or hydrophilic.

6.3 Physical aspects of solvation

Solvation is the interaction of a solvent with dissolved molecules (the solute). Ionized and uncharged molecules interact strongly with solvent, and the strength and nature of this interaction influences many properties of the solute, including solubility, reactivity, and color, as well as influencing the properties of the solvent such as the viscosity and density. In the process of solvation, solute ions are surrounded by a concentric shell of solvent molecules and, thus, solvation is the process of reorganizing solvent and solute molecules into solvation complexes. Solvation involves bond formation—typically, hydrogen bonding or van der Waals forces—and, in the case of water, the process is referred to as *hydration*.

The logarithm of the solubility of a solute is proportional to the thermodynamic work, or hydration free energy necessary to transfer it into water from a reference solvent (here cyclohexane). High water solubility corresponds to a negative (favorable) hydration free energy, low solubility to a positive hydration free energy (work must be performed to dissolve the solute). Hydration free energy is directly related to the properties of the solute, the water, and the strength of interactions between water and solvent. It is here that the high surface tension and dielectric constant of water are crucial. The surface tension is the work necessary to create a unit area of water-vacuum interface (units of force per unit length are equivalent to energy per unit area). Work is necessary since interactions must be broken to bring water from the interior to the surface. Hydrating a solute can be divided into two steps: (i) creation of a solute-shaped cavity in water, which requires work to be done against the surface tension of water, and (ii) placing the solute in the cavity, which involves interactions of the solute with water molecules and restructuring of the water.

The first step always opposes dissolution of any solutes. If the interactions between the solute and water are weak, as they are for non-polar solutes and groups, the cavity term dominates and the solubility will be low. The cavity term drives aggregation of non-polar molecules to reduce the surface area in contact with solvent. This is known as the hydrophobic effect. In contrast, when a polar or ionic solute is dissolved in water the electric field from the partial atomic charges of the solute induces a large polarization (reorientation) of the water dipoles resulting in an attractive electrostatic field (the reaction field) back at the solute. This results in a high solubility—a consequence of high dielectric constant of water, and the reason that water can dissolve a wide range of ionic solutes and polar solutes.

In summary, the solubility is determined by two major contributions: (i) the cavity contribution, which is unfavorable and approximately proportional to area of the solute or solute group(s) exposed to water, and (ii) the electrostatic contribution, which depends on the strength of the reaction field induced in water. This in turn depends on the magnitude of the partial atomic charge, the dielectric constant of water, and how near the atomic charge is to the water (i.e. the radius of the atom, and whether atom is buried or exposed to solvent).

6.4 Role of solvation

Solvation is the interaction of a solvent with the dissolved solute—in the case of water, solvation is often referred to as *hydration*. Solvent polarity is the most important factor in determining how well it solvates a particular solute. Polar solvents have molecular dipoles, meaning that part of the solvent molecule has more electron density than another part of the molecule. The part with more electron density will experience a partial negative charge while the part with less electron density will experience a partial positive charge. Polar solvent molecules can solvate polar solutes and ions because they can orient the appropriate partially charged portion of the molecule towards the solute through electrostatic attraction. This stabilizes the system and creates a solvation shell (or hydration shell in the case of water) around each particle of solute (Adreev et al., 2018).

In the solvation process, ionized and uncharged molecules interact strongly with solvent, and the strength and nature of this interaction influences many properties of the solute, including solubility, reactivity, and color, as well as influencing the properties of the solvent such as the viscosity and density. Solvation involves bond formation, hydrogen bonding, and van der Waals forces. Thus solvation is an interaction of a solute with the solvent which leads to stabilization of the solute species in the solution. Solvation is, in concept, distinct from solubility.

Many biological macromolecules, such as proteins, nucleic acids and lipids, contain both hydrophilic and hydrophobic groups. The differential ability of water to solvate the different groups produces a driving force for them to adopt structures or self-assemble in ways where the hydrophilic groups are exposed to water and the hydrophobic groups are sequestered from water. This is a major factor in the folding, assembly and maintenance of precise, complex three-dimensional structures of proteins, membranes, nucleic acids and protein-nucleic acid assemblies.

For example, the hydrophobic effect promotes: (i) the formation of a buried non-polar core of amino acids in protein folding, (ii) helix formation in nucleic acids through base stacking, (iii) the formation of lipid membranes with an non-polar lipid tail region, (iv) the formation of macromolecular complexes such as multimeric proteins, protein-nucleic acid assemblies and membrane protein-lipid assemblies, and (v) the specific binding and recognition of molecules with complementary non-polar surface groups.

Solvation of polar groups acts in a reverse fashion to the hydrophobic effect in the above processes: there is a strong driving force to keep the ionic and polar portions of proteins, lipids and nucleic acids on the surface in contact with water. This is also the reason that the low dielectric lipid tail region of membranes is impervious to ions, a key property of biological membranes. The delicate balance between polar and non-polar solvation forces contributes to a remarkable fidelity and accuracy of self-assembly.

The ability of water to solvate different molecular types produces a driving force for them to adopt structures or self-assemble in ways where the hydrophilic groups are exposed to water and the hydrophobic groups are sequestered from water. Solvation of polar groups acts in a reverse fashion to the hydrophobic effect in the above processes: there is a strong driving force to keep the ionic and polar portions of molecules in contact with water. This is also the reason that the low dielectric lipid tail region of membranes is impervious to ions, a key property of biological membranes. The delicate balance between polar and non-polar solvation forces contributes to a remarkable fidelity and accuracy of self-assembly.

Solvation also affects host-guest complexation since many host molecules have a hydrophobic pore that readily encapsulates a hydrophobic guest. These interactions can be used in applications such as drug delivery, such that a hydrophobic drug molecule can be delivered in a biological system without needing to covalently modify the drug in order to solubilize it.

References

Adreev, M., De Pable, J., Chremos, A., Douglas, J.F., 2018. Influence of ion solvation on the properties of electrolyte solutions. J. Phys. Chem. B 122 (14), 4029–4034.

Akpor, O., Muchie, M., 2010. Remediation of heavy metals in drinking water and wastewater treatment systems: Processes and applications. Int. J. Phys. Sci. 5 (12), 1807–1817.

Birkle, P., Torres Alvarado, I.S., 2010. Water-Rock Interaction XIII. CRC Press, Taylor & Francis Group, Boca Raton, Florida.

Botta, O., Bada, J.L., 2002. Extraterrestrial organic compounds in meteorites. Surv. Geophys. 23, 411–467.

Brantley, S., Kubicki, J., White, A. (Eds.), 2008. Kinetics of Water-Rock Interaction. Springer, New York.

Dragon, K., Marciniak, M., 2010. Chemical composition of groundwater and surface water in the Arctic environment (Petuniabukta region, Central Spitsbergen). J. Hydrol. 386, 160–172.

Drever, J.I., 1997. The Geochemistry of Natural Waters: Surface and Groundwater Environments, third ed. Prentice Hall, Englewood Cliffs, New Jersey.

Drever, J.I., Zobrist, J., 1992. Chemical weathering of silicate rocks as a function of elevation in the southern Swiss Alps. Geochim. Cosmochim. Acta 56, 3209–3216.

Guseva, N., 2016. The origin of the natural water chemical composition in the permafrost region of the eastern slope of the Polar Urals. Water 8, 594–613.
Hinman, N.W., 2013. Water-rock interaction and life. Proc. Earth Planet. Sci. 7, 354–359.
Ibrahim, R.G.L., Korany, E.A., Tempel, R.N., Gomaa, M.A., 2019. Processes of water–rock interactions and their impacts upon the groundwater composition in Assiut area, Egypt: Applications of Hydrogeochemical and multivariate analysis. J. African Earth Sci. 149, 72–83.
Millot, R., Gaillardet, J., Dupré, B., Allègre, C.J., 2002. The global control of silicate weathering rates and the coupling with physical Erosion: New insights from Rivers of the Canadian shield. Earth Planet. Sci. Lett. 196, 83–98.
Morel, F.M.M., Hering, J.G., 1993. Principles and Applications of Aquatic Chemistry. John Wiley and Sons Inc, Hoboken, New Jersey.
Morgan, B., Lahav, O., 2009. The effect of pH on the kinetics of spontaneous Fe(II) oxidation by O2 in aqueous solution – Basic principles and a simple heuristic description. Geochim. Cosmochim. Acta 73, 6631–6677.
Rapaport, D.C., 1983. Hydrogen bonds in water. Mol. Phys. 50 (5), 1151.
Reynolds Jr., R.C., Johnson, N.M., 1972. Chemical weathering in the temperate glacial environment of the Northern Cascade Mountains. Geochim. Cosmochim. Acta 36, 537–554.
Speight, J.G., 2014. The Chemistry and Technology of Petroleum, fifth ed. CRC Press, Taylor & Francis Group, Boca Raton, FL.
Speight, J.G., 2017. Handbook of Petroleum Refining. CRC Press, Taylor & Francis Group, Boca Raton, FL.
Spencer, R.G.M., Aiken, G.R., Wickland, K.P., Striegl, R.G., Hernes, P.J., 2008. Seasonal and spatial variability in dissolved organic matter quantity and composition from the Yukon River basin, Alaska. Global Biogeochem. Cycles. Page 22.
Stober, I., Bucher, K. (Eds.), 2002. Water-Rock Interaction. Springer Netherlands, Rotterdam, Netherlands.
Stumm, W., Morgan, J., 1996. Aquatic Chemistry. Chemical Equilibria and Rates in Natural Waters, third ed. John Wiley & Sons, Inc, Hoboken, New Jersey.
Urben, P.G., 2007. Bretherick's Handbook of Reactive Chemical Hazards, seventh ed. Academic Press, New York.
Wisniak, J., 2004. The nature and composition of water. Ind. J. Chem. Technol. 11, 434–444.

Further reading

Bandura, A.V., Lvov, S.N., 2006. The ionization constant of water over wide ranges of temperature and density. J. Phys. Chem. Ref. Data Monogr. 35 (1), 15–30.
Geissler, P.L., Dellago, C., Chandler, D., Hutter, J., Parrinello, M., 2001. Autoionization in liquid water. Science 291 (5511), 2121–2124.

Thermodynamics of water

4

Chapter outline

1 Introduction

Energy occurs in various forms in the natural universe and may, for example, have the form of radiation, heat, electricity, motion, or chemical interaction. The principle of conservation of energy states, however, that although its form may change, the total amount of energy in the universe remains constant. This principle is also known as the first law of thermodynamics. A second broad principle, based on experience and observation, states that energy transfers occur only along favorable potential gradients. For example, water flows down slopes, heat passes from hot objects to cooler ones, and electrical currents flow from points of high potential to points of lower potential. This general principle also implies that energy in any closed system tends to become evenly distributed. It is known as the second law of thermodynamics.

Achieving the goal of understanding these processes, and being able to make quantitative statements about the processes, requires the application of theoretical analysis to develop tentative models (conceptual models) which can be quantified and tested using experimental data. Thermodynamic principles, applied to chemical energy transfers, form a basis for evaluating quantitatively the feasibility of various possible

Natural Water Remediation. https://doi.org/10.1016/B978-0-12-803810-9.00004-8

chemical processes in natural water systems, for predicting the direction in which chemical reactions may proceed, and in many instances for predicting the actual dissolved concentrations of reaction products that should be present in the water.

The fundamental concepts relating to chemical processes that are most useful in developing a unified approach to the chemistry of natural water are mainly related to chemical thermodynamics and to reaction mechanisms and rates. Thermodynamics also offers a unified way of viewing chemical and physical processes occurring in natural systems, but it has not been applied this way in hydrology to any significant degree. The total energy in a groundwater system, for example, includes components of gravitational, thermal, and chemical energy, but generalized thermodynamic treatments of hydrologic systems including all three parameters are rare. The term *system* as used here refers to a body of correlating chemical processes with biological or physical water, its dissolved material, and the potentially inter-processes. However, for many environmental effects it is acting solids and gases that are in contact with the water.

Systems at the surface of the Earth are continually responding to energy inputs derived from solar radiation or from the radiogenic heat in the interior. These energy inputs drive plate movements and erosion, exposing metastable mineral phases at the surface and, in addition, these energy fluxes are harvested and transformed by living organisms. As long as these processes persist, chemical disequilibrium at the surface of the Earth will be perpetuated. Chemical disequilibrium is also driven by human activities related to production of food, extraction of water and energy resources, and burial of wastes. To understand how the surface of the Earth will change over time, it is necessary to understand the rates at which reactions occur and the chemical feedbacks that relate these reactions across extreme temporal and spatial scales. This book addresses fundamental and applied questions concerning the rates of water-rock interactions driven by tectonic, climatic, and anthropogenic forces.

The present Chapter aims to present the current understanding the thermodynamics of water through an understanding of hydrogen bonding in liquid water (Xu and Goddard III, 2004).

2 The states of water

Water is by far, the most common liquid on Earth and, like all liquids, is difficult to compress but that conforms to the shape of its container and retains a (nearly) constant volume independent of pressure. As such, it is one of the four fundamental states of matter (the others being gas, solid, and plasma), and is the only state with a definite volume but no fixed shape.

Water is made up of small vibrating particles of matter held together by intermolecular bonds. Water able to flow and take the shape of the container but unlike a gas, water does not disperse to fill every space of a container, and maintains a fairly constant density. A distinctive property of the water in the liquid state is surface tension, leading to the phenomenon of wetting. In addition, the density of water liquid is usually close to the density of ice but and much higher than the density of steam (Chapter 2) (Vedamuthu et al., 1994).

In addition, the mechanisms of atomic/molecular diffusion (or particle displacement) in liquids are closely related to the mechanisms of viscous flow and solidification in liquids. In fact, descriptions of viscosity in terms of molecular free space within the liquid were modified as needed in order to account for liquids, such as water, in which the molecules are known to be associated in the liquid state at ambient temperature. When various molecules combine together to form an associated molecule, they enclose within a semi-rigid system a certain amount of space which before was available as free space for mobile molecules. Thus, increase in viscosity upon cooling due to the tendency of most substances to become *associated* on cooling. Similar arguments could be used to describe the effects of pressure on viscosity, where it may be assumed that the viscosity is chiefly a function of the volume for liquids with a finite compressibility. An increasing viscosity with rise of pressure is therefore expected. In addition, if the volume is expanded by heat but reduced again by pressure, the viscosity remains the same.

The tendency to orientation of molecules in small groups allows the liquid a certain degree of association. This association results in a considerable internal pressure within a liquid, which is due almost entirely to those molecules which, on account of the temporary low velocity have coalesced with other molecules. The internal pressure between several such molecules might correspond to that between a groups of molecules in the solid form. In addition, water has been shown to have exhibit bilinear behavior in the 0–100 °C (32–212 °F) that defines a crossover temperature at 50 ± 10 °C (120 ± 18 °F) which supports the hypothesis that there are two states of liquid water. These two states play an important role in the thermal and optical properties of nanomedical systems and also the structure of liquid water strongly influences the stability of hydrogen bonding in proteins (Breiten et al., 2013; Weldon et al., 2014).

However, water molecules in water vapor have few hydrogen bonds and more space between them, making vapor light and less dense than water or ice. While the water molecules are closer together in liquid water than in solid ice, there are fewer hydrogen bonds in liquid water than in the rigid lattice of ice (Némethy and Scheraga, 1964; Abraham and Acree Jr, 2012). Therefore, water is fluid whereas ice is solid.

In thermodynamics, a state of matter is one of the distinct forms in which matter can exist. A state of matter is also characterized by phase transitions—the phase transition indicates a change in structure and can be recognized by an abrupt change in properties. As an example, when ice melts to form liquid water at 0 °C (32 °F), the thermal conductivity changes from 2.18 W/(m·K) to 0.56 W/(m·K). A distinct state of matter can be defined as any set of states distinguished from any other set of states by a phase transition, such as the properties (Table 4.1). All matter can change from one **state** to another—in some cases it might require extreme temperatures or extreme pressures, but it can be done and under the appropriate conditions each of the four states of matter—viz.: (i) gas, (ii) liquid, (iii) and (iv) plasma—are interchangeable (Table 4.2) with distinct properties for each of the states of water (Table 4.3).

Matter in the gaseous state (also called the *vapor state*) has both variable volume and shape, adapting both to fit its container and the particles are neither close together nor fixed in place. A gas is a compressible fluid—not only will a gas conform to the shape of its container but it will also expand to fill the container. In a gas, the

Table 4.1 Common properties of water (H_2O)

IUPAC name	**Water (rarely used: oxidane)**
Other (uncommon) names	Rarely used: hydrogen hydroxide, hydrogen oxide, dihydrogen monoxide, hydrogen monoxide, dihydrogen oxide, hydric acid, hydrohydroxic acid, hydroxic acid, hydrol
Molar mass	18.01528(33) g/mol
Appearance	White crystalline solid, colorless liquid
Odor	None
Crystal structure	Hexagonal
Molecular shape	Angular (bent)
Dipole moment	1.8546 D
Conjugate acid	Hydronium, H_3O^+
Conjugate base	Hydroxide, OH^-
Density	Liquid: 0.9998396 g/mL at 0 °C, 0.9970474 g/mL at 25 °C, 0.961893 g/mL at 95 °C; Solid: 0.9167 g/ml at 0 °C
Melting point	0 °C (32 °F; 273 K)
Boiling point	100 °C (212 °F; 373.13 K)
Vapor pressure	3.1690 kilopascals, 0.031276 atm, 044 psi
Solubility	Poorly soluble in haloalkane derivatives, aliphatic and aromatic hydrocarbon derivatives, ether derivatives. Miscible with methanol, ethanol, propanol, isopropanol, acetone, glycerol, 1,4-dioxane, tetrahydrofuran, sulfolane, acetaldehyde, dimethylformamide, dimethoxyethane, dimethyl sulfoxide, acetonitrile. Partially miscible with diethyl ether, methyl ethyl ketone, dichloromethane, ethyl acetate, bromine.
Viscosity	0.890 cP (centipoise)
Refractive index(n_D)	1.3330 (20 °C)
Surface tension	Water-air: 72.86 ± 0.05 mN m^{-1} at 20 °C
Heat capacity (C)	75.385 ± 0.05 J/(mol·K)
Thermal conductivity	0.6089 W/(m·K)
Standard molar entropy ($S^o{}_{298}$)	69.95 ± 0.03 J/(mol·K)
Standard enthalpy of formation($\Delta_f H^o{}_{298}$)	−285.83 ± 0.04 kJ/mol
Gibbs free energy($\Delta_f G$)	−237.24 kJ/mol

molecules have sufficient kinetic energy so that the effect of intermolecular forces is small (or zero for an ideal gas) and the typical distance between neighboring molecules is much greater than the molecular size. A gas has no definite shape or volume, but occupies the entire container in which it is confined. At temperatures below the critical temperature, a gas can be liquefied by compression alone without cooling. A gas can exist in equilibrium with a liquid (or solid), in which case the gas pressure is equal to the vapor pressure of the liquid (or solid). A supercritical fluid (SCF) is a gas in which the temperature and pressure are above the critical temperature and critical

Table 4.2 Interrelationship of the various states of matter of water

Original state	Process	Product state
Gas	Condensation	Liquid
Gas	Deposition	Solid
Gas	Ionization	Plasma
Liquid	Vaporization	Gas
Liquid	Freezing	Solid
Solid	Sublimation	Gas
Solid	Melting	Liquid
Plasma	Recombination	Gas

Table 4.3 Distinction between the gas, liquid, and solid states of matter

	Solid	Liquid	Gas
Shape	Definite shape	Indefinite shape	Indefinite shape
Volume	Definite volume	Definite volume	Indefinite volume
Inter particular forces	Strong inter particular forces	Weaker	Inter-particular forces are negligible
Inter-particular space	Negligible	Comparatively large inter particular space	Very large inter particular space
Particular motion	Vibratory motion	Particle motion is slow	Particle motion is very rapid and random
Packing of particles	Closely packed	Particles are loosely packed	Particles are very loosely packed
Compressibility	Incompressible	Compressible	Highly compressible
Density	Very high density	Low density	Very low density

pressure, respectively. In this state, the distinction between liquid and gas disappears insofar as a supercritical fluid has the physical properties of a gas but the high density of the supercritical fluid confers solvent properties in some cases.

Compared to the other states of matter, gases have low density and viscosity. Pressure and temperature influence the particles within a certain volume to produce a variation in particle separation and speed (referred to as *compressibility*) which influences the optical properties and, finally, gas particles spread apart or diffuse in order to homogeneously distribute themselves throughout any container.

Water vapor is the gaseous phase of water and is produced from the evaporation or boiling of liquid water of from the sublimation of ice. Unlike other forms of water, water vapor is invisible—it is less dense than air and initiates convection currents that can lead to clouds. Being a component of the hydrosphere hydrologic cycle of the Earth, it is particularly abundant in atmosphere where it is also a potent greenhouse gas along with other gases such as carbon dioxide and methane. Use of water vapor, as steam, has been important to humans for cooking and as a major component in energy production and transport systems since the Industrial Revolution.

Matter in the liquid state maintains a fixed volume, but has a variable shape that adapts to fit its container and the particles are still close together but move freely. Typically, a liquid is a difficult to compress (most liquids are nearly incompressible) to a shape that conforms to the shape of the container but retains a (nearly) constant volume independent of pressure. The volume is definite if the temperature and pressure are constant. However, this means that the shape of a liquid is not definite but is determined by its container. The volume is usually greater than that of the corresponding solid, the best known exception being water. The highest temperature at which a liquid can exist is the critical temperature of the liquid. A liquid may be converted to a gas by heating at constant pressure to the boiling point, or else by reducing the pressure at constant temperature.

Water in the liquid state is nearly incompressible fluid that conforms to the shape of the container but retains a (nearly) constant volume independent of pressure. As such, it is the only state with a definite volume but no fixed shape. Water, like most liquids, resists compression, although others can be compressed. Unlike a gas, a liquid does not disperse to fill every space of a container, and maintains a fairly constant density. A distinctive property of the liquid state is the surface tension which is related to the phenomenon of the liquid being able or unable) to wet a surface.

The density of the liquid water is usually close to that of ice and much higher than the density of the gas. Therefore, liquid water and solid water (ice) are both termed condensed mater. Although liquid water is abundant on Earth, this state of matter is actually the least common in the known universe, because liquids require a relatively narrow temperature/pressure range to exist.

Matter in the solid state maintains a fixed volume and shape, with component particles (atoms, molecules or ions) close together and fixed into place. In a crystalline solid, such as ice, the particles (atoms, molecules, or ions) are packed in a regularly ordered, repeating pattern. There are various different crystal structures for ice—the same substance can have more than one structure (or solid phase).

Ice molecules can exhibit at least fifteen phases (packing geometries) that depend on temperature and pressure. When water is cooled rapidly (quenching), up to three different types of amorphous ice can form depending on the pressure history and the temperature history. Amorphous ice is an ice lacking crystal structure and exists in three forms: low-density amorphous ice (LDA ice) formed at atmospheric pressure, or below, high density amorphous ice (HAD ice) and very high density amorphous ice (VHDA ice), forming at higher pressures. Low-density amorphous ice forms by extremely quick cooling of liquid water (hyper-quenched glassy water, HGW), by depositing water vapor on very cold substrates (amorphous solid water, ASW) or by heating high density forms of ice at ambient pressure (LDA ice).

Ice may also be deposited directly by water vapor, as happens in the formation of frost. When ice melts, it absorbs as much energy as it would take to heat an equivalent mass of water by 80 °C (144 °F). During the melting process, the temperature remains constant at 0 °C (32 °F). While melting, any energy added breaks the hydrogen bonds between ice (water) molecules. Energy becomes available to increase the thermal energy (temperature) only after enough hydrogen bonds are broken that the ice can be considered liquid water—the energy consumed in breaking hydrogen bonds in the transition from ice to water is the heat of fusion.

Matter in the plasma state has variable volume and shape, but as well as neutral atoms, it contains a significant number of ions and electrons, both of which can move around freely. Like a gas, a plasma does not have definite shape or volume but unlike a gas a plasma is (i) electrically conductive, produces a magnetic fields and an electric current, and (ii) responds strongly to electromagnetic forces. Positively charged nuclei exist in the midst freely-moving disassociated electrons.

A gas is usually converted to a plasma in one of two ways such as (i) from a huge voltage difference between two points, or (ii) by exposing the gas to extremely high temperatures. Heating matter to high temperatures causes electrons to leave the atoms, resulting in the presence of free electrons which creates a partially ionized plasma. At very high temperatures, it is assumed that essentially all electrons are free and that a very high-energy plasma is essentially non-electron nuclei existing in a sea of electrons which forms a fully ionized plasma. The plasma state does not exist under normal conditions on Earth but is commonly generated by either lightning, electric sparks, fluorescent lights, and neon lights.

3 Thermodynamics

Thermodynamics is the branch of physics that deals with the relationships between heat and other forms of energy. In particular, thermodynamics can be used to describe the methods by which thermal energy is converted to and from other forms of energy and how it affects matter. The science of thermodynamics has been developed over centuries, and its principles apply to nearly every device ever invented. Its importance in modern technology, especially in the behavior of water in various systems, cannot be overstated (Kühne et al., 2006).

Thermodynamics is concerned with the properties of matter—as the name suggest, foremost among these is heat. Heat is energy transferred between substances or systems due to a temperature difference between them. As a form of energy, heat is conserved insofar as it cannot be created or destroyed. However, heat can be transferred from one place to another (from system to system) and can also be converted to and from other forms of energy. The amount of heat transferred by a substance or system depends on the speed and number of atoms or molecules in motion—the faster the atoms or molecules move, the higher the temperature, and the more atoms or molecules that are in motion, the greater the quantity of heat they transfer.

All properties of matter are either extensive or intensive and either physical or chemical. Extensive properties, such as mass and volume, depend on the amount of matter that is being measured. Intensive properties, such as density and color, do not depend on the amount of matter. Both extensive and intensive properties are physical properties, which means they can be measured without changing the chemical identity of a substance. For example, the freezing point of a substance is a physical property—when water freezes, it is still water—the formula (H_2O) has not changed but the physical state has changed.

Knowledge of the thermodynamic aspects of water science is an important part of developing an understanding of and predicting phase and chemical equilibria in industrial and natural aqueous systems at elevated temperatures and pressures. Such systems contain a variety of organic and inorganic solutes ranging from non-polar nonelectrolytes to strong electrolytes. Temperature and pressure strongly affect speciation of solutes that are encountered in molecular or ionic forms, or as ion pairs or complexes. Properties such as the thermodynamic equilibrium constants of hydrothermal reactions and activity coefficients of aqueous species, are required for practical use by geologists, power-cycle chemists and process engineers.

Derivative properties (such as enthalpy, heat capacity and volume are useful in extrapolations when calculating the Gibbs energy at conditions remote from ambient. They also indicate evolution in molecular interactions with changing temperature and pressure. In the thermodynamics of hydrothermal solutions, the unsymmetrical standard-state convention is generally used; in this case, the standard thermodynamic properties (STP) of a solute reflect its interaction with the solvent (water), and the excess properties, related to activity coefficients, correspond to solute-solute interactions. For dilute and moderately concentrated solutions, the standard-state functions have a dominant role and can be used as a reasonable approximation for semi-quantitative modeling. The solute-solvent interactions particularly prevail at near-critical conditions, where all the STP of solutes undergo rapid variations. The standard derivative properties of a solute scale with the thermal expansivity and isothermal compressibility of the solvent and diverge at the solvent critical point. In addition, this extreme behavior strongly affects the properties of solutions in a relatively wide range of conditions below and above the critical point.

Temperature is a measure of the average kinetic energy of the particles in a sample of matter, expressed in terms of units or degrees designated on a standard scale. The most commonly used temperature scale is Celsius (often referred to as the Centigrade scale), which is based on the freezing and boiling points of water, assigning respective values of 0 degrees Celsius (0 °C) and 100 degrees Celsius (100 °C). The Fahrenheit scale is also based on the freezing and boiling points of water which have assigned values of 32 degrees Fahrenheit (32 °F) and 212 degrees Fahrenheit (212 °F), respectively. Another scale that is often used in thermodynamics is the Kelvin scale (shown as K with no degree sign) which uses the same increment as the Celsius scale, i.e., a temperature change of 1 °C is equal to 1 K. However, the Kelvin scale starts at absolute zero, the temperature at which there is a total absence of heat energy and all molecular motion stops. Thus, a temperature of 0 K is equal to 273.15 °C (−459.67 °F).

Thus, the thermal energy is the energy a substance or system has due to its temperature, i.e., the energy of moving or vibrating molecules and thermodynamics involves measuring this energy. A system, such as water, consists of large numbers of molecules interacting in complicated ways are at equilibrium, the system can be described using criteria such as (i) the mass of the system, (ii) the pressure of the system, and (iii) the volume of the system. In addition, the word system is very commonly used in thermodynamics. A certain quantity of matter or the space which is under thermodynamic study or analysis is called as system. The system is covered by the boundary and the area beyond the boundary is called as universe or surroundings. The boundary of the system can be fixed or it can be movable. Between the system and surrounding the exchange of mass or energy or both can occur. There are three mains types of system: (i) an open system, (ii) a closed system, and (iii) an isolated system.

An *open system* is a system in which the transfer of mass as well as energy can take place across its boundary is called as an open system. An example of such a system is an engine in which case fuel is provided to the engine which then produces power—there is an exchange of mass as well as energy. The engine also emits heat which is exchanged with the surroundings.

A *closed system* is a system in which the transfer of energy takes place across the boundary of the system with the surrounding, but no transfer of mass takes place. The closed system is a fixed-mass system. Using the engine as the example, the gas that is compressed in the piston and cylinder arrangement is an example of the closed system. In this case the mass of the gas remains constant but it can get heated or cooled.

The *isolated system* is a system in which neither the transfer of mass nor that of energy takes place across its boundary with the surroundings. For example—again using the engine as the example—if the piston and cylinder arrangement in which the gas is being compressed or expanded is insulated it becomes isolated system. Here there will neither transfer of mass nor that of energy.

Thermal energy is the energy a substance or system has due to its temperature, i.e., the energy of moving or vibrating molecules. Thus, thermodynamics is concerned with several properties of matter—foremost among these is heat. Heat is energy transferred between substances or systems due to a temperature difference between them. As a form of energy, heat is conserved, i.e., it cannot be created or destroyed. It can, however, be transferred from one place to another. Heat can also be converted to and from other forms of energy. For example, a steam turbine can convert heat to kinetic energy to run a generator that converts kinetic energy to electrical energy.

Specific heat (also called the heat capacity) is the amount of heat required to increase the temperature of a certain mass of a substance by a specific amount. The conventional unit for this is calories per gram per kelvin. The calorie is defined as the amount of heat energy required to raise the temperature of 1 g of water at 4C by 1°. The specific heat of a metal depends almost entirely on the number of atoms in the sample, not its mass. The specific heat of a gas is more complex and depends on whether it is measured at constant pressure or constant volume.

The thermal conductivity (k) is the rate at which heat passes through a specified material, expressed as the amount of heat that flows per unit time through a unit area with a temperature gradient of one degree per unit distance. The unit for k is watts (W) per meter (m) per kelvin (K). Values of k for metals such as copper and silver are relatively high at 401 and 428 W/m·K, respectively. This property makes these materials useful for automobile radiators and cooling fins for computer chips because they can carry away heat quickly and exchange it with the environment. The highest value of k for any natural substance is diamond at 2200 W/m·K.

Heat transfer is the transfer of heat from one body to another or between a body and the environment by three different means: conduction, convection and radiation. Conduction is the transfer of energy through a solid material. Conduction between bodies occurs when they are in direct contact, and molecules transfer their energy across the interface. Convection is the transfer of heat to or from a fluid medium. Molecules in a gaseous state or in a liquid state that are in contact with a solid body transmit or absorb heat to or from that body and then move away, allowing other molecules to move into place and repeat the process. Efficiency can be improved by increasing the surface area to be heated or cooled, as with a radiator, and by forcing the fluid to move over the surface, as with a fan.

Entropy is another property that is part of thermodynamics and all thermodynamic systems generate waste heat. This waste results in an increase in entropy, which for a closed system is "a quantitative measure of the amount of thermal energy not available to do work. Entropy in any closed system always increases and never decreases. Additionally, moving parts produce waste heat due to friction, and radiative heat inevitably leaks from the system. Entropy is also defined as a measure of the disorder or randomness in a closed system, which also increases and is impossible to prevent. While some processes appear to be completely reversible, in practice, none actually are. Entropy, therefore, provides us with an arrow of time: forward is the direction of increasing entropy.

The fundamental principles of thermodynamics were originally expressed in three laws. Later, it was determined that a more fundamental law had been neglected, apparently because it had seemed so obvious that it did not need to be stated explicitly. To form a complete set of rules, scientists decided this most fundamental law needed to be included. The issue was that the first three laws of thermodynamics had already been established and were well known by their assigned numbers. When faced with the prospect of renumbering the existing laws, which would cause considerable confusion, or placing the pre-eminent law at the end of the list, which would make no logical sense and, hence, the Zeroth Law came into being. The Zeroth Law of thermodynamics states that if two bodies are in thermal equilibrium with some third body, then they are also in equilibrium with each other. This law establishes temperature as a fundamental and measurable property of matter.

The First Law of thermodynamics states that the total increase in the energy of a system is equal to the increase in thermal energy plus the work done on the system. Thus, heat is a form of energy and is therefore subject to the principle of conservation. The Second Law of thermodynamics states that heat energy cannot be transferred from a body at a lower temperature to a body at a higher temperature without the addition of energy. The Third Law of thermodynamics states that the entropy of a pure crystal at absolute zero is zero.

As explained above, entropy is sometimes referred to as waste energy, i.e., energy that is unable to do work, and since there is no heat energy whatsoever at absolute zero, there can be no waste energy. Entropy is also a measure of the molecular disorder in a system, and while a perfect crystal is by definition perfectly ordered, any positive value of temperature means there is motion within the crystal, which causes disorder. For these reasons, there can be no physical system with lower entropy and, thus, entropy always has a positive value.

Water is the only substance on Earth that naturally occurs in three physical states: solid, liquid, and gas. Depending on temperature and atmospheric pressure, water can change from one state to another, a process called physical phase change.

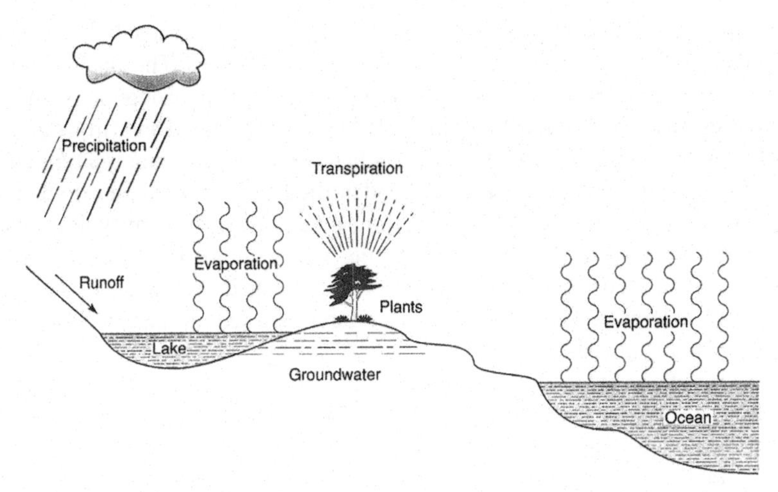

Fig. 4.1 The water cycle.

Because of this, some geographical regions of the world experience humidity, rain, snow, or even a combination of all three. Hydrogen bonds are again the key since the number of bonds between molecules determines whether water will be a solid, liquid, or gas. In the solid state, water molecules have the maximum number of hydrogen bonds (4 per molecule), giving water the rigid characteristic of ice. In its liquid state, water has fewer hydrogen bonds, which accounts for its less-structured, fluid character.

As water changes from solid to liquid to gas, hydrogen bonds are broken, giving water molecules more freedom of movement. This cycling of water through its states is the basis for the water cycle, also now as the hydrologic cycle (Fig. 4.1), that is essential for life on earth, purifying water and distributing it across land masses.

Water has more than sixty anomalies, such as the increase of density upon increasing temperature or its extraordinary large capacity of absorbing heat, essential for regulating the body temperature of humans (Vedamuthu et al., 1994). The heat capacity of water, contrarily to most of the liquids, increases at low temperatures, where other anomalies appear. For example, water can stay liquid at very low temperature in a metastable super-cooled state: down to −47 °C (−53 °F) in plants and −92 °C (−134 °F) (Kiss and Baranyai, 2014; Nilsson and Pettersson, 2015).

However, to gain fundamental understanding of the origin of these anomalies, it is necessary to address the instantaneous local structure of the liquid at various thermodynamic state points and establish how this structure couples to the dynamics of the molecular motion. Many different plausible explanations exist for the unusual properties of water where maybe both homogeneous and heterogeneous models could be viable and sophisticated structural and dynamical experimental data are needed to determine their validity.

3.1 Properties of a pure substance

A pure substance is a substance in which the chemical composition does not change during thermodynamic processes—water is an example of a pure substances. As it pertains to mixtures and to differentiate a mixture from water, a substance that has uniform thermodynamic properties throughout is homogeneous. The characteristics of a homogeneous mixture are; (i) mixtures, which are the same throughout with identical properties everywhere in the mixture, (ii) not easily separated and (iii) is generally referred to as a solution. On the other hand, a heterogeneous mixture is a type of mixture in which the composition can easily be identified and, often, there is two or more phases present. Each substance in the mixture retains its own identifying properties and mixtures have: (i) have different properties when sampled from different areas, (ii) the individual components can be seen with the naked eye, (iii) the components can be easily separated. The properties of mixtures also require understanding of the properties of pure substances.

In the current context, water is a substance of prime importance that exists in three states namely: (i) water vapor, (ii) liquid water, and (iii) solid ice and undergoes transformation from one state to another (Reddy et al., 2016; Yagasaki et al., 2016). A pure substance does not have to be of a single element or compound—a mixture of two or more phases of a pure substance is still a pure substance as long as the chemical composition of all phases is the same.

Water, as a pure substance may exist in different phases: (i) solid, (ii) liquid, and (iii) gas. Each phase has a having a distinct molecular arrangement that is homogenous throughout and separated from others (if any) by easily identifiable boundary surfaces. On the other hand, there are substances that have several phases within a principal phase, each with a different molecular structure. For example, carbon may exist as graphite or diamond in the solid phase, and ice may exist in seven different phases at high pressure. In the case of water, the various form of ice may also fit this description.

For water, the molecular bonds are the strongest in the solid phase and weakest in the gas phase—in the solid phase the molecules are arranged in a three-dimensional pattern (lattice) throughout the solid and cannot move relative to each other but may continually oscillate about their equilibrium position. In the liquid phase, the molecular spacing in liquid phase is not much different from that of the solid phase (generally slightly higher), except the molecules are no longer at fixed positions relative to each other. In the gas phase, the molecules are far apart from each other and molecular order does not exist. Thus, the molecules in steam move randomly and continually collide with each other and the walls of the container. Molecules in the gas phase are at a considerably higher energy level than they are in liquids or solid phases.

If a liquid (pure substance) is heated at constant pressure, the temperature at which it boils is called saturation temperature. This temperature will remain constant during heating until all of the liquid boils off and, at this temperature, the liquid and the associated vapor are in equilibrium and are the saturated liquid and vapor respectively. The saturation temperature of a pure substance is a function of pressure only. At atmospheric pressure, the saturation temperature is the normal boiling point. Similarly, if the vapor of a pure substance is cooled at constant pressure, the temperature at which the condensation starts, is called dew point temperature.

For a pure substance, dew point and boiling point are same at a given pressure and, similarly, when a solid is heated it melts at a definite temperature (the melting point) and when cooled, the liquid become a solid (at the freezing point). The melting point

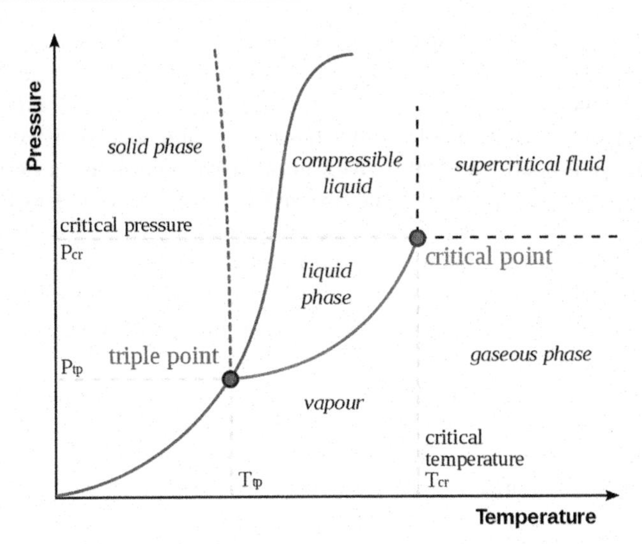

Fig. 4.2 Illustration of the triple point.

and freezing point are same at same pressure for a pure substance and the solid and liquid are in equilibrium at this temperature. For all pure substances there is a temperature at which all the three phases exist in equilibrium (the triple point) (Fig. 4.2). In the case of water, a most important aspect of the substance is the ability to exit in three phases leading to the value of the triple point (Velasco and Fernández-Pineda, 2007; Kiss et al., 2012).

3.2 Intensive and extensive properties

A thermodynamic property is a point function that is used to define the state of a system and it is independent of the path followed. As example, pressure, temperature, and specific volume are thermodynamic properties. Furthermore, there are two properties that deserve recognition and definition: (i) an intensive property and (ii) an extensive property, both of which can affect the outcome of any study on thermodynamic properties.

An intensive property is a physical quantity where the value does not depend on the amount of the substance for which it is measured. For example, the temperature of a system in thermal equilibrium is the same as the temperature of any part of it and if the system is divided, the temperature of each subsystem is identical. The same rationale applies to the density of a homogeneous system and if the system is divided into two equal parts, the mass and the volume are both divided in half and the density remains unchanged. Additionally, the boiling point of a substance is another example of an intensive property—for example, the boiling point of water is 100 °C (212 °F) at a pressure of 14.7 psi (one atmosphere), which remains true regardless of the quantity of the water. However, some intensive properties do not apply at very small sizes. For example, viscosity is a macroscopic quantity and is not relevant for extremely small systems.

On the other hand, an extensive property is a physical quantity where the value is (i) proportional to the size of the system that the property describes or (ii) to the quantity of matter in the system. For example, the mass of a sample is an extensive quantity since it depends on the amount of substance. Dividing one extensive property by another extensive property generally gives an intensive value—as an example: mass (extensive property) divided by volume (extensive property) gives density (intensive property).

The distinction between intensive and extensive properties has some theoretical uses as, for example, in thermodynamics, where the state of a simple compressible system is completely specified by two independent, intensive properties. Other intensive properties are derived from those two variables.

The ratio of two extensive properties of the same object or system is an intensive property. For example, the ratio of the mass and the volume of an object, which are two extensive properties, is density, which is an intensive property. More generally properties can be combined to give new properties, which may be called derived or composite properties. For example, the base quantities mass and volume can be combined to give the derived quantity density.

A specific property is the intensive property obtained by dividing an extensive property of a system by its mass. For example, the heat capacity of water is an extensive property of a system and dividing heat capacity, C_p, by the mass of the system yields the specific heat capacity, c_p, which is an intensive property. By way of explanation and by common convention, when the extensive property is represented by an upper-case letter, the symbol for the corresponding intensive property is typically represented by a lower-case letter (as illustrated above).

If the amount of substance in moles can be determined, then each of the thermodynamic properties may be expressed on a molar basis, and their name may be qualified with the adjective *molar*, yielding terms such as molar volume, molar internal energy, molar enthalpy, and molar entropy. The symbol for molar quantities may be indicated by adding a subscript *m* to the corresponding extensive property. For example, molar enthalpy is shown as Hm. Briefly, enthalpy is a thermodynamic quantity equivalent to the total heat content of a system and is equal to the internal energy of the system plus the product of pressure and volume. Entropy is a thermodynamic quantity representing the unavailability of the thermal energy of a system for conversion into mechanical work, which is often interpreted as the degree of disorder or randomness in the system.

For the characterization of substances or reactions, the data are usually reported so that the molar properties are referred to a standard state and, in that case, an additional superscript ° is added to the symbol. For example: (i) $V_m{}^\circ = 22.41$ l/mol is the molar volume of an ideal gas at standard conditions of temperature and pressure, (ii) $C_{P,m}{}^\circ$ is the standard molar heat capacity of a substance at constant pressure, (iii) $\Delta_r H_m{}^\circ$ is the standard enthalpy variation of a reaction with subcases such as, for example, formation enthalpy and combustion enthalpy, and (iv) E° is the standard reduction potential of a redox couple i.e. Gibbs energy over charge, which is measured in volts.

In thermodynamics, the Gibbs energy (more commonly known as the Gibbs free energy—although IUPAC does recommend the name: Gibbs energy or Gibbs function and also known as free enthalpy to distinguish it from Helmholtz free energy—is a thermodynamic potential that can be used to calculate the maximum of reversible work that may be performed by a thermodynamic system at constant temperature (iso thermal system) and constant pressure (isobaric system). The Gibbs free energy is given by the equation:

$$\Delta G^{\circ} = \Delta H^{\circ} - T\Delta S^{\circ} \left(\text{J is in SI units}\right)$$

This equation represents the *maximum* amount of non-expansion work that can be extracted from a thermodynamically closed system—a system that can exchange heat and work with its surroundings, but not matter. This maximum can be attained only in a completely reversible process. When a system transforms reversibly from an initial state to a final state, the decrease in Gibbs free energy equals the work done by the system to its surroundings, minus the work of the pressure forces.

3.3 Triple point

In thermodynamics, the triple point of a substance is the temperature and pressure at which the three phases (gas, liquid, solid) of that substance coexist in thermodynamic equilbrium. It is that temperature and pressure at which the sublimation curve, the fusion curve, and the vaporization curve meet (Fig. 4.2). In addition to the triple point for solid, liquid, and gas phases, a triple point may involve more than one solid phase, for substances with multiple polymorphs (such as the various crystal form of ice).

The single combination of pressure and temperature at which water vapor, liquid water, and solid ice can coexist in a stable equilibrium occurs and, at that point, it is possible to change all of the substance to ice, water, or vapor by making arbitrarily small changes in pressure and temperature. Even if the total pressure of a system is well above the triple point of water, the system can still be brought to the triple point of water. More specifically, the surfaces separating the different phases should also be perfectly flat, to negate the effects of surface tension.

Thus, the gas–liquid–solid triple point of water corresponds to the minimum pressure at which liquid water can exist. At pressures below the triple point solid ice when heated at constant pressure is converted directly into water vapor (sublimation). Above the triple point, solid ice when heated at constant pressure first melts to form liquid water, and then evaporates or boils to form vapor at a higher temperature. For most substances the gas–liquid–solid triple point is also the minimum temperature at which the liquid can exist. For water, however, this is not true because the melting point of ordinary ice decreases as a function of pressure. At temperatures just below the triple point, compression at constant temperature transforms water vapor first to solid and then to liquid water (water ice has lower density than liquid water, so increasing pressure leads to liquefaction.

Thus, water below the critical pressure when heated first becomes a mixture of liquid and vapor and then becomes saturated vapor and finally a superheated vapor. At the critical point there is no distinction between liquid state and vapor state; these two merge together. At constant pressure greater than critical pressure, when the liquid is heated in supercritical region, there is no distinction between liquid and vapor; as a result if heating is done in a transparent tube, the meniscus of liquid and vapor does not appear as transformation from liquid to vapor takes place. At pressures below critical pressure, when a liquid is heated there is a clear-cut meniscus between liquid and vapor, until all the liquid evaporates (Galkin and Lunin, 2005; Brunner, 2009).

3.4 Equations of state

In a closed system, the pressure, volume, and temperature may be easily measured. If the volume (V) is set at an arbitrary value and the temperature (T) is maintained at a specific value, then the pressure (P) will be fixed at a definite value. Once the volume and temperature are chosen, then the value of the pressure at equilibrium is fixed. That is, of the three thermodynamic coordinates—pressure, volume, and temperature—only two of the coordinates are independent variables. There exists an equation of equilibrium which connects the thermodynamic coordinates and which robs one of them of its independence. Such an equation (an equation of state) is a mathematical function relating the appropriate thermodynamic coordinates of a system in equilibrium.

Every thermodynamic system has its own equation of state, although in some cases the relation may be so complicated that it cannot be expressed in terms of simple mathematical functions. For a closed system, the equation of state relates the temperature to two other thermodynamic variables. An equation of state expresses the individual peculiarities of one system as compared with another system and must, therefore, be determined either by experiment or by molecular theory. A general theory like thermodynamics, based on general laws of nature, is incapable of generating an equation of state for any substance. An equation of state is not a theoretical deduction from the theory of thermodynamics, but is an experimentally derived law of behavior for any given pure substance. The equation of state expresses the results of experiments in which the thermodynamic coordinates of a system were measured as accurately as possible, over a range of values.

An equation of state is only as accurate as the experiments that led to its formulation, and holds only within the range of values measured. As soon as this range is exceeded, a different form of equation of state may be required. In two-phase regions, including the borders of the regions, specifying temperature alone will set the pressure and vice versa. Giving both pressure and temperature will not define the volume because we will need to know the relative proportion of the two phases present. The mass fraction of the vapor in a two-phase liquid-vapor region is called the quality.

3.5 Thermodynamic anomalies

Life on Earth depends on water, yet many unique properties of water still present a puzzle for which different interpretations have been proposed over the past several decades. As liquid water is so common-place on the Earth, it is often regarded as a typical liquid. Nevertheless, in reality, water is most atypical as a liquid, behaving as a quite different material at low temperatures to that when it is hot, with a division temperature of approximately 50 °C (122 °F).

In fact, water is in a class by itself with exceptional properties when compared with other materials. The anomalous properties of water are those where the behavior of liquid water is entirely different from what is found with other liquids. The hydrogen bonds also produce and control the local tetrahedral arrangement of the water molecules (Chapter 2). The strength and directionality of the hydrogen bonds control thermodynamic and dynamic behavior of liquid water. If hydrogen-bonding did not exist, water would behave non-anomalously as expected from similar molecules. No other material is commonly found as solid (ice), liquid (water), or gas (steam). Frozen water (ice) also shows anomalies when compared with other solids. Although it is an apparently simple molecule, water has a highly complex and anomalous character due to its inter-molecular hydrogen bonding (Stokely et al., 2010). For example): (i) as a gas, water is one of lightest known, (ii) as a liquid, water is much denser than expected, and (iii) as a solid, it is much lighter than expected when compared with the liquid form.

In fact, in terms of properties and behavior, water has many anomalies (Table 4.4), such as the increase of density upon increasing temperature or its extraordinary large capacity of absorbing heat, essential for regulating the human body temperature (Vedamuthu et al., 1994). The heat capacity of water, contrary to the heat capacity of many liquids, increases at low temperatures.

Density is, perhaps the most widely known anomaly of water. Unlike other simple liquids which expand upon heating, water expand upon cooling below 4 °C (39 °F) at ambient pressure. Because of anomalous property, ice floats on water and fish species can survive in warm waters below a layer of ice at temperatures well below 0 °C (32 °F).

Furthermore, water is in a class by itself with exceptional properties when compared with other materials. The anomalous properties of water are those where the behavior of liquid water is entirely different from what is found with other liquids. These hydrogen bonds also produce and control the local tetrahedral arrangement of the water molecules. The strength and directional of the hydrogen bonds control thermodynamic and dynamic behavior of liquid water. If hydrogen-bonding did not exist, water would behave non-anomalously as expected from similar molecules. No other material is commonly found as solid (ice), liquid (water), or gas (steam). Frozen water (ice) also shows anomalies when compared with other solids.

Although it is an apparently simple molecule (H_2O), it has a highly complex and anomalous character due to its inter-molecular hydrogen bonding (Chapter 2) (Grabowski, 2001; Chen et al., 2003; Govender et al., 2003; Lu et al., 2008; Stokely et al., 2010). The high cohesion between the molecules gives water a high freezing and melting point. The high heat capacity, high thermal conductivity, and high water content in organisms contribute to thermal regulation and prevent local temperature fluctuations, thus allowing humans to control the body temperature more easily. The high latent heat of evaporation gives resistance to dehydration and considerable evaporative cooling. Water self-ionizes and allows easy proton exchange between molecules, so contributing to the richness of the ionic interactions in biology. Also, water is an excellent solvent due to its polarity, high relative permittivity (dielectric constant) and small size, particularly for polar and ionic compounds and salts.

At 4 °C (39 °F), water expands on heating or cooling and this density maximum, together with the low ice density, results in (i) the necessity that all of a body of fresh water (not just its surface) is close to 4 °C (39 °F before any freezing can occur,

Table 4.4 Anomalous properties of water

1.	The density of ice increases on heating (up to 70 K).
2.	Water shrinks on melting.
3.	Pressure reduces the melting point if ice.
4.	Liquid water has a high-density that increases on heating (up to 4 °C, 39 °F).
5.	The surface of water is denser than the bulk.
6.	Pressure reduces the temperature of maximum density.
7.	There is a minimum in the density of supercooled water.
8.	Water has a low coefficient of expansion (thermal expansivity).
9.	The thermal expansion reduces (becoming negative) at low temperatures.
10.	The thermal expansion increases with increased pressure.
11.	Water has unusually low compressibility.
12.	The compressibility drops as temperature increases up to 46.5 °C (116 °F).
13.	The refractive index of water has a maximum value at just below 0 °C (32 °F).
14.	Unusually high surface tension.
15.	Thermal conductivity of water is high and rises to a maximum at 130 °C (266 °F).
16.	The dielectric constants of water and ice are high.
17.	The electrical conductivity of water rises to a maximum at about 230 °C 446 °F).
18.	The electrical conductivity of water rises considerably with frequency.
Bulk anomalies	
1.	No aqueous solution is ideal.
2.	D_2O and T_2O differ significantly from H_2O in their physical properties.
3.	Liquid H_2O and D_2O differ significantly in their phase behavior.
4.	H_2O and D_2O ices differ significantly in their quantum behavior.
5.	Water has unusually high viscosity.
6.	Viscosity decreases with pressure below 33 °C (91 °F).
7.	Solutes have varying effects on properties such as density and viscosity.
8.	Acidity constants of weak acids show temperature minima.
Thermodynamic anomalies	
1.	The heat of fusion of water with temperature exhibits a maximum at −17 °C (1 °F).
2.	Liquid water has over twice the specific heat capacity of ice or steam.
3.	High specific heat capacity.
4.	Specific heat capacity (C_P): minimum at 36 °C (97 °F), maximum at −45 °C (−49 °F).
5.	The specific heat capacity (C_P) has a minimum with respect to pressure.
6.	High heat of vaporization.
7.	High heat of sublimation.
8.	High entropy of vaporization.
9.	The heat of fusion of water with temperature exhibits a maximum at −17 °C (1 °F).
Phase anomalies	
1.	Water has an unusually high melting point.
2.	Water has an unusually high boiling point.
3.	Water has an unusually high critical point.
4.	Solid water exists in a wide variety of crystal and amorphous structures.
5.	The thermal conductivity of ice reduce with increasing pressure.
6.	The shear modulus of ice reduce with increasing pressure.
7.	The transverse sound velocity of ice reduce with increasing pressure.
8.	The structure of liquid water changes at high pressure.
9.	Liquid water is easily supercooled.
10.	Supercooled water has two phases and a second critical point at −91 °C (−132 °F).

(ii) the freezing of rivers, lakes, and oceans is from the top down, so permitting survival of the bottom ecology, insulating the water from further freezing, reflecting back sunlight into space, and allowing rapid thawing, and (iii) density driven thermal convection causing seasonal mixing in deeper temperate waters carrying life-providing oxygen into the depths. The high heat capacity of the oceans and seas allows them to act as heat reservoirs such that sea temperatures vary only a third as much as land temperatures and so moderate the climate of the Earth (for example, the Gulf stream carries tropical warmth to northwestern Europe otherwise the British Isles might be ice-bound). The compressibility of water reduces the sea level by about 40 m giving us 5% more land and the high surface tension of water as well as the expansion of water on freezing promotes the erosion of rocks and provides soil for agriculture.

Water has the high specific heat—as water is heated, the increased movement of water causes the hydrogen bonds to bend and break. As the energy absorbed in these processes is not available to increase the kinetic energy of the water, it takes considerable heat to raise the temperature of. Additionally, as water is a light molecule, there are more molecules per gram than most similar molecules to absorb this energy. Heat absorbed is given out on cooling, so allowing water to act as a heat reservoir for the surface of the Earth, buffering against changes in temperature. Thus, the water in our oceans stores vast amounts of energy, so moderating the climate of the climate of the Earth.

Water has the highest heat of vaporization per gram of any molecular liquid and, hence, a very low volatility but high cohesive energy density. There is still considerable hydrogen-bonding in water at 100 °C (212 °F). As effectively all these bonds need to be broken (very few indeed remaining in the gas phase), there is a great deal of energy required to convert the water to gas, where the water molecules are effectively separated. The increased hydrogen-bonding at lower temperatures causes higher heats of vaporization.

Liquid metals aside, water has the highest thermal conductivity of any liquid. For most liquids, the thermal conductivity (the rate at which energy is transferred down a temperature gradient) falls with increasing temperature, but this occurs only above 130 °C (266 °F) in liquid water (Kell, 1972; Ramires et al., 1995; Maestro et al., 2016).

As the temperature of water is lowered, the rate at which energy is transferred is reduced to an ever-increasing extent. Instead of the energy being transferred between molecules, it is stored in the hydrogen-bonding fluctuations within the increasingly large clusters that occur at lower temperatures. There is a minimum in the thermal conductivity-temperature behavior just below −37 °C as the amount of fully expanded network increases and in line with that indicated by the much higher value one form of ice (ice Ih) (Johari and Andersson, 2007). If the density is kept constant, the thermal conductivity is proportional to the square root of the absolute temperature, between 100 °C (212 °F) and 400 °C (750 °F) (Abramson, 2001).

The high dielectric constant of water constant makes it a good solvent for polar and ionized groups, and thus helps account for such fundamental phenomena as the solubility of salts, the ionization of many acids and bases in water, and the tendency of polar residues to lie at protein surfaces. It also greatly weakens charge-charge interactions relative to what they would be in vacuum. The hydrophobic effect drives the association of nonpolar surfaces in water, and is typically interpreted as resulting from the tendency of water to form structures that maintain energetically favorable water-water hydrogen bonds at hydrophobic surfaces. These central solvent effects are illustrated by continuum solvation models to explain the electrostatic solvation effects at modest computational cost, and the hydrophobic effect may be approximated through surface area.

4 The hydrogen bond

The structure and bonding arrangements in water has been dealt with in detail elsewhere in this text (Chapter 2) but it is necessary to consider this aspect of water science as it pertains to the thermodynamics of water systems (Urquidi et al., 1999; Suresh and Naik, 2000). In fact, the hydrogen bonding phenomenon in liquid water continues to be one of the most challenging topics to understand. Each water molecule possesses two proton donors and two proton acceptors (lone pairs of electrons) that contribute to hydrogen bonding.

In the simplest sense, the thermodynamics of water centers on the occurrence of intermolecular hydrogen bonding which is an attractive interaction between a group X-H and an atom or group of atoms Y in the same or different molecule(s), where there is evidence of bond formation (Rastogi et al., 2011). The energy of the hydrogen bond (approximately 5 Kcal/mol) is intermediate between those of Van der Waals interaction (approximately 0.3 Kcal/mol) and a covalent chemical bond (approximately 100 Kcal/mol) chemical bonds (Stillinger, 1980).

Since the energy of the hydrogen bond is a relatively weak bond, thermal energy constantly acts to disrupt hydrogen bonds. One can thus consider the energetics to drive formation of hydrogen bonds, and entropic factors arising from thermal energy to break hydrogen bonds. The result is a time-varying distribution of hydrogen bonds among the different donor-acceptor pairs in the system. In addition, combination of femtosecond two-dimensional infrared spectroscopy and molecular dynamics simulations demonstrated that the vast majority of average numbers of hydrogen bonds are part of a hydrogen-bonded well of attraction and virtually all molecules return to a hydrogen bonding partner within a short time (Eaves et al., 2005). Despite this continuous dynamics, fluctuation in the total number of hydrogen bonds in a system containing a large number of molecules is quite small. Most simulation models suggest that a given hydrogen atom in water is hydrogen bonded for the majority (85–90%) of the time (Bakker and Skinner, 2010).

The concept of the hydrogen bond is a century old but youthful because of its vital role in so many branches of science and because of continued advances in experiment, theory and simulation. The significance of hydrogen bonds can be best understood by comparing the physical state of water and methane, both of similar size; at room temperature, while methane is supercritical, water exists in liquid state, making it possible for life to sustain on earth. The anomalous expansion of water at 4 °C (39 °F) makes it possible for marine life to exist. The high dielectric constant of water opens up the entire field of electrochemistry. The internal structure of water is largely responsible for self-assembly of surfactants, leading to a wide array of liquid crystalline phases.

Hydrogen bonds are largely responsible for preserving the structure/conformation of several life-supporting biological molecules such as deoxyribonucleic acid (DNA), ribonucleic acid (RNA), and proteins in aqueous solution.

The structure and orientation of water molecules adjacent to charged surfaces play an important role in surface science, electrochemistry, geochemistry and biology. Several force fields are operational in such situations. The presence of ions (H+, OH^-) in the liquid phase further induces formation of an electrical double layer within which the electric field decays with distance from the surface. Moreover, the length of oxygen-hydrogen bond (of water) near the electrode surface is smaller, indicating a weaker hydrogen bond.

Also, it is not yet clear whether the hydrogen bond structure near a charged electrode surface is disrupted or not. The additional presence of stabilized (hydrogen-bonded) water molecules with their dipoles lying perpendicular to that of field. The fraction of such molecules is relatively small; for every thousand molecules with their dipoles aligned in the direction of field, roughly one was found aligned perpendicular to that of the field. Nevertheless, the role of these changes in alignment on the transport or solvation properties of water and whether ions induce long-range changes in the structure of water is still an open question (Gulliver, 2007; Bakker, 2008).

The thermodynamic properties of confined water are generally considered to be different from those of bulk water; however, what gives rise to these differences is still an open question (Wagner and Pruss, 2002; Mallamace et al., 2014a, 2014b; Sippola and Taskinen, 2018). Thus, hydrogen bonding in water, together with the tendency of the water molecules to form to form open tetrahedral networks at low temperatures, gives rise to the characteristic properties of water which differ from those of other similar liquids (Chapter 2). Such properties are often described as anomalous although it can be argued that water possesses exactly those properties that can be predicted from the structure of water (Chaplin, 2007). An important feature of the hydrogen bond is that it possesses direction and the cohesion of water due to hydrogen bonding is responsible for water being a liquid over the range of temperatures on Earth. However, it is the clustering of the water, due to the directed characteristics of the hydrogen bonding that is responsible for the properties of water that allow water to act in diverse ways under a variety of conditions.

Finally, it is worthy of note that the consequences of strengthening or weakening the hydrogen bonding need to be considered (Silverstein et al., 2000; Chaplin, 2007). If the hydrogen bond strength was slightly different from the natural value, at the extremes water would not be liquid on the surface of Earth at typical surface temperatures. The temperature of maximum density (approximately 4 °C, 39 °F) would disappear. Also, the ionization of water would be much less evident and the important alkali metal ions sodium (Na^+) and potassium (K^+) would lose their distinctive properties (Hribar et al., 2004). Stronger hydrogen bonding in water leads to water molecules existing in clusters and are not available for biomolecular hydration.

5 Electronic structure

By definition, a molecule is an aggregation of atomic nuclei and electrons that is sufficiently stable to possess observable properties. In water, each hydrogen nucleus is bound to the central oxygen atom by a pair of electrons that are shared between them (a covalent chemical bond). In water, only two of the six outer-shell electrons of oxygen are used for this purpose, leaving four electrons which are organized into two non-bonding pairs. The four electron pairs surrounding the oxygen tend to arrange themselves as far from each other as possible in order to minimize repulsions between these clouds of negative charge. This would typically, result in a tetrahedral geometry in which the angle between electron pairs (and therefore the hydrogen-oxygen-hydrogen, H-O-H, bond angle) is 109.5°. However, because the two non-bonding pairs remain closer to the oxygen atom, these exert a stronger repulsion against the two covalent bonding pairs, effectively pushing the two hydrogen atoms closer together. The result is a distorted tetrahedral configuration in which the hydrogen-oxygen-hydrogen bond angle is 104.45°.

Although the water molecule carries no net electric charge (Chapter 2), the eight electrons are not distributed uniformly—there is a somewhat higher negative charge at the oxygen end of the molecule, and a compensating positive charge at the hydrogen end (Chapter 2) and the resulting polarity is largely responsible for the unique chemistry and properties of water (Chapter 3). Thus, water is a unique substance that is a major component of all living things (Chapter 1).

Water continues to remain incompletely understood primarily because water is anomalous in many of its physical and chemical properties (Chapter 2). Some of the unique properties of water make it (water) essential for life, while other properties have profound effects on the size and shape of living organisms, how they work, and the physical limits or constraints within which they must operate.

Water has long been known to exhibit many physical properties that distinguish it from other small molecules of comparable size (Table 4.5) and these properties are often referred to as the anomalous properties of water, but they are by no means mysterious. All of the proeprties are entirely predictable consequences of the way the size and nuclear charge of the oxygen atom conspire to distort the electronic charge clouds of the atoms of other elements. When the charge clouds are chemically bonded to the oxygen, this is where an understanding of the thermodynamic aspects of water chemistry need to be clearly understood.

Table 4.5 Boiling points and freezing points of the hydrides related to water

Compound	Formula	Molar mass	Boiling point[a]		Freezing point[a]	
			°C	°F	°C	°F
Hydrogen telluride	H_2Te	129.6	−4	25	−49	−56
Hydrogen selenide	H_2Se	81	−42	−44	−64	−83
Hydrogen sulfide	H_2S	34	−62	−80	−84	−119
Water	H_2O	18	100	212	0	32

[a] Rounded to the nearest degree.

5.1 Structure and properties

Water is one of the few known substances whose solid form is less dense than the liquid. The plot at the right shows how the volume of water varies with the temperature; the large increase (about 9%) on freezing shows why ice floats on water and why pipes burst when they freeze. The expansion between −4° (25 °F) and 0° (32 °F) is due to the formation of larger hydrogen-bonded aggregates. Above 4 °C (39 °F), thermal expansion sets in as vibrations of the oxygen-hydrogen bonds becomes more vigorous, tending to shove the molecules farther apart. The other widely-cited anomalous property of water is its high boiling point. From molecular structure studies, it might be predicted that water should boil at a temperature on the order of around −90 °C (130 °F)—water would exist in the world as a gas rather than a liquid if hydrogen-bonding was absent.

In water, each hydrogen nucleus is bound to the central oxygen atom by a pair of electrons that are shared between them; chemists call this shared electron pair a covalent chemical bound. In water, only two of the six outer-shell electrons of oxygen are used for this purpose, leaving four electrons which are organized into two non-bonding pairs. The four electron pairs surrounding the oxygen tend to arrange themselves as far from each other as possible in order to minimize repulsions between these clouds of negative charge. This would ordinarily result in a tetrahedral geometry in which the angle between electron pairs (and therefore the H-O-H bond angle, as per regular tetrahedral geometry) is 109.5°. However, because the two non-bonding pairs remain closer to the oxygen atom, these exert a stronger repulsion against the two covalent bonding pairs, effectively pushing the two hydrogen atoms closer together. The result is a molecular tetrahedral arrangement in which the hydrogen-oxygen-hydrogen (H-O-H) angle is distorted to 104.45°. Nevertheless, because of this tetrahedral or near-tetrahedral structure, water was often referred to (by some observers) as *tetrahedral soup*!

More to the point, the water molecule is electrically neutral, but the positive and negative charges are not distributed uniformly. This is illustrated by the gradation in color in the schematic diagram here. The electronic (negative) charge is concentrated at the oxygen end of the molecule, owing partly to the nonbonding electrons (solid blue circles), and to the high nuclear charge of oxygen which exerts stronger attractions on the electrons. This charge displacement constitutes an electric dipole which is often considered to be the electrical image (or electrical fingerprint) of a water molecule.

The geometry of the water molecule (Chapter 2) consists of two oxygen-hydrogen bonds of length 0.096 pm at an angle of 104.45 (Fig. 4.3) and is often considered to be an angular (or bent) molecule. Other basic properties of water are its size, shape and polarity. Atoms that are not bonded will repel each other strongly if brought close enough that their electron orbitals overlap. At larger distances two atoms attract each other weakly due to an induced dipole-induced dipole (London dispersion) force. The combination of repulsive and attractive interactions is termed the van der Waals interaction. The point at which the repulsive and attractive forces balance is commonly used to define the diameter of an atom, which for oxygen and hydrogen are 0.32 nm and 0.16 nm (nm = nanometers, I nanometer = 0.000000001 m = 1×10^{-9} m), respectively.

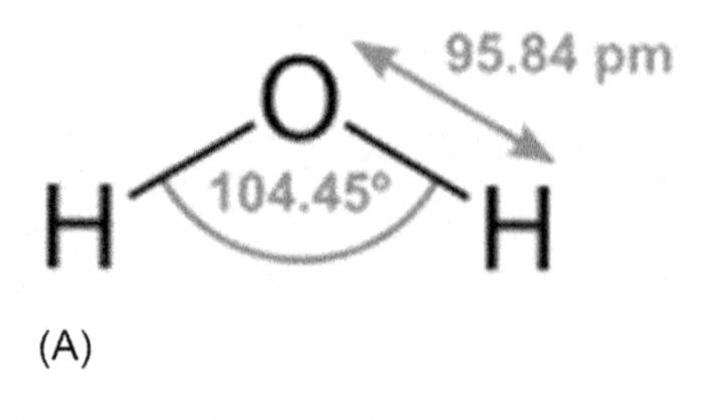

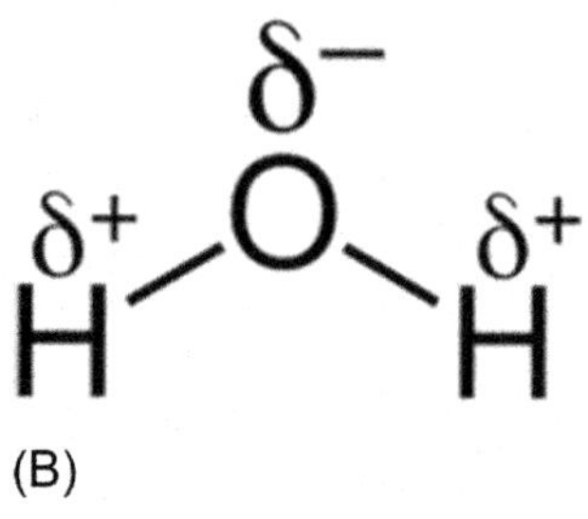

Fig. 4.3 (A) The structure of water also showing (B) the relative negativity of the hydrogen and oxygen atoms.

The water molecule is approximately spherical and is electrically neutral but, because the electronegativity of oxygen is much greater than that of hydrogen, the electron distribution is concentrated more around the former, i.e. water is electrically polarized, having a permanent dipole moment of 6×10^{-30} in the gas phase. The dipole moment is even larger (8×10^{-30}) in liquid and ice because neighboring water dipoles mutually polarize each other. A useful way to represent the polarity of a molecule is to assign a partial charge to each atom, so as to reproduce the net charge on a molecule as well as the dipole moment, and possibly higher-order electrical moments (Chapter 2).

The magnitude of the partial charge on an atom is a measure of its polarity. For water there is approximately +0.5 on each hydrogen, and a charge of opposite sign and twice this magnitude on the oxygen. In contrast, the hydrogens of a non-polar molecule such as methane have a partial charge on the order of 0.1, and the dipole moment of methane is zero. Thus, water is a very polar molecule with the ability to make strong electrostatic interactions with itself, other molecules and ions.

5.2 Surface tension and wetting

Surface tension is the elastic tendency of a fluid surface which makes it acquire the least possible surface area possible. At liquid–air interfaces, surface tension results from the greater attraction of liquid molecules to each other (due to cohesion) than to the molecules in the air (due to adhesion). The net effect is an inward force at its surface that causes the liquid to behave as if its surface were covered with a stretched elastic membrane. Surface tension has the dimension of force per unit length or of energy per unit area. Both are equivalent, but it is more common or conventional, when referring to energy per unit of area, to use the term surface energy which is a more general term insofar as it also applies to solids.

When a liquid is in contact with a solid surface, its behavior depends on the relative magnitudes of the surface tension forces and the attractive forces between the molecules of the liquid and of those comprising the surface. If a water molecule is more strongly attracted to its own kind, then surface tension will dominate, increasing the curvature of the interface. This is what happens at the interface between water and a hydrophobic surface such as a plastic item. By contract, a clean glass surface has hydroxyl groups on the surface that readily interact with and attach to water molecules through hydrogen bonding. As a result, this causes the water to spread out evenly over the surface, or to wet it. A liquid will wet a surface if the angle at which it contacts the surface is more than 90°. The value of this contact angle can be predicted from the properties of the liquid and solid separately.

The water surface acts like an elastic film that resists deformation when a small weight is placed on the surface which is all due to the surface tension of the water (Sharp, 2001). A molecule within the bulk of a liquid experiences attractions to neighboring molecules in all directions, but since these average out to zero, there is no net force on the molecule. For a molecule that finds itself at the surface, the situation is quite different; it experiences forces only sideways and downward, and this is what creates the stretched-membrane effect.

The distinction between molecules located at the surface and those deep inside is especially prominent in water, owing to the strong hydrogen-bonding forces. The difference between the forces experienced by a molecule at the surface and one in the bulk liquid gives rise to the surface tension of the liquid.

If water is to wet a surface that is not ordinarily wettable, a detergent is added to the water to reduce the surface tension. A detergent is a special kind of molecule in which one end is attracted to water molecules but the other end is not, so these ends stick out above the surface and repel each other, canceling out the surface tension forces due to the water molecules alone.

Because of the relatively high attraction of water molecules to each other through a network of hydrogen bonds, water has a higher surface tension (72.86 m-Newtons per meter for the water-air interface at 20 °C, 68 °F) than most other liquids. In addtion, surface tension is an important factor in the phenomenon of capillarity.

5.3 *Liquid water*

Liquid water is a transparent, tasteless, odorless, and colorless chemical substance which is the main constituent of rivers, lakes, and oceans of the Earth (Chapter 1) and is a vital constituent of the fluids of most living organisms. In addition to the occurrence of water in land forms of water (i.e. rivers, lakes, and oceans), water also occurs as precipitation in the form of rain, hail, and snow which contribute to the land form of water. Also, clouds are formed from suspended droplets of water and ice. Water moves continually through the water cycle of evaporation, transpiration, condensation, precipitation, and runoff, the terminus of which is the sea.

The nature of liquid water and how the water molecules within it are organized and interact with (attract) each other through the special type of dipole-dipole interaction known as hydrogen bonding (forming a hydrogen-bonded cluster) in which four water molecules are located at the corners of a tetrahedral structure is an especially favorable (low-potential energy) configuration, but the molecules undergo rapid thermal motions on a time scale of picoseconds (one second × 10^{-12}) and, as a result, the lifetime of any specific cluster configuration will be fleetingly brief.

A variety of techniques including infrared absorption, neutron scattering, and nuclear magnetic resonance have been used to probe the microscopic structure of water. The information gathered from these experiments and from theoretical calculations has led to the development of approximately twenty models that attempt to explain the structure and behavior of water. More recently, computer simulations of various kinds have been employed to explore how well these models are able to predict the observed physical properties of water. Whether or not these models have advanced the science and technology of water is, according to some observers, debatable.

The possible locations of neighboring molecules around a given water are limited by energetic and geometric considerations, thus giving rise to a certain amount of structural order (or structural definition) within any small volume element. It is not clear, however, to what extent these structures interact as the size of the volume element is enlarged. The view first developed in the 1950s that water is a collection of transient clusters (tetrahedron) of varying sizes has gradually been abandoned as being unable to account for many of the observed properties of the liquid (Peters, 1995; Keutsch and Saykally, 2001).

5.4 Local structures and water clusters

Chemically, a *water cluster* is a discrete hydrogen-bonded assembly or cluster of water molecules (Keutsch and Saykally, 2001; Ludwig, 2001; Tokmachev et al., 2010). There is a general commentary that water manifests itself as clusters rather than an isotropic collection may help explain many anomalous water characteristics such as its highly unusual density temperature dependence. Water clusters are also implicated in the stabilization of certain supramolecular structures.

It is quite likely that over very small volumes, any localized polymeric clusters of water may have a fleeting existence, and many theoretical calculations have been made showing that some combinations are more stable than others. While this might prolong their lifetimes, it does not appear that they remain intact long enough to detect as directly observable entities in ordinary bulk water at normal pressures.

Theoretical models suggest that the average cluster may encompass as many as 90 water molecules at 0 °C (32 °F), so that very cold water can be thought of as a collection of ever-changing ice-like structures. At 70 °C (158 °F), the average cluster size is believed to be less than 25. It must be emphasized that no stable clustered unit or arrangement has ever been isolated or identified in pure bulk liquid water. Another view is that small clusters of four water molecules may come together to form water bicyclo-octamers (Bagchi, 2012; Skinner et al., 2014).

Water clusters are of considerable interest as models for the study of water and water surfaces, and many articles on them are published every year. One view is that the majority (on the order of 80%) of the water molecules are bound in chain-like fashion to only two other molecules at room temperature, thus supporting the prevailing view of a dynamically-changing, disordered water structure.

5.5 Liquid and solid water

Ice, like all solids, has a well-defined structure; each water molecule is surrounded by four neighboring water molecules. Two of these are hydrogen-bonded to the oxygen atom on the central water molecule, and each of the two hydrogen atoms is similarly bonded to another neighboring water. The hydrogen bonds are represented by the dashed lines in this two-dimensional schematic diagram. In reality, the four bonds from each oxygen atom point toward the four corners of a tetrahedron centered on the oxygen atom. This basic assembly repeats itself in three dimensions to build the ice crystal.

When ice melts to form liquid water, the uniform three-dimensional tetrahedral organization of the solid breaks down as thermal motions disrupt, distort, and occasionally break hydrogen bonds. The methods used to determine the positions of molecules in a solid do not work with liquids, so there is no unambiguous way of determining the detailed structure of water. The illustration here is probably typical of the arrangement of neighbors around any particular water molecule, but very little is known about the extent to which an arrangement like this gets propagated to more distant molecules.

To a chemist, the term *pure* has meaning only in the context of a particular application or process. The distilled or de-ionized water often contains dissolved atmospheric gases and occasionally some silica, but the small amounts of these gases and relative inertness make these impurities insignificant for most purposes. When water of the highest obtainable purity is required for certain types of exacting measurements, it is commonly filtered, de-ionized, and triple-vacuum distilled. But even this chemically pure water is a mixture of isotopic species: there are two stable isotopes of both hydrogen (^{1}H and ^{2}H, the latter often denoted by D) and oxygen (^{16}O and ^{18}O) which give rise to combinations such as $^{1}H_2^{18}O$ and $^{2}H_2^{16}O$, all of which are readily identifiable in the infrared spectra of water vapor. Furthermore, the two hydrogen atoms in the water molecule contain protons whose magnetic moments can be parallel or antiparallel, giving rise to ortho-water and para-water, respectively.

5.6 Ionic hydration shells

The term *ionic hydration shell* relates to the number of molecules of water that are bound to the ions and their distribution and coordination. Separately, the strength of binding and their residence times may be determined. The ionic charge, radial distance of the inner hydration water, and width of the inner hydration layer together determine these residence times; with the width of the inner hydration layer playing a particularly important part for otherwise similar ions.

Concentration is particularly important as a salt may form water-separated ion-pairs and direct ion-pairs, particularly at higher concentrations where there is competition from solute and solvent for the remaining water molecules, with both reducing the total number of water molecules involved in inner-shell hydration. In concentrated solution, all ions slow down the rotation of the water molecules, regardless of whether they weaken or strengthen the water hydrogen-bonding network at lower concentrations.

Water molecules interact strongly with ions, which are electrically-charged atoms or molecules. Dissolution of ordinary salt (NaCl) in water yields a solution containing the sodium (Na^+) and chloride (Cl^-) ion. Owing to its high polarity, the water molecules closest to the dissolved ion are strongly attached to it, forming what is known as the inner or primary hydration shell (Smiechowski and Stangret, 2007; Tan et al., 2014; Comez et al., 2016). Positively-charged ions such as sodium (Na^+) attract the negative (oxygen) ends of the water molecules, as shown in the diagram below. The ordered structure within the primary shell creates, through hydrogen-bonding, a region in which the surrounding waters are also somewhat ordered—this is the outer hydration shell, or cybotactic region which is that part of a solution in the vicinity of a solute molecule in which the ordering of the solvent molecules is modified by the presence of the solute molecule; the term solvent cosphere of the solute has also been used.

5.7 Environmental systems

It has long been known that the intracellular water very close to any membrane or organelle (sometimes called vicinal water) is organized very differently from bulk water, and that this structured water plays a significant role in governing the shape (and thus biological activity) of large folded biopolymers. It is important to bear in mind, however, that the structure of the water in these regions is imposed solely by the geometry of the surrounding hydrogen bonding sites.

Water can hydrogen-bond not only to itself, but also to any other molecules that have hydroxyl units (-OH) or amino units (-NH2) in accessible locations. This includes simple molecules such as alcohol derivatives, surfaces such as glass, and macromolecules such as proteins. The biological activity of proteins (of which enzymes are an important subset) is critically dependent not only on their composition but also on the way these huge molecules are folded; this folding involves hydrogen-bonded interactions with water, and also between different parts of the molecule itself. Anything that disrupts the intramolecular hydrogen bonding arrangement will denature the protein and destroy the structure of the protein and cause it to lose the biological activity. This is essentially the process that occurs when an egg is boiled—the protein is denatured because the bonds that hold the egg-white protein in its compact folded arrangement break apart so that the molecules unfold into a tangled, insoluble mass which cannot be restored to the original form.

As a further note, hydrogen-bonding need not always involve water. In fact, the two parts of the DNA double helix are held together by hydrogen-nitrogen-hydrogen (H-N-H) hydrogen bonds (Watson and Crick, 1953; Bansal, 2003).

6 Adsorption and desorption

The term *sorption* is an all-inclusive term that includes the physical and chemical interactions involved in absorption, adsorption, and desorption—desorption is the opposite effect of sorption.

Adsorption and absorption are important processes that occur in chemistry and biology. It is important to understand both processes and the differences between them when considering separation protocols, particularly in gas and liquid chromatography. The major difference between adsorption and absorption is that one is a surface process and the other a bulk process: (i) adsorption takes place on the surface of a substrate, (ii) absorption occurs when one substance enters the bulk, or volume, of another substance e.g. a gas absorbed by a liquid.

More specifically, adsorption is a surface process, the accumulation of a gas or liquid on a liquid or solid. Adsorption can be defined further based on the strength of the interaction between the adsorbent (the substrate onto which chemicals will attach) and the adsorbed molecules. **Physisorption occurs when** Van der Waals interactions between substrate and adsorbate (the molecule that is adsorbed) while **chemisorption occurs when** chemical bonds involved (covalent bonds usually) in sticking the adsorbate to the adsorbent. Chemisorption involves more energy than physisorption. The difference between the two processes is loosely based on the binding energy of the interaction.

Adsorption has importance for industries which work with natural gas, crude oil, air purification, and water purification (Mokhatab et al., 2006; Speight, 2014, 2017, 2019). Adsorption is applied for purifications of organics and sulfur dioxide (SO_2) from the gas phase. Also water can be extracted from oxygen, methane, and nitrogen and, additionally, nitrogen oxides can be extracted from nitrogen. Adsorption is also used for gas separations, such as nitrogen from oxygen, acetone (CH_3COCH_3) and acetylene ($CH{\equiv}CH$) from vent stream, and carbon monoxide, methane, carbon dioxide, nitrogen, and argon from hydrogen. In the liquid phase, adsorption is applied, for example, for organic and inorganic removal, and decolorization.

On the other hand, absorption is a phenomenon involving the bulk properties of a solid, liquid or gas. It involves atoms or molecules crossing the surface and entering the volume of the material. As in adsorption, there can be physical and chemical absorption. **Physical absorption occurs in a** non-reactive process e.g. when oxygen present in air dissolves in water. The process depends on the liquid and the gas, and on physical properties like solubility, temperature and pressure. **Chemical absorption occurs when a** chemical reaction takes place when the atoms or molecules are absorbed. An example is when hydrogen sulphide is removed from biogas streams and converted into solid sulfur.

Desorption is the release of one substance from another, either from the surface or through the surface. Desorption can occur when an equilibrium situation is altered. Imagine a tank of water in equilibrium with its surroundings. The amount of oxygen entering and leaving the water from the air will be the same — and the oxygen concentration in the water will be constant. If the water temperature increases, the equilibrium and solubility are changed, and the oxygen will desorb from the water — lowering the oxygen content.

As an example of the application of thermodynamics to the adsorption of chemical on clay in the environment, the thermodynamic quantities to be evaluated for the reaction between water and clay (such as montmorillonite) are the free energy, heat, and entropy. The precise nature of the reaction may be represented as follows:

$$n_m \text{ clay}(\text{dry at P} = 0) + n_w \text{ H}_2\text{O}(\text{at P}_o) \leftrightarrow n_m \text{ clay n}, \text{H}_2\text{O}(\text{at P})$$

In this equation, n_m is the grams or moles of clay, n_w is the grams or moles of water, P_o and P = vapor pressure of water, which is the pressure at which water vapor is in thermodynamic equilibrium with the condensed state of water—at higher pressures, the water would condense. The water vapor pressure is the partial pressure water vapor in any gas mixture in equilibrium with solid or liquid water. As for other substances, water vapor pressure is a function of temperature. The reaction proceeding from left to right is the sorption reaction whereas the reaction proceeding from right to left is the desorption reaction. In the reaction as written it is also indicated that the thermodynamic quantities to be determined are for change in state of free water and dry clay as the standard states to the combined state at a given vapor pressure P.

Clay minerals are important natural adsorbents as they pertain to water chemistry and thermodynamics. Typically, clay minerals consist of hydrated aluminum and silicon oxides and are formed by weathering and other processes acting on primary rocks. The general formulas of some common clay minerals can be repressed by chemical formulas but these formulas are subject to variance. The clay mineral families are distinguished from each other by general chemical formula, structure, and chemical and physical properties:

Clay	Approximate formula
Kaolinite	$Al_2(OH)_4Si_2O_5$
Montmorillonite	$Al_2(OH)_2Si_4O_{10}$
Nontronite	$Fe_2(OH)_2Si_4O_{10}$
Hydrous mica	$KAl_2(OH)_2(AlSi_3)O_{10}$

Clay minerals are characterized by layered structures consisting of sheets of silicon oxide alternating with sheets of aluminum oxide. Units of two or three sheets constitute *unit layers*. Some clay minerals, particularly the montmorillonites, may absorb large quantities of water between unit layers, a process accompanied by swelling of the clay. Clay minerals are usually (but not necessarily) ultrafine-grained (normally considered to be less than 2 μm (2 μm, 2×10^{-6} m) in size, using the standard particle size classification).

All the thermodynamic quantities for the reaction between water and clay indicate that the magnitude of change in these values due to the interaction of the water molecules with the exchangeable ions is much greater than that due to the interaction of the water with the oxygen surfaces. In fact, the magnitude of change in the thermodynamic values due to moving the exchangeable cations out of the hexagonal cavities is much greater than that due to the parting of the oxygen sheets during interlayer expansion.

References

Abraham, M.H., Acree Jr., W.E., 2012. The hydrogen bond properties of water from 273 K to 573 K; equations for the prediction of gas-water partition coefficients. Phys. Chem. Chem. Phys. 14, 7433–7440.

Bagchi, B., 2012. From anomalies in neat liquid to structure, dynamics and function in the biological world. Chem. Phys. Lett. 9, 1–9.

Bakker, H.J., 2008. Ultrafast energy equilibration in hydrogen-bonded liquids. Chem. Rev. 108, 1456–1473.

Bakker, H.J., Skinner, J.L., 2010. Vibrational spectroscopy as a probe structure and dynamics in liquid water. Chem. Rev. 110, 1498–1517.

Bansal, M., 2003. DNA structure: revisiting the Watson-Crick double helix. Curr. Sci. 85 (11), 1556–1563.

Breiten, B., Lockett, M.R., Sherman, W., Al-Sayah, M., Lange, H., Bowers, C.M., Heroux, A., Krilov, G., Whitesides, G.M., 2013. Water networks contribute to enthalpy/entropy compensation in protein-ligand binding. J. Am. Chem. Soc. 135 (41), 15579–15584.

Brunner, G., 2009. Near, supercritical water. Part II. Oxidative processes. J. Supercrit. Fluids 47, 382–390.

Chaplin, M., 2007. Water's Hydrogen bond Strength. https://arxiv.org/ftp/arxiv/papers/0706/0706.1355.pdf.

Chen, B., Ivanov, I., Klein, M.L., Parrinello, M., 2003. Hydrogen bonding in water. Phys. Rev. Lett. 91, 215503.

Comez, L., Paolantoni, M., Sassi, P., Corezzi, S., Morresi, A., Fioretto, D., 2016. Molecular properties of aqueous solutions: a focus on the collective dynamics of hydration water. Soft Matter 12, 5501–5514.

Eaves, J.D., Loparo, J.J., Fecko, C.J., Roberts, S.T., Tokmakoff, A., Geissler, P.L., 2005. Hydrogen bonds in liquid water are broken only fleetingly. Proc. Natl. Acad. Sci. 102, 13019–13022.

Galkin, A.A., Lunin, V.V., 2005. Subcritical and supercritical water: a universal medium for chemical reactions. Russ. Chem. Rev. 74, 21–35.

Govender, M.G., Rootman, S.M., Ford, T.A., 2003. An ab initio study of the properties of some hydride dimers. Cryst. Eng. 6, 263–286.

Grabowski, S.J., 2001. A new measure of hydrogen bonding strength—ab initio and atoms in molecules studies. Chem. Phys. Lett. 338, 361–366.

Gulliver, J.S., 2007. Introduction to Chemical Transport in the Environment. Cambridge University Press, Cambridge, United Kingdom.

Hribar, B., Southall, N.T., Vlachy, V., Dill, K.A., 2004. How ions affect the structure of water. J. Am. Chem. Soc. 124, 12302–12311.

Johari, G.P., Andersson, O., 2007. Vibrational and relaxational properties of crystalline and amorphous ices. Thermochim. Acta 461, 14–43.

Kell, G.S., 1972. Thermodynamic and transport properties of fluid water. In: Franks, F. (Ed.), Water: A Comprehensive Treatise. Vol. 1. Plenum Press, New York, pp. 363–412.

Keutsch, F.N., Saykally, R.J., 2001. Water clusters: untangling the mysteries of the liquid, one molecule at a time. Proc. US Natl. Acad. Sci. 98, 10533–10540.

Kiss, P.T., Baranyai, A., 2014. Anomalous properties of water predicted by the BK3 model. J. Chem. Phys. 140, 154505.

Kiss, P.T., Bertsyk, P., Baranyai, A., 2012. Testing recent charge-on-spring type polarizable water models. I. Melting temperature and ice properties. J. Chem. Phys. 137, 194102.

Kühne, T.D., Krack, M., Parrinello, M., 2006. Static and dynamical properties of liquid water from first principles by a novel car-parrinello-like approach. J. Chem. Theory Comput. 5, 235–241.

Lu, H., Wang, Y., Wu, Y., Yang, P., Li, L., Li, S., 2008. Hydrogen-bond network and local structure of liquid water: an atoms-in-molecules perspective. J. Chem. Phys. 129, 124512.

Ludwig, R., 2001. Water: from clusters to the bulk. Angew. Chem. Int. Ed. 40 (10), 1808–1827.

Maestro, L.M., Marqués, M.I., Camarillo, E., Jaque, D., García Solé, J., Gonzalo, J.A., Jaque, J., Del Valle, J.C., Mallamace, F., Stanley, H.E., 2016. On the existence of two states in

liquid water: impact on biological and nanoscopic systems. Int. J. Nanotechnol. 13 (8/9), 667–677.
Mallamace, F., Corsaro, C., Mallamace, D., Vasi, S., Vasi, C., Stanley, H.E., 2014a. Thermodynamic properties of bulk and confined water. J. Chem. Phys. 141, 18C504.
Mallamace, F., Corsaro, C., Mallamace, D., Vasi, S., Vasi, C., Stanley, H.E., 2014b. Erratum: thermodynamic properties of bulk and confined water. J. Chem. Phys. 141, 249903.
Mokhatab, S., Poe, W.A., Speight, J.G., 2006. Handbook of Natural Gas Transmission and Processing. Elsevier, Amsterdam, Netherlands.
Némethy, G., Scheraga, H.A., 1964. Structure of water and hydrophobic bonding in proteins. IV. The thermodynamic properties of liquid deuterium oxide. J. Chem. Phys. 41, 680–689.
Nilsson, A., Pettersson, L.G.M., 2015. The structural origin of anomalous properties of liquid water. Nat. Commun. 6, 8998.
Peters, D., 1995. Hydrogen bonds in small water clusters: a theoretical point of view. J. Mol. Liq. 67, 49–61.
Ramires, M.L.V., Nieto de Castro, C.A., Nagasaka, Y., Nagashima, A., Assael, M.J., Wakeham, W.A., 1995. Standard reference data for the thermal conductivity of water. J. Phys. Chem. Ref. Data 24, 1377–1381.
Rastogi, A., Ghosh, A.K., and Suresh, S.J. 2011. Hydrogen bond interactions between water molecules in bulk liquid, near electrode surfaces and around ions, Thermodynamics—Physical Chemistry of Aqueous Systems, Dr. Juan Carlos Moreno (Editor). InTech, University Campus STeP Ri, Slavka Krautzeka, Rijeka, Croatia.
Reddy, S.K., Straight, S.C., Bajaj, P., Pham, C.H., Riera, M., Moberg, D.R., Morales, M.A., Knight, C., Götz, A.W., Paesani, F., 2016. On the accuracy of the MB-pol many-body potential for water: interaction energies, vibrational frequencies, and classical thermodynamic and dynamical properties from clusters to liquid water and ice. J. Chem. Phys. 145, 194504.
Sharp, K.A., 2001. Water: structure and properties. In: Encyclopedia of Life Sciences. John Wiley & Sons Inc., Hoboken, NJ.
Silverstein, K.A.T., Haymet, A.D.J., Dill, K.A., 2000. The strength of hydrogen bonds in liquid water and around nonpolar solutes. J. Am. Chem. Soc. 122, 8037–8041.
Sippola, H., Taskinen, P., 2018. Activity of supercooled water on the ice curve and other thermodynamic properties of liquid water up to the boiling point at standard pressure. J. Chem. Eng. Data 63, 2986–2998.
Skinner, L.B., Benmore, C.J., Neuefeind, J.C., Parise, J.B., 2014. The structure of water around the compressibility minimum. J. Chem. Phys. 141, 214507.
Smiechowski, M., Stangret, J., 2007. Hydroxide ion hydration in aqueous solutions. J. Phys. Chem. A 111, 2889–2897.
Speight, J.G., 2014. The Chemistry and Technology of Petroleum, fifth ed. CRC Press, Taylor & Francis Group, Boca Raton, FL.
Speight, J.G., 2017. Handbook of Petroleum Refining. CRC Press, Taylor & Francis Group, Boca Raton, FL.
Speight, J.G., 2019. Natural Gas: A Basic Handbook, second ed. Gulf Publishing Company, Elsevier, Cambridge, MA.
Stillinger, F.H., 1980. Water revisited. Science 209, 451–457.
Stokely, K., Mazzaa, M.G., Stanley, H.E., Franzese, G., 2010. Effect of hydrogen bond cooperativity on the behavior of water. Proc. Natl. Acad. Sci. 107, 1301–1306.
Suresh, S.J., Naik, V.M., 2000. Hydrogen bond thermodynamic properties of water from dielectric constant data. J. Chem. Phys. 113, 9727–9732.
Tan, M.L., Cendagorta, J.R., Ichiye, T., 2014. The molecular charge distribution, the hydration shell, and the unique properties of liquid water. J. Chem. Phys. 141, 244504.

Tokmachev, A.M., Tchougreeff, A.L., Dronskowski, R., 2010. Hydrogen-bond networks in water clusters: an exhaustive quantum-chemical. Eur. J. Chem. Phys. Phys. Chem. 11 (2), 384–388.
Urquidi, J., Cho, C.H., Singh, S., Robinson, G.W., 1999. Temperature and pressure effects on the structure of liquid water. J. Mol. Struct. 485-486, 363–371.
Vedamuthu, M., Singh, S., Robinson, G.W., 1994. Properties of liquid water: origin of the density anomalies. J. Phys. Chem. 98, 2222–2230.
Velasco, S., Fernández-Pineda, C., 2007. Thermodynamics of a pure substance at the triple point. Am. J. Physiol. 75 (12), 1086–1091.
Wagner, W., Pruss, A., 2002. The IAPWS formulation 1995 for the thermodynamic properties of ordinary water substance for general and scientific use. J. Phys. Chem. Ref. Data 31, 387–535.
Watson, J., Crick, F., 1953. A structure for deoxyribose nucleic acid. Nature 171 (4356), 737–738.
Weldon, D.J., Chittiboyina, A.G., Sheri, A., Chada, R.R., Gut, J., Rosenthal, P.J., Shivakumar, D., Sherman, W., Desai, P., Jung, J.C., Avery, M.A., 2014. Synthesis, biological evaluation, hydration site thermodynamics, and chemical reactivity analysis of α-keto substituted peptidomimetics for the inhibition of plasmodium falciparum. Bio-Org. Med. Chem. Lett. 24 (5), 1274–1279.
Xu, X., Goddard III, W.A., 2004. Bonding properties of the water dimer: a comparative study of density functional theories. J. Phys. Chem. A 108, 2305–2313.
Yagasaki, T., Matsumoto, M., Tanaka, H., 2016. Anomalous thermodynamic properties of ice XVI and metastable hydrates. Phys. Rev. B 93, 054118.

Further reading

Chandra, A., 2000. Effects of ion atmosphere on hydrogen-bond dynamics in aqueous solutions. Phys. Rev. Lett. 85, 768–771.

Sources of water pollution

Chapter outline

1 Introduction

Water contamination is usually the result of human activity. In areas where population density is high and human use of the land is intensive, groundwater is especially vulnerable. Any activity whereby chemicals or domestic and industrial wastes are released to the environment, either intentionally or accidentally, will pollute a water system. When the water becomes contaminated, it is difficult (sometimes impossible) and expensive to clean up. To begin to address pollution prevention or remediation, it is necessary to understand how water systems interrelate.

Generally, water systems are interconnected and can be fully understood and intelligently managed only when that fact is acknowledged. For example, if there is a water supply (such as a well) near a source of contamination, that well water is in imminent danger of becoming contaminated and, following from this, the potential to contaminate a water system (such as a river, stream, or lake, and eventually the ocean)

Natural Water Remediation. https://doi.org/10.1016/B978-0-12-803810-9.00005-X

also have a high risk of being polluted by the flow of contaminants from the well into one or more of these systems.

In addition, groundwater can become contaminated in many ways such as from contaminated surface water that recharges an aquifer. The contaminated groundwater can then affect the quality of surface water at discharge areas. Groundwater can also be contaminated when hazardous substances (either in liquid form, in solution, or as leachate) travel through the soil into groundwater. Contaminants that dissolve in groundwater will move along with the water to wells that are used to provide drinking water. If there is a continuous source of contamination entering moving groundwater, an area of contaminated groundwater (the *plume*) will form. Thus, a combination of moving groundwater and a continuous source of contamination can add large volumes of pollutants.

Some hazardous substances dissolve very slowly in water and when these substances seep into groundwater faster than they can dissolve, some of the contaminants will stay in the soil or in the aquifer overburden). If these contaminants are in liquid form and if the contaminants are less dense than water (i.e. density < 1.00), the contaminant will float on top of the water table. These types of pollutants are referred to as light non-aqueous phase liquids (LNAPLs). On the other hand, if the contaminant is in liquid form and is more dense than water (i.e. density > 1.00), these types of contaminants are referred to as dense non-aqueous phase liquids (DNAPLs) which can form pools at the bottom of an aquifer. These pools of dense non-aqueous phase will continue to contaminate the aquifer as they slowly dissolve (or suspend in the water) and are carried away by moving groundwater. As the dense non-aqueous phase flow downward through an aquifer, tiny globs of liquid become trapped in the spaces between soil particles leading to the phenomenon known as *residual contamination*. In fact, there are several processes can affect the manner by which contamination spreads and the ultimate path of a contaminant in the groundwater. One or more of these processes can make the contaminant more harmful, less harmful, or even toxic. Thus, the possible processes that can influence the transport of chemicals (i) advection, (ii) sorption, and (iii) biological degradation.

Advection occurs when contaminants move with the groundwater. This is the main form of contaminant migration in groundwater. Thus, in the current context, advection is the transport of a chemical by bulk motion and the properties of that chemical substance remain intact during the process. An example of advection is the transport of or silt in a river by bulk water flow downstream from the entry point of the chemical. Advection is sometimes confused with the more encompassing process of *convection* which is the combination of advective transport and diffusive transport.

Convection is the transfer due to the bulk movement of molecules within fluids such as occurs in gases and liquids. Convection includes sub-mechanisms of advection (directional bulk-flow transfer), and diffusion (non-directional transfer of energy or mass particles along a concentration gradient). Convection cannot take place in most solids because neither bulk current flows nor significant diffusion of matter can take place. Diffusion of heat takes place in rigid solids (in which it is referred to as heat conduction. Convection, additionally may take place in soft solids or mixtures where solid particles can move past each other.

In fact, while the term *advection* often serves as a synonym for *convection*, more technically, convection applies to the movement of a fluid (which is often due to density gradients created by thermal gradients), whereas advection is the movement of some material by the velocity of the fluid. Thus, it is technically correct to think of momentum being adverted by the velocity field, although the resulting motion would be considered to be convection. Because of the specific use of the term convection to indicate transport in association with thermal gradients, it is more correct to use the term advection if there is any uncertainty about which terminology best describes their particular system.

Sorption occurs when contaminants attach themselves to soil particles. Sorption slows the movement of contaminants in groundwater, but also makes it harder to clean up contamination. Thus, sorption is a physical and chemical process by which a chemical becomes attached to another chemical or substance. Specific cases of sorption are treated in the following articles: (i) absorption, which is the incorporation of a chemical in one state into another chemical or substance of a different state such as when a gas is absorbed into a liquid or solid or when a liquid is absorbed into a (porous) solid, (ii) adsorption, which is the physical adherence or bonding of ions or molecules on to the surface of another phase such as the adsorption of a chemical on to the surface of a solid catalyst, and (iii) ion exchange, which is an exchange of ions between two electrolytes. The reverse of sorption is desorption. Biological degradation occurs when microorganisms, such as bacteria and fungi, use hazardous substances as a food and energy source. In the process, contaminants are decomposed (break down or degraded into less complex products) and often become less harmful.

In spite of the availability of such process, the decontamination of contaminated water—especially in the case of a contaminated groundwater system—often takes longer than expected because groundwater systems are complicated and the contaminants not always visible, in contact to the typical contaminants on the surface. This makes it more difficult to find contaminants and to design a treatment system that either destroys the contaminants in the ground or takes them to the surface for cleanup.

Depending on its physical, chemical, and biological properties, a contaminant that has been released into the environment may move within an aquifer. Some contaminants, because of their physical or chemical properties, do not always follow the water flow.) It is possible to predict, to some degree, the transport within an aquifer of those substances that move along with water flow. For example, both water and certain contaminants flow in the direction of the topography from a recharge area to a discharge area. Soils that are porous and permeable tend to transmit water and certain types of contaminants with relative ease to an aquifer below. Just as water generally moves slowly, so do contaminants in a water system. Because of this slow movement, contaminants tend to remain concentrated in the form of a plume that flows along the same path as the water flow. The size and speed of the plume depend on the amount and type of contaminant, its solubility and density, and the velocity of the surrounding water.

Water contaminants can move rapidly through fractures in rocks. Fractured rock presents a unique problem in locating and controlling contaminants because the fractures are generally randomly spaced and do not follow the contours of the land surface or the hydraulic gradient. Contaminants can also move into the water system through macropores, such as root systems, animal burrows, abandoned wells, and other systems of holes and cracks that supply pathways for contaminants. In areas surrounding active pumping wells, the potential for contamination increases because water from the zone of contribution, a land area larger than the original recharge area, is drawn into the well and the surrounding aquifer. Some wells that provide water for drinking draw water from nearby streams, lakes, or rivers. Contaminants present in these surface waters can contribute contamination to the water system. Some wells rely on artificial recharge to increase the amount of water infiltrating an aquifer, often using water from storm runoff, irrigation, industrial processes, or treated sewage.

In several cases, this practice has resulted in increased concentrations of nitrates, metals, microbes, or synthetic chemicals in the water. Under certain conditions, pumping can also cause the water (and associated contaminants) from another aquifer to enter the one being pumped. This phenomenon is called inter-aquifer leakage. Thus, properly identifying and protecting the areas affected by well pumping is important to maintain water quality. Generally, the greater the distance between a source of contamination and a water source, the more likely that natural processes will reduce the impacts of contamination. Processes such as oxidation, biological degradation (which sometimes renders contaminants less toxic), and adsorption (binding of materials to soil particles) may take place in the soil layers of the unsaturated zone and reduce the concentration of a contaminant before it reaches a water system. Even contaminants that reach water directly, without passing through the unsaturated zone, can become less concentrated by dilution (mixing) with the water. However, because some waters move slowly, contaminants generally undergo less dilution than when in surface water.

In addition, land-based activities (industry, agriculture) and run-off from land are major sources of coastal water pollution. Other issues which are equally or even more dangerous than oil pollution also threaten the marine world, for example, alien species migration, hydrocarbon derivatives in ballast water, biocides, pollution from ship-building, pollution from ship repair, pollution from ship scrapping, and noise pollution that affects sea mammals.

Thus, there are many sources of water pollutants for which practical solutions are required to minimize the effects of these pollutants are recent entrants into the environment as well as those pollutants that require remediation (cleaning up) past problems. In the end, there are many choices on the personal level and the societal level that must be made (consciously or not) to decrease the amount of pollution. Within the context of this book, the effects of water pollution are varied and include (i) poisonous drinking water, (ii) unbalanced river and lake ecosystems that can no longer support full biological diversity, as well as (iii) deforestation from acid rain. These effects are, of course, specific to the various contaminants. This chapter presents the various methods by which water systems can be polluted.

2 Sources

Water pollution can occur from two sources: (i) point source and (ii) non-point source. Point sources of pollution are those which have direct identifiable source as, for example, the pipe attached to a factory, oil spill from a tanker, effluents coming out from various industrial operations. Thus, point sources of pollution include wastewater effluent (both municipal and industrial) and storm sewer discharge and affect mostly the area near it.

On the other hand, non-point sources of pollution (also often referred to as diffuse sources of pollution) are those source of pollutants in which the pollutants are released from different sources of origin (Table 5.1). Examples include runoff from agricultural land and urban areas. This type of pollution can be difficult to identify and control. Typically, in such cases, there is a number of ways by which contaminants enter into a water system or into the surface water (such as the groundwater system) and arrive in the environment from different non-identifiable sources.

In some systems of nomenclature, pollutants are referred to as being (i) pollutants from (i) direct sources or (ii) pollutants from indirect sources. Direct sources of pollutants include effluent outfalls from factories, refineries, and waste treatment plants that emit fluids of varying quality directly into urban water supplies. Pollutants from indirect sources include contaminants that enter the water supply from soil systems and groundwater systems and from the atmosphere via rain water. Soil systems and groundwater systems contain the residue of human agricultural practices (such as fertilizers and pesticides) and improperly disposed of industrial wastes. Atmospheric contaminants are also derived from human practices (such as gaseous emissions from automobiles, factories, and even from bakeries).

The types of nonpoint source pollutants entering a water system vary depending on the type of human activity in the watershed. Runoff from agricultural regions typically contains elevated concentrations of suspended solids, dissolved salts and nutrients from fertilizers, biodegradable organic matter, pesticides, and pathogens from animal wastes. Activities that disrupt the vegetation cover or soil surface, such as construction and silvaculture, will contribute suspended sediments and sediment-bound nutrients

Table 5.1 Various sources of contaminants in urban water systems[a]

Contaminant	Typical sources
Metals	Automobiles, bridges, atmospheric deposition, industrial areas, soil erosion, corroding metal surfaces, combustion processes
Microbial pathogens	Lawns, roads, leaky sanitary sewer lines, sanitary sewer cross-connections, animal waste, septic systems
Nutrients (N and P)	Lawn fertilizer, atmospheric deposition, automobile exhaust, soil erosion, animal waste, detergents
Oil/grease/hydrocarbons	Roads, driveways, parking lots, vehicle maintenance areas, gas stations, illicit dumping to storm drains
Organic materials	Residential lawns and gardens, commercial landscaping, animal wastes
Pesticides/herbicides	Residential lawns and gardens, roadsides, utility right-of ways, commercial and industrial landscaped areas, soil wash-off
Sediment/floatables	Streets, lawns, driveways, roads, construction activities, atmospheric deposition, and drainage channel erosion

[a] Listed alphabetically rather than by any order of preference or importance.

such **as** phosphorus to surface runoff. Runoff from silvaculture sites (sites for the care and cultivation of forest trees) may also contain herbicides that were applied to control the growth of undesirable plants. Urban runoff, one of the worst sources of nonpoint source pollution, often contains high concentrations of (i) suspended and dissolved solids, (ii) nutrients and pesticides from landscaped areas (iii) toxic metals, oil and grease, and hydrocarbon derivatives from road asphalt, (iv) pathogens from pet wastes and leaking septic tanks, (v) and synthetic organics such as detergents, degreasers, chemical solvents, and other compounds that accumulate on impervious surfaces or are carelessly poured down storm drains. There are also cases where pollutants enter the environment in one place and here is no noticeable effect until the pollutant has been transported over a great distance—sometimes hundreds of miles or even thousands of miles to another site (*transboundary pollution*). One example of transboundary pollution is the release of radioactive waste that travels through the oceans from nuclear reprocessing plants to nearby countries.

Water can become contaminated from natural sources or numerous types of human activities. Some substances found naturally in rocks or soils, such as iron, manganese, arsenic, chlorides, fluorides, sulfates, or radionuclides, can become dissolved in water. Thus, contaminants may reach water from activities on the land surface, such as releases or spills from stored industrial wastes; from sources below the land surface but above the water table, such as septic systems or leaking underground crude oil storage systems; from structures beneath the water table, such as wells; or from contaminated recharge water.

Other naturally occurring substances, such as decaying organic matter, can move in water as particles. Whether any of these substances appears in water depends on local conditions. Some substances may pose a health threat if consumed in excessive quantities; others may produce an undesirable esthetics in the form of an obnoxious odor or taste or color. Water that contains unacceptable concentrations of these substances is not used for drinking water or other domestic water uses unless it is treated to remove these contaminants.

Water pollutants may be (i) organic water pollutants and (ii) inorganic water pollutants. Organic water pollutants comprise of insecticides and herbicides, organo-halide derivatives and other forms of chemicals; bacteria from sewage and livestock farming; food processing wastes; pathogens; and volatile organic compounds (often referred to as VOCs). Inorganic water pollutants may arise from heavy metals from acid mine drainage; silt from surface run-off, logging, slash and burning practices and land filling; fertilizers from agricultural run-off which include nitrates and phosphates as well as chemical waste from industrial effluents.

Some of the important sources of water pollution are (listed alphabetically rather than by preference or by relative importance: (i) acid rain pollution, (ii) alien species, (iii) climate change, (iv) disruption of sediments, (v) industrial chemicals and agro-chemical wastes, (vi) nutrient enrichment, (vii) oil spillage, (viii) radioactive waste, (ix) sewage and other oxygen demanding wastes, (x) thermal; pollution, and (xi) urbanization.

2.1 Acid rain

Acids and buses from industrial and mining activities can alter the water quality in a stream or lake to the extent that it kills the aquatic organisms living there, or prevents them from reproducing. Acid mine drainage has polluted surface waters since the beginning of ore mining. Sulfur-laden water leached from mines, including old and abandoned mines as well as active ones, contains compounds that oxidize to sulfuric acid on contact with air.

More pertinent to this section, the term *acid rain* refers to the deposition of a mixture from wet (rain, snow, sleet, fog, cloud water, and dew) and dry (acidifying particles and gases) acidic components. Distilled water, once the carbon dioxide is removed, has a neutral pH of 7. Liquids with a pH less than 7 are acidic, and those with a pH greater than 7 are alkaline. Acid rain is a rain or any other form of precipitation that is unusually acidic insofar as the rain has elevated levels of hydrogen ions (low pH). It can have harmful effects on plants, aquatic animals and infrastructure. Acid rain is caused by emissions of sulfur dioxide and nitrogen oxide, which react with the water in the atmosphere to produce acids. Nitrogen oxides can also be produced naturally by lightning strikes and sulfur dioxide is produced by volcanic eruptions.

Water pollution that alters the surrounding pH level, such as due to acid rain, can harm or kill many aquatic species. Atmospheric sulfur dioxide and nitrogen dioxide emitted from natural and human-made sources like volcanic activity and burning fossil fuels\interact with atmospheric chemicals, including hydrogen and oxygen, to form sulfuric and nitric acids in the air. These acids fall down to earth through precipitation in the form of rain or snow. Once acid rain reaches the ground, it flows into waterways that carry its acidic compounds into water bodies. Acid rain that collects in aquatic environments lowers water pH levels and affects the aquatic biota.

2.2 Agricultural wastes

Agricultural wastes that flow directly into surface waters are typically high in nutrients (phosphorus and nitrogen), biodegradable organic carbon, pesticide residues, and fecal coliform bacteria (bacteria that normally live in the intestinal tract of warm-blooded animals and indicate contamination by animal wastes). Feedlots where large numbers of animals are contained into relatively small pens provide an efficient way to raise animals for food.

Nitrogen in the form of ammonia (NH_3) and nitrate derivatives ($-NO_3^-$) form part of the plant nutrients that can lead to eutrophication. Nitrate enrichment through sewage contamination and fertilizer runoff is not as critical as it is with phosphates because aquatic ecosystems are not as sensitive to increases in nitrate levels. Nitrogen normally occurs in a form that plants cannot use (i.e. nitrogen gas), however, it may be used in the decomposition of dead water plants and by blue-green algae which can convert nitrogen in the air into ammonia and nitrates that plants can use. This subtle interdependence illustrates the complexity of the relationships within an aquatic ecosystem. The best way to stop this kind of pollution is to prevent human sewage, cattle excrement and fertilizers from washing into rivers.

Feedlot drainage (and drainage from intensive poultry cultivation) creates an extremely high potential for water pollution. Aquaculture has a similar problem because wastes are concentrated in a relatively small space. Even relatively low densities of animals can significantly degrade water quality if the animals are allowed to trample the water system bank, or runoff from manure-holding ponds is allowed to overflow into nearby waterways.

Both surface and groundwater pollution are common in agricultural regions because of the extensiveness of fertilizer and pesticide application.

2.3 Alien species

In the context of this book, an alien species (also referred to as an *invasive species* or an *introduced species*) is a species that is not native to a specific location and which has a tendency to spread to a degree believed to cause damage to the environment, human economy or human health. The term as most often used applies to introduced species (also called non-indigenous species or non-native species) that adversely affect the habitats amend bioregions that they invade economically, environmentally, or ecologically. Such invasive species may be either plants or animals and may disrupt by dominating a region, wilderness area, particular habitat, or wildland-urban interface land from loss of natural controls (such as predators or herbivores).

An introduced species might become invasive if it can outcompete native species for resources such as nutrients, light, physical space, water, or food. If these species evolved under great competition or predation, then the new environment may host fewer able competitors, allowing the invader to proliferate quickly. Ecosystems which are being used to their fullest capacity by native species can be modeled as zero-sum systems in which any gain for the invader is a loss for the native. However, such unilateral competitive superiority (and extinction of native species with increased populations of the invader) is not the rule. Invasive species often coexist with native species for an extended time, and gradually, the superior competitive ability of an invasive species becomes apparent as its population grows larger and denser and it adapts to its new location. An invasive species might be able to use resources that were previously unavailable to native species, such as deep water sources accessed by a long taproot, or an ability to live on previously uninhabited soil types.

Thus, ecosystems that have been invaded by alien species may not have the natural predators and competitors present in its native environment that would normally control their populations. Native ecosystems that have undergone human-induced disturbance are often more prone to alien invasions because there is less competition from native species.

In some parts of the world, alien species also known as invasive species are a major problem of water pollution. Outside their normal environment, they have no natural predators, so they rapidly spread and dominate the animals or plants that thrive there. Common examples of alien species include zebra mussels in the Great Lakes of the USA, which were carried there from Europe by ballast water (waste water flushed from ships). The Mediterranean Sea has been invaded by a kind of alien algae called *Caulerpa taxifolia*. In the Black Sea, an alien jellyfish called *Mnemiopsis leidyi* reduced fish stocks by 90% after arriving in ballast water. In San Francisco Bay, Asian clams called *Potamocorbula amurensis*, also introduced by ballast water, have dramatically altered the ecosystem.

2.4 Climate change

Climate change is a change in the statistical distribution of weather patterns when that change lasts for an extended period of time (i.e., decades to millions of years). Climate change is caused by factors such as biotic processes, variations in solar radiation received by Earth, plate tectonics, volcanic eruptions, and certain human activities have been identified as primary causes of ongoing climate change, often referred to as global warming. There is no general agreement in scientific, media or policy documents as to the precise term to be used to refer to anthropogenic forced change; either global warming or climate change may be used.

The term *climate change* (*global warming*) is often used to refer specifically to anthropogenic climate change. Anthropogenic climate change is caused by human activity, as opposed to changes in climate that may have resulted as part of natural processes of the Earth. In this sense, especially in the context of environmental policy, the term climate change has become synonymous with anthropogenic global warming. Within scientific journals, global warming refers to surface temperature increases while climate change includes global warming and everything else that increasing greenhouse gas levels affect, including the temperature of the Earth (Fig. 5.1) although other natural phenomena also effect the temperature of the Earth and the increase in temperature and climate change are due to other natural effects (Speight and Islam, 2016; Radovanović and Speight, 2018).

Global warming has also an impact on water resources through enhanced evaporation, geographical changes in precipitation intensity, duration and frequency (together affecting the average runoff), soil moisture, and the frequency and severity of droughts and floods. Future projections using climate models pointed out that there will be an increase in the monsoon rainfall in most parts of India, with increasing greenhouse gases and sulphate aerosols. Relatively small climatic changes can have huge impact on water resources, particularly in arid and semiarid regions such as North-West India. This will have impacts on agriculture, drinking water, and on generation of hydroelectric power, resulting in limited water supply and land

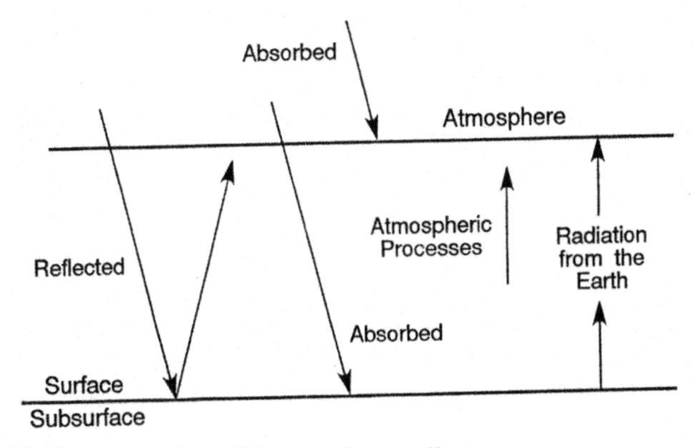

Fig. 5.1 Simplified representation of the greenhouse effect.

degradation. Apart from monsoon rains, India uses perennial rivers which originate in the Hindukush and Himalayan ranges and depend on glacial melt-waters. Since the melting season coincides with the summer monsoon season, any intensification of the monsoon is likely to contribute to flood disasters in the Himalayan catchment. Rising temperatures will also contribute to a rise in the snowline, reducing the capacity of these natural reservoirs, and increasing the risk of flash floods during the wet season. Increase in temperatures can lead to increased eutrophication in wetlands and fresh water supplies.

Briefly, eutrophication is the process through which a body of water becomes enriched with chemicals such as nitrates and phosphates. Algae and other aquatic plants then feed on these nutrients leading to excess growth. This leads to a reduction in the amount of dissolved oxygen available as algal blooms on the surface restrict the amount of sunlight penetrating the water limiting photosynthesis which causes the death and decomposition of plant life underwater. The lack of dissolved oxygen also kills all animal life in the water body.

2.5 Disruption of sediments

Sediments are a natural part of a stream, lake, or river, and the type and amount found in water systems are influenced by the geology of the surrounding area. The deposition of sediment is the result of a combination of physical, chemical, and biological processes. In these processes, sedimentary material may be carried into a body of water by erosion or through sloughing (caving in) of the shore and, as a result, clay minerals, sand, and organic matter are carried by water flow into a lake and settle out as layers of sediment.

Sediments consist of mostly inorganic material washed into a water system as a result of land cultivation, construction, demolition, and mining operations. Sediments can cover gravel beds and block light penetration, making it more difficult for aquatic species (such as fish) to find food. Sediments can also damage gill structures directly, smothering aquatic insects and fishes. Organic sediments can deplete the water of oxygen, creating anaerobic (without oxygen) conditions, and may create unsightly conditions and cause unpleasant odors. Furthermore, suspended particulate matter affects the mobility of organic compounds sorbed to particles. Also, sorbed organic matter undergoes chemical degradation and biodegradation at different rates and by different pathways compared to organic matter in solution.

Natural processes that add to sediments in waterways include in-stream scouring of the river bed and banks and erosion of sediments from the surrounding catchment from natural slips and any exposed soils. Sediments can enter water systems from alongside a reach or from upstream via the myriad of smaller interconnecting water systems that form a river network within a catchment area. However, if the level of the sediment becomes too high (though natural or anthropogenic processes), it can disrupt the ecosystem. Excess sediment in a water system can cause damage by blocking light that allows algae (an important food source) to grow, harming fish gills, filling up important habitats, and stopping fish from seeing well enough to move around or feed.

While sediment movement is a natural part of a functioning freshwater ecosystem, human activities around a waterway can greatly increase the amount of sediment that enters the system. This can have considerable effects on water quality and the plants and animals that live there. The addition of sediment to rivers and streams above normal levels is a serious issue. Sediments in waterways travel downstream in suspension when water velocity is high or turbulent. When there is a decrease in velocity, especially in pools and deep areas of a water system, sediments will eventually settle and can be seen as deposits of fine material or by the formation of sand bars on the bed of the water system.

Deposition of silt in water bodies occurs as a result of erosion carrying silt laden water and due to flood. It increases the turbidity of water and reduces light penetration in deep water causing decline in abundance of submerged plants. Siltation inhibits the growth of aquatic plants. The abundance of phytoplankton is affected due to reduction in surface exchange of gases and nutrients. Plants that are tolerant to turbidity are abundant followed by those that are intermediate and the least tolerant species.

Sediments in suspension can have a significant impact on the water quality of a waterway because sediments decrease water clarity, which reduces visibility. Water clarity is usually measured as turbidity. Turbid waters prevent the growth of aquatic plants and algae (because plants need light for photosynthesis) and decrease the ability of fish to find food or to detect predators and prey, thereby increasing stress. Sediments may smother invertebrates in a water system which are an important food source for fish.

Excessive sediment deposits on the river/stream bed can significantly alter and degrade habitat. Some animals are dependent on the rocky bottoms of water systems, while others live in deep sandy pools or around woody debris. Sediments fill the spaces between stones that invertebrates live in, and in extreme cases can bury woody debris, stony substrates (gravels and cobbles), and root mats, and fill pools and channels. This reduces the amount of invertebrate habitat and cover and spawning grounds (a place to lay eggs) for fish. An increase in the amount of sediment deposited on the river/stream bed can also significantly change the flow and depth of rivers or streams over time and infill lakes and estuaries. Natural cleaning processes—where the water flows through the gravel bed of a water system and interacts with the microbes living on stone surfaces, removing nutrients and some pollutants—can also be short-circuited by excessive sediment deposits.

Thus, the impacts of sediment on a water system are (i) decreased water clarity—increased sediment loading into a water system will decrease water clarity and reduce visibility for fish seeking food and places to live, (ii) damage to fish gills and filter feeding apparatus of invertebrates, (iii) changes to the benthic (bottom) structure of the stream/river bedcoarse substrates such as gravels and boulders are replaced/smothered by sand and silt, (iv) decreased numbers of invertebrate species from smothering of habitat—invertebrates are a food source to some fish and diverse invertebrate communities are also an indicator of healthy water systems, (v) decreased algal food supply at base of food chain—sediments can scour algae from rocks, make algae unpalatable, or reduce light to levels where algae cannot grow, because plants need light to for photosynthetic processes, and (vi) increased contaminants from surrounding land—sediments can transport attached pollutants such as nutrients, bacteria, and toxic chemicals from agriculture and horticulture into water systems.

The construction of dams for hydroelectric power or water reservoirs can reduce the sediment flow affecting adversely the formation of beaches, increases coastal erosion and reduces the flow of nutrients from rivers into seas (potentially reducing coastal fish stocks). Increased sediment flow can also create a problem. During construction work, soil, rock, and other fine powders sometimes enter nearby rivers in large quantities, causing water to become turbid (muddy or silted).

Active and abandoned mines can contribute to water contamination. Precipitation can leach soluble minerals from the mine wastes (known as spoils or tailings) into the water below. These wastes often contain metals, acid, minerals, and sulfides. Abandoned mines are often used as wells and waste pits, sometimes simultaneously. In addition, mines are sometimes pumped to keep them dry; the pumping can cause an upward migration of contaminated water, which may be intercepted by a well.

2.6 Industrial waste and agrochemicals

Industrial waste is the waste produced by industrial activity which includes any material that is rendered useless during a manufacturing process such as that of factories, industries, mills, and mining operations. Some examples of industrial wastes are chemical solvents, pigments, sludge, metals, ash, paints, sandpaper, paper products, industrial byproducts, and radioactive waste. Toxic waste, chemical waste, industrial solid waste, and municipal solid waste are designations of industrial wastes.

Effluents from industries contain various organic and inorganic waste products. Synthetic organic chemicals and pesticides can adversely affect aquatic ecosystems as well as making the water unusable for human contact or consumption. These compounds may come from point source industrial effluents or from nonpoint source agricultural and urban runoff.

Fly ash form thick floating cover over the water thereby reducing the penetration of light into deeper layers of water bodies. Fly ash increases the alkalinity of water and cause reduced uptake of essential bases leading to death of aquatic plants. Liquid organic effluents change the pH of water and the specific toxicity effects on the aquatic plants vary depending on their chemical composition. There may be synergistic, additive or antagonistic interactions between metals with respect to their effects on plants however these effects are reduced in hard and buffered freshwater bodies.

Agrochemicals (sometimes spelled *agrichemical*) are chemicals used in agriculture such as pesticides, insecticides fungicides, and nematicides (types of chemical pesticides used to kill plant-parasitic roundworms). The term agrochemicals may also include synthetic fertilizers, hormones, and other chemical growth agents as well as raw animal manure with high concentrations of undigested insecticides.

Insecticides are chemicals that are sprayed onto crops to kill the insects that eat crops. One of the more controversial insecticides is DDT which was used to control the malaria mosquito.

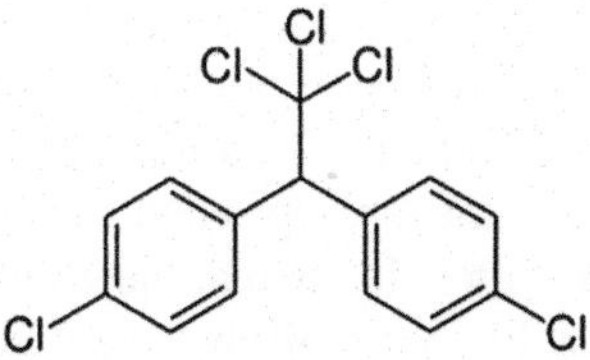

DDT

DDT (dichlorodiphenyltrichloroethane) is a colorless, tasteless, and almost odorless crystalline chemical compound, an organochlorine, originally developed as an insecticide, and ultimately becoming infamous for its environmental impacts. At certain levels (10 mg per kg) DDT can cause human poisoning with the following symptoms: dizziness, vomiting and convulsions. At lower levels, DDT can affect the development of babies and has been associated with cancer. Another group of controversial insecticides are the organo-phosphates. These insecticides have been used in the place of DDT since it was banned. Some observers believe that these insecticides have caused even greater environmental damage than DDT and that they are even more toxic to humans and other mammals.

Insecticides are easily washed by the rain into streams and groundwater where they poison fish and domestic animals. Many insecticides are stored for a long time in the bodies of animals and can end up in the meat, fish, egg and milk that you eat. Fruits and vegetables that have been sprayed with insecticides also remain poisonous for many days afterwards and must be washed very well before eating.

Many agrochemicals are toxic and such chemicals that are in bulk storage can pose significant risks to the environment and to human health, particularly when an accidental spill occurs. In many countries, use of agrichemicals is highly regulated. On farms, proper storage facilities and labeling, emergency clean-up equipment and procedures, and safety equipment and procedures for handling, application, and disposal are often subject to mandatory standards and regulations. Usually, the regulations are carried out through the registration process.

Many of the industries are situated along the banks of river such as the steel industry and the pulp and paper industry for their requirement of huge amounts of water in manufacturing processes and finally their wastes containing acid derivatives, alkali derivatives, dyestuffs, and other chemicals are dumped and poured into rivers as effluents. Chemical industries concerning with manufacture of aluminum tend to release large amount of fluoride through their emissions to air and effluents to water bodies. The fertilizer industry generates huge amount of ammonia whereas steel plants generate cyanide. Chromium salts are used in industrial process for the production of sodium dichromate and other compounds containing chromium. All such discharges finally arrive at a water system in the form of effluents affecting human health and the organism living in the water system.

Underground and aboveground storage tanks are commonly used to store crude oil products and other chemical substances. For example, many homes have underground heating oil tanks. Many businesses and municipal highway departments also store gasoline, diesel fuel, fuel oil, or chemicals in on-site tanks. Industries use storage tanks to hold chemicals used in industrial processes or to store hazardous wastes for pickup by a licensed hauler. Over the years, the contents of many of these tanks have leaked and spilled into the environment. If an underground storage tank develops a leak, which commonly occurs as the tank ages and corrodes, its contents can migrate through the soil and reach the water. Tanks that meet federal/state standards for new and upgraded systems are less likely to fail, but they are not foolproof. Abandoned underground tanks pose another problem because their location is often unknown. Aboveground storage tanks can also pose a threat to water if a spill or leak occurs and adequate barriers are not in place. Improper chemical storage, sloppy materials handling, and poor-quality containers can be major threats to water. Tanker trucks and train cars pose another chemical storage hazard. Each year, chemical spills occur from trucks, trains, and storage tanks, often when materials are being transferred. At the site of an accidental spill, the chemicals are often diluted with water and then washed into the soil, increasing the possibility of water contamination.

In the agricultural sector, water and electricity for irrigation are subsidized for political reasons. This leads to wasteful flood irrigation rather than adoption of more optimal practices such as sprinkler and drip irrigation. Cropping patterns and farming practices also do not necessarily encourage the judicious use of water. There are losses of water due to breaches and seepage resulting in water logging and salinity. Agro-chemical wastes include fertilizers, pesticides which may be herbicides and insecticides widely used in crop fields to enhance productivity. Improper disposal of pesticides from field farms and agricultural activities contributes a lot of pollutants to water bodies and soils.

Pesticides reach water bodies through surface runoff from agricultural fields, drifting from spraying, washing down of precipitation and direct dusting and spraying of pesticides in low lying areas polluting the water quality. Most of them are non-biodegradable and persistent in the environment for long period of time. These chemicals may reach human through food chain leading to biomagnification. In fact, chemicals from fertilizers, pesticides, insecticides, and herbicides that have been applied in excess to crops are washed away with rainwater as runoff, then enter into soil and finally arrive at the water bodies. Chemicals from fertilizers result in eutrophication by enrichments of nutrients. Ammonium from fertilizers is acidic in nature causing acidification of water. Similarly pesticides, herbicides and insecticides also cause change in pH of the water bodies. Most common effect of these substances is the reduction in photosynthetic rate. Some may uncouple oxidative phosphorylation or inhibit nitrate reductase enzyme. The uptake and bioaccumulation capacities of these substances are great in macrophytic plants due to their low solubility in water.

Bioaccumulation is the build-up of toxic substances in a food chain. A common example in aquatic systems is the accumulation of heavy metals such as mercury (Hg) in fish.

Heavy metals such as nickel, molybdenum, zinc, cadmium and lead are mined and processed by the mining and ore-smelting industries and these metals are easily washed into streams and groundwater (Bibi et al., 2016). Copper and mercury are another two heavy metals, which are found in fungicides, which are also sprayed on crops and easily washed into rivers. These heavy metals are toxic to biological life including the people who may have to drink from the polluted rivers. Crops that have been irrigated with polluted water can also be dangerous. In a similar way to DDT, heavy metals can also build up in the body causing symptoms of poisoning.

At the start of the food chain, mercury is absorbed by algae in the form of methyl mercury (CH_3Hg^+). Fish then eat the algae and absorb the methylmercury and since they are absorbing it at a faster rate than it can be excreted, it accumulates in the body of the fish. Further up the food chain, predatory fish and birds then absorb the mercury from the fish they consume, which then accumulates in their bodies leading to a higher concentration of the mercury in their own bodies than in the species they have eaten (known as *biomagnification*). This process can be dangerous to humans if fish which have bioaccumulated mercury are eaten and absorbed into human tissues thereby causing health problems such as damage to the central nervous system.

2.7 Nutrient enrichment

A nutrient is a chemical used by an organism to survive, grow, and reproduce. Different types of organism have different essential nutrients. Nutrients may be organic or inorganic: organic compounds include most compounds containing carbon, while all other chemicals are inorganic. Inorganic nutrients include nutrients such as iron, selenium, and zinc, while organic nutrients include, among many others, energy-providing compounds and vitamins. Plant nutrients consist of more than a dozen minerals absorbed through roots, plus carbon dioxide and oxygen absorbed or released through leaves. All organisms obtain all their nutrients from the surrounding environment.

Nutrients, especially nutrients containing nitrogen and phosphorus, can promote accelerated eutrophication, or the rapid biological aging of lakes, streams, and estuaries. Phosphorus and nitrogen are common pollutants in residential and agricultural runoff, and are usually associated with plant debris, animal wastes, or fertilizer. Phosphorus and nitrogen are also common pollutants in municipal wastewater discharges, even if the wastewater has received conventional treatment. Phosphorus adheres to inorganic sediments and is transported with sediments in storm runoff. Nitrogen tends to move with organic matter or is leached from soils and moves with groundwater.

The sources of nutrients in surface water can be divided broadly into natural and anthropogenic types. Contribution to pollution by natural source is low due to balance established by the natural system between the production and consumption of nutrients over the course of time. Anthropogenic sources of contaminants are contributed from agriculture, domestic and industrial wastes. Nutrient concentrations in water systems have been strongly correlated with human land use and disturbance gradients. The enrichment of both nitrogen and phosphorus have links with the agricultural and urban land uses in the watershed. Fertilizers containing nitrogen are the main sources of nitrogen in water systems. Similarly, the nutrient enrichment of aquatic systems from anthropogenic sources includes point and nonpoint sources. In contrast to point sources of nutrients that are relatively easy to monitor and regulate, nonpoint sources such as livestock, crop fertilizers, and urban runoff exhibit more variability.

Nutrient enrichment in aquatic water bodies leads to eutrophication which is a process whereby water bodies receive excess inorganic nutrients, especially N and P, stimulating excessive growth of plants and algae. Eutrophication can happen naturally in the course of normal succession of some freshwater ecosystems. However, when the nutrient enrichment is due to the activities of humans, it is referred to as "cultural eutrophication", where the rate of nutrient enrichment is greatly intensified. Eutrophication was recognized as a pollution problem in North American lakes and reservoirs in the mid-20th Century. Plants must take in nutrients from the surrounding environment in order to grow. Nitrogen and phosphorous, in particular, encourage growth because they stimulate photosynthesis. This is why they are common ingredients in plant fertilizers. When agricultural runoff pollutes waterways with nitrogen and phosphorous rich fertilizers, the nutrient-enriched waters often paves way to algal bloom leading to eutrophication. The result is oxygen depletion and dying of fishes due to suffocation.

2.8 Crude oil spills

Oil spillage (an oil spill) refers to the spills of crude oil itself or a crude oil product (or products) and is the release of liquid crude oil hydrocarbon derivatives into the environment, especially the marine ecosystem, due to human activity. The term is usually applied to marine oil spills, where oil is released into the ocean or into coastal offshore platforms, drilling rigs, and water, but the term is also applied to oil spills that occur on land. The spillage of crude oil may be due to releases of crude oil from tankers offshore platforms and drilling rigs as well as spills of refined crude oil products (such as naphtha or and kerosene or diesel fuel) and their by-products, heavier fuel oil as used by large ships such as bunker fuel oil, or the spill of any oily refuse or waste oil.

Oil spills into the ocean are generally much more damaging than those on land, since they can spread for hundreds of nautical miles in a thin oil slick which can cover beaches with a thin coating of oil. These can kill seabirds, mammals, shellfish and other organisms they coat. Oil spills on land are more readily containable if a makeshift earth dam can be rapidly bulldozed around the spill site before most of the oil escapes, and land animals can avoid the oil more easily.

However, cleanup and recovery from an oil spill is difficult and depends upon many factors, including the type of oil spilled, the temperature of the water (affecting evaporation and biodegradation), and the types of shorelines and beaches involved. Spills may take weeks, months or even years to clean up and can have disastrous consequences for society; economically, environmentally, and socially. As a result, oil spill accidents have initiated intense media attention and political uproar, bringing many together in a political struggle concerning government response to oil spills and what actions can best prevent them from happening.

Oil discharge into the surface of sea by way of accident or leakage from cargo tankers carrying naphtha (or gasoline) or kerosene (or diesel fuel) and their derivatives pollute sea water to a great extent. Exploration of oil from offshore also lead to oil pollution in water. The residual oil spreads over the water surface forming a thin layer of water-in-oil emulsion.

Oil spillage:

Oil pollution due to spillage of oil tankers and storage containers prevents oxygenation of water and depletes the oxygen content of the water body by reducing light transmission inhibiting the growth of planktons and photosynthesis in macrophytes.

The behavior and fate of oil spills in the marine environment is described by complex processes of oil transformations, which depend on the composition or other properties of the oil itself, as well as on the parameters of oil spills or environmental conditions. Most of the processes (such as physical transport, evaporation, dissolution, natural degradation, oxidation, and sedimentation) are natural and do not require commercialized equipment or products. The following two processes may require advanced research: biodegradation and emulsification.

Burning of oil in situ is strictly regulated and requires prior approval from governmental bodies. In-situ burning of oil spills usually has a negative environmental impact, but it can be used when oil needs to be removed quickly to prevent the spread of contamination or further environmental damage or when other cleanup options prove ineffective or might be more harmful to the environment, and as the only alternative when spill locations have restricted access due to terrain, weather or other factors. In some cases, oil and debris can be placed in a landfill. Governments strictly regulate the disposal of such materials. Landfill, land farming and similar methods have important environmental implications and its practice is less and less used, so research should focus mainly on recycling (oil), incineration or on finding oil spill debris clean-up methods, e. g., thermal desorption.

2.9 Radioactive waste

Radioactive waste is waste that contains radioactive material and is usually a byproduct of nuclear power generation and other applications of nuclear technology. Radioactivity naturally decays over time, so radioactive waste has to be isolated and confined in appropriate disposal facilities for a sufficient period until it no longer poses a threat. The time radioactive waste must be stored for depends on the type of waste and radioactive isotopes. Current approaches to managing radioactive waste have been segregation and storage for short-lived waste, near-surface disposal for low and some intermediate level waste, and deep burial or partitioning for the high-level waste.

There are two broad classifications: high-level or low-level waste. High-level waste is primarily spent fuel removed from reactors after producing electricity. Low-level waste comes from reactor operations and from medical, academic, industrial, and other commercial uses of radioactive materials. Both forms of radioactive waste are deadly to life forms.

High-level radioactive waste primarily is uranium fuel that has been used in a nuclear power reactor and is spent, or no longer efficient in producing electricity. Spent fuel is thermally hot as well as highly radioactive and requires remote handling and shielding. Nuclear reactor fuel contains ceramic pellets of uranium-235 (^{235}U) inside of metal rods. Before these fuel rods are used, they are often less radioactive and may be handled without special shielding.

High-level wastes are hazardous because they produce fatal radiation doses during short periods of direct exposure. For example, 10 years after removal from a reactor, the surface dose rate for a typical spent fuel assembly exceeds 10,000 rem/h—far greater than the fatal whole-body dose for humans of about 500 rem received all at once. If isotopes these high-level wastes get into groundwater or rivers, they may enter food chains. The dose produced through this indirect exposure would be much smaller than a direct-exposure dose, but a much larger population could be exposed.

Low-level wastes, generally defined as radioactive wastes other than high-level and wastes from uranium recovery operations, are commonly disposed of in near-surface facilities rather than in a geologic repository. There is no intent to recover the wastes once they are disposed of. Low-level waste includes items that have become contaminated with radioactive material or have become radioactive through exposure to neutron radiation. This waste typically consists of contaminated protective shoe covers and clothing, wiping rags, mops, filters, reactor water treatment residues, equipment and tools, luminous dials, medical tubes, swabs, injection needles, syringes, and laboratory animal carcasses and tissues. The radioactivity can range from just above background levels found in nature to much higher levels in certain cases such as parts from inside the reactor vessel in a nuclear power plant. Low-level waste is typically stored on-site by licensees, either until it has decayed away and can be disposed of as ordinary trash, or until sufficient amounts are available for transport to a low-level waste disposal facility.

Thus, radioactive pollution is caused by the presence of radioactive materials in water such contaminants are classified as (i) small doses, which temporary stimulate the metabolism and (ii) large doses, which gradually damage the organism causing genetic mutation. Source may be from radioactive sediment, waters used in nuclear atomic plants, radioactive minerals exploitation, nuclear power plants and use of radioisotopes in medical and research purposes.

2.10 Sewage and oxygen demanding wastes

Sewage (also referred to as domestic wastewater or municipal wastewater) is a type of wastewater that is produced by a human community and is characterized by produced volume, rate of flow, physical condition, chemical and toxic constituents, and the character and quantity of organisms that it contains consists mostly of gray water (or greywater). Sewage consists of organic matter, inorganic salts, heavy metals, bacteria, viruses, and nitrogen.

By definition, gray water is water from sinks, household baths, showers, dishwashers, and clothes washers. On the other hand black water (also written as blackwater) is the water used to flush toilets (combined with the human waste that it carries away from the home), soaps, and detergents. Sewage usually travels from the plumbing system of a building either into a sewer, which will carry it elsewhere, or into a sewage treatment facility.

Management of sewage-type waste is not successful due to huge volumes of organic and non-biodegradable wastes generated daily. As a consequence, garbage in some countries is unscientifically disposed and ultimately leads to increase in the pollutant load of surface and water courses. Sewage can be a fertilizer as it releases important nutrients to the environment such as nitrogen and phosphorus which plants and animals need for growth. Chemical fertilizers used by farmers also add nutrients to the soil, which drain into rivers and seas and add to the fertilizing effect of the sewage. Together, sewage and fertilizers can cause a massive increase in the growth of algae or plankton that can exist in oceans, lakes, or rivers thereby creating a condition known as algal bloom which reduces the dissolved oxygen content of water and killing other forms of life like fish.

One of the main causes of water contamination in many countries is the effluent (outflow) from septic tanks. Although each individual system may release a relatively small amount of waste into the ground, the large number and widespread use of these systems makes them a serious contamination source. Sewer pipes carrying wastes sometimes leak fluids into the surrounding soil and water. Other pipelines carrying industrial chemicals and oil brine have also been known to leak, especially when the materials transported through the pipes are corrosive.

Septic systems that are improperly sited, designed, constructed, or maintained can contaminate water with bacteria, viruses, nitrates, detergents, oils, and chemicals. Along with these contaminants are the commercially available septic system cleaners containing synthetic organic chemicals (such as 1,1,1- trichloroethane, CH_3CCl_3, or methylene chloride, CH_2Cl_2). These chemicals can contaminate water supply wells and interfere with natural decomposition processes in septic systems. Most, if not all, state and local or national regulations require specific separation distances between septic systems and drinking water wells to reduce (hopefully to mitigate completely) the risk of contamination of the water destined from human use and consumption.

Detergents from domestic and industrial uses wash down into water bodies causing serious effects on plants. Detergents contain high phosphates which results in phosphate-enrichment of water. Phosphorus is an essential element for life, both as a nutrient for plant life and as a key element in the metabolic processes of all living things. The normal low phosphate (PO_4) level in water inhibits the growth of plants but a small increase of phosphates can result in a rapid increase in plant growth such as blue-green algae and water hyacinth, especially in dams. As a result, the water plants become overcrowded and die. When they die the decomposing bacteria uses up more oxygen and affects other forms of life badly—for example, fish suffocate. This process (eutrophication) can be increased by human activities, such as by injection of domestic effluents (especially soapy water), farm and lawn fertilizers, industrial effluent and the destruction of wetlands.

Phosphate derivatives enter the plants through roots or surface absorption causing retarded growth of plants, elongation of roots, carbon dioxide fixation, photosynthesis, cation uptake, pollen germination and growth of pollen tubes, destruction of chlorophylls and cell membranes and denaturation of proteins causing enzyme inhibition in various metabolic processes.

2.11 Thermal pollution

Thermal pollution is the degradation of water quality by any process that changes the ambient water temperature. Heat is classified as a water pollutant when it is caused by heated industrial effluents or from anthropogenic (human) alterations of stream bank vegetation that increase the water system temperatures due to solar radiation. A common cause of thermal pollution is the use of water as a coolant by power plants and industrial manufacturers. When water used as a coolant is returned to the natural environment at a higher temperature, the sudden change in temperature decreases oxygen supply and affects the ecosystem. Fish and other organisms adapted to particular temperature range can be killed by an abrupt change in water temperature (either a rapid increase or decrease in the temperature of the water known as *thermal shock*).

Heated discharges may into a stream or lake may drastically alter the ecology of the water system. Although localized heating can have beneficial effects like freeing harbors from ice, the ecological effects are generally deleterious. Heated effluents lower the solubility of oxygen in the water because gas solubility in water is inversely proportional to temperature, thereby reducing the amount of dissolved oxygen available to aerobic (oxygen-dependent) species. Heat also increases the metabolic rate of aquatic organisms (unless the water temperature gets too high and kills the organism), which further reduces the amount of dissolved oxygen because respiration increases.

The release of heated water into water bodies from the thermal power plants has an adverse effect on the aquatic life. It reduces the activity of aerobic decomposers due to oxygen depletion because of high temperature. If there is a decrease in the decompositon of organic matter, the availability of nutrients in the water bodies is jeopardized. Aquatic plants show reduced photosynthesis rate due to inhibition of enzyme activity with increased temperature. Primary productivity and diversity of aquatic plant species decline because of increased temperature of water bodies as a result of thermal pollution. Thus, when a power plant first opens or shuts down for repair or other causes, fish and other organisms adapted to particular temperature range can be killed by the abrupt change in water temperature.

The elevated temperature generally decreases the level of dissolved oxygen in the water—typically gases are less soluble in hotter liquids. This can harm aquatic animals such as fish, amphibians and other aquatic organisms. Thermal pollution may also increase the metabolic rate of aquatic animals, as enzyme activity which results in these organisms consuming more food in a shorter time than if their environment were not changed. An increased metabolic rate may result in fewer resources; the more adapted organisms moving in may have an advantage over organisms that are not used to the warmer temperature. As a result, the food of the old environment and the new environments may be compromised. In addition, the higher temperature of the water limits oxygen dispersion into deeper waters, contributing to anaerobic conditions which can lead to increased levels of bacterial species when there is ample food supply. Also, many aquatic species will fail to reproduce at elevated temperatures.

Primary producers (such as plants and cyanobacteria) are affected by warm water because higher water temperature increases plant growth rates, resulting in a shorter lifespan and overpopulation of the species which can cause an algal bloom which, in turn, reduces the oxygen levels. In limited cases, warm water has little deleterious effect and may even lead to improved function of the receiving aquatic ecosystem. This phenomenon is seen especially in seasonal waters (called thermal enrichment).

The release of unnaturally cold water from a reservoir can dramatically change the fish and macroinvertebrate fauna of rivers, and reduce river productivity.

This can cause the elimination of indigenous fish species and the alteration of macro-invertebrate fauna populations. This may be mitigated by designing the dam to release warmer surface waters instead of the colder water at the bottom of the reservoir.

Thus, changes in water temperature adversely affect water quality and aquatic biota. Majority of the thermal pollution in water is caused due to human activities. Some of the important sources of thermal pollution are nuclear power and electric power plants, crude oil refineries, steel melting factories, coal fire power plant, boiler from industries which release large amount of heat to the water bodies leading to change in the physical, chemical and biological characteristics of the receiving water bodies. High temperature declines the oxygen content of water; disturbs the reproductive cycles, respiratory and digestive rates and other physiological changes causing difficulties for the aquatic life.

2.12 Urbanization

In the current context, urbanization refers to the population shift from rural to urban residency, the gradual increase in the proportion of people living in urban areas, and the ways in which each society adapts to this change. It is predominantly the process by which towns and cities are formed and become larger as more people begin living and working in central areas. Although the two concepts are sometimes used interchangeably, urbanization should be distinguished from urban growth: (i) urbanization is the proportion of the total national population living in areas classed as urban, (ii) urban growth refers to the absolute number of people living in areas classed as urban.

In general, urbanization leads to higher phosphorus concentrations in urban catchments. Increasing imperviousness (i.e. not allowing liquid to go through) increased runoff from urbanized surfaces, and increased municipal and industrial discharges all result in increased loadings of chemicals (some of which are nutrients) to urban water systems. This makes urbanization second only to agriculture as the major cause of impairment of a water system. Urban wastewater (also referred to as municipal wastewater) often contains high concentrations of organic carbon, phosphorus, and nitrogen, and may contain pesticides, toxic chemicals, salts, inorganic solids (e.g., silt), and pathogenic bacteria and viruses.

Waste should always be disposed of properly, typically by a qualified hazardous waste handler or through municipal hazardous waste collection days. Many chemicals should not be disposed of in household septic systems, including oils (e.g., cooking, motor), lawn and garden chemicals, paints and paint thinners, disinfectants, medicines, photographic chemicals, and swimming pool chemicals. Similarly, many substances used in industrial processes should not be disposed of in drains at the workplace because they could contaminate a drinking water source. Companies should train employees in the proper use and disposal of all chemicals used on site. The many different types and the large quantities of chemicals used at industrial locations make proper disposal of wastes especially important for water protection.

Solid waste is disposed of in thousands of municipal and industrial landfills throughout the country. Chemicals that should be disposed of in hazardous waste landfills sometimes end up in municipal landfills. In addition, the disposal of many household wastes is not regulated. Once in the landfill, chemicals can leach into the water by means of precipitation and surface runoff. New landfills are required to have clay or synthetic liners and leachate (liquid from a landfill containing contaminants) collection systems to protect water. Most of the older landfills, however, do not have these safeguards. Older landfills were often sited over aquifers or close to surface waters and in permeable soils with shallow water tables, enhancing the potential for leachate to contaminate water. Closed landfills can continue to pose a water contamination threat if they are not capped with an impermeable material (such as clay) before closure to prevent the leaching of contaminants by precipitation.

A surface impoundment is a natural topographic depression, artificial excavation, or diked area formed primarily of earthen materials but which may be lined with artificial materials. Thus, surface impoundments are relatively shallow ponds or lagoons used by industries and municipalities to store, treat, and dispose of liquid wastes. Like landfills, most surface impoundment facilities are required to have liners, but even these liners sometimes leak.

Typically, a surface impoundment is constructed in on-channel and off-channel areas. An off-channel surface impoundment is an excavated disposal pits that is constructed with sloping sides and bordered by elevated berms. Usually the impoundment is constructed nearby production areas to minimize costs of piping the water from the coalbed gas wells. An on-channel impoundments is situated within a designated water features (e.g. perennial and ephemeral streams, creeks, lakes, ponds, and dry washes) or within the alluvial floodplain of the water features. An off-channel impoundment is not situated within the water feature (e.g. perennial and ephemeral streams, dry washes, lakes, and ponds) and is constructed >500 ft from the outermost alluvial floodplain deposits of any stream, or river. An impoundment pit mainly serves as an evaporating and/or infiltrating unit depending on whether it is lined or unlined as well as watering for livestock. In addition, an impoundment pit is commonly used to minimize or eliminate co-produced water discharge or disposal into surface waters. The impoundment pit can also be used to manage co-produced water flow from the coalbed gas wells to an outfall.

Water impoundments are used to store water for preparing fluids to be used for hydraulic fracturing (Chapter 7) and can also can be used to store flow-back water before and after treatment. In arid regions, the industry uses open pits and tanks to evaporate liquid from the solid pollutants. Full evaporation ultimately leaves precipitated solids that must be disposed in a landfill. However, there is some concern that, if not properly managed, an evaporative pit may allow air emissions of volatile organic compounds and other pollutants. Birds and wildlife, and sometimes domesticated animals like cattle, can mistake these pits for freshwater sources.

The industry is increasingly replacing open pits with closed-loop fluid systems that keep fluids within a series of pipes and watertight tanks inside secondary containment. Additional measures include establishing setback requirements for open pits, determining the composition of wastewater stored in evaporative ponds for appropriate disposal or treatment, since contaminants can become more concentrated as water evaporates, and placing a fence around open pits to keep out animals.

Drainage wells are used in wet areas to help drain water and transport it to deeper soils. These wells may contain agricultural chemicals and bacteria. Injection Wells/

Floor Drains Injection wells are used to collect storm water runoff, collect spilled liquids, dispose of wastewater, and dispose of industrial, commercial, and utility wastes. Disposal wells that pose threats to drinking water supplies are prohibited and must be closed, connected to a public sewage system, or connected to a storage tank. Moreover, problems associated with improperly constructed wells can result in water contamination when contaminated surface or water is introduced into the well.

Also, improperly abandoned wells can act as a conduit through which contaminants can reach an aquifer if the well casing has been removed, as is often done, or if the casing is corroded. In addition, some people use abandoned wells to dispose of wastes such as used motor oil. These wells may reach into an aquifer that serves drinking supply wells. Abandoned exploratory wells (e.g., for gas, oil, or coal) or test hole wells are usually uncovered and are also a potential conduit for contaminants.

Active drinking water supply wells that are poorly (constructed) can result in water contamination. Issues such as faulty casings, inadequate covers, or lack of concrete pads, allow outside water and any accompanying contaminants to flow into the well. Sources of such contaminants can be surface runoff or wastes from farm animals or septic systems. Contaminated fill packed around a well can also degrade well water quality. Well construction problems are more likely to occur in older wells that were in place prior to the establishment of well construction standards and in domestic and livestock wells.

Also, poorly constructed irrigation wells can allow contaminants to enter water. Often pesticides and fertilizers are applied in the immediate vicinity of wells on agricultural land.

3 Effects on specific water systems

As described elsewhere (Chapter 1), the existence of a water system is due to a collection of factors. For example, the pore structure of the soil and the sediment are central influences on groundwater movement. Hydrologists quantify this influence primarily in terms of (i) porosity and (ii) permeability.

As a recap, water pollution is usually caused by human activities. Different human sources add to the pollution of water. Also, water pollution is any chemical, physical or biological change in the quality of water that has a harmful effect on any living thing that drinks or uses or lives (in) it. When humans drink polluted water it often has serious effects on their health. Water pollution can also make water unsuited for the desired use. There are several classes of water pollutants.

The first are disease-causing agents. These are bacteria, viruses, protozoa and parasitic worms that enter sewage systems and untreated waste. A second category of water pollutants is oxygen-demanding wastes; wastes that can be decomposed by oxygen-requiring bacteria. When large populations of decomposing bacteria are converting these wastes it can deplete oxygen levels in the water. This causes other organisms in the water, such as fish, to die. A third class of water pollutants is water-soluble inorganic pollutants, such as acids, salts and toxic metals. Large quantities of these compounds will make water unfit to drink and will cause the death of aquatic life. Another class of water pollutants are nutrients; they are water-soluble nitrates and phosphates that cause excessive growth of algae and other water plants, which deplete the water's oxygen supply. This kills fish and, when found in drinking water, can kill young children. Water can also be polluted by a number of organic compounds such as oil, plastics and pesticides, which are harmful to humans and all plants and animals in the water. A very dangerous category is suspended sediment, because it causes depletion in the water's light absorption and the particles spread dangerous compounds such as pesticides through the water.

Finally, water-soluble radioactive compounds can cause cancer, birth defects and genetic damage and are thus very dangerous water pollutants.

Water is a complex system of chemical species that has received considerable study because of a series of environmental issues that have been raised (Eglinton, 1975; Melchior and Bassett, 1990; Easterbrook, 1995). In fact, throughout history, the quality and quantity of water available to humans have been vital factors in determining not only the quality of life but also the existence of life (Boyd, 2015; Laws, 2017). However, in many parts of the world, water pollution is a major issue.

3.1 Rivers, lakes, and streams

The effect of pollution on lakes differs in several respects from the effect on water systems. Water movement in lakes is slower than in rivers and streams, so reaeration is more of an issue in lakes. Because of the slow movement of water in a lake, sediments, and pollutants bound to sediments, tend to settle out of the water column rather than being transported downstream.

Light has an important influences on a lake—light is the source of energy in the photosynthetic reaction and, this, the penetration of light into the lake water determines the amount of photosynthesis that can occur at various depths in the lake decreasing with depth in the lake. Because of this, algal growth is concentrated near the surface of **a** lake, in the photic zone, which is limited to the maximum depth where there is still enough light to support photosynthesis.

Temperature and heat often have a profound effect on a lake. Water is at maximum density at 4 °C (39 °F)—warmer or colder water (including ice) is less dense, and will float. Water is also a poor conductor of heat and retains heat quite well. During the winter, if the lake does not freeze, the temperature is relatively constant with depth. As the weather warms in the spring the top layers of water begin to warm. Since warner water is less dense, and water is a poor conductor of heat, the lake eventually stratifies into a warm, less dense, surface layer (the *epilimnion*) and a cooler, denser, bottom layer (the *hypolimnion*). A third thermal gradient (the *metalimnion*) exists between the epilimnion and the hypolimnion (Fig. 5.2). The metalimnion is sometimes referred to as the *thermocline* because it represents the temperature gradient between the epilimnion and the hypolimnion.

Circulation of water occurs only within a stratum, and thus there is only limited transfer of biological or chemical material (including dissolved oxygen) between the epilimnion and the hypolimnion. As colder weather approaches, the top layer cools, becomes denser, and sinks. This creates circulation within the lake, known as the autumn turnover (fall turnover, in the United States). If the lake freezes over in the winter, the lake surface temperature will be <4 °C (39 °F), and the ice will float on

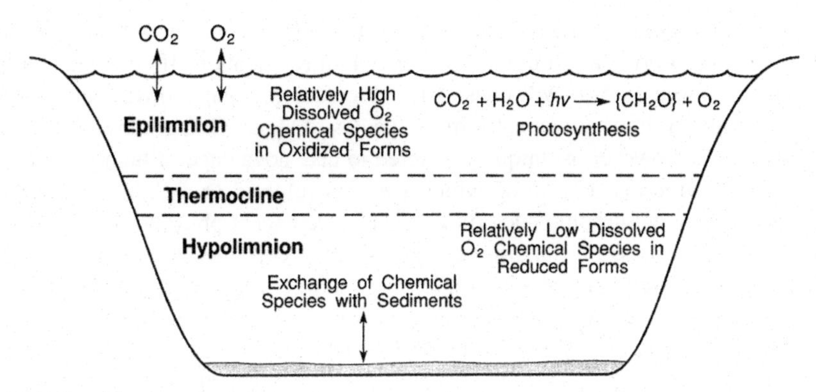

Fig. 5.2 Stratification of a Lake.

top of the slightly denser underlying water. When spring comes, the lake surface will warm slightly and there will be a spring turnover as the ice thaws.

In terms of the biochemistry of the lake, a river feeding the lake would contribute carbon, phosphorus, and nitrogen, either as high-energy organics or as low-energy compounds. The phytoplankton (free-floating algae) take carbon, phosphorus, and nitrogen, and, using sunlight as an energy source, make high-energy compounds. Algae are eaten by zooplankton (tiny aquatic animals), which are in turn eaten by larger aquatic life such as fish. All of these life forms defecate or excrete waste products, contributing a pool of dissolved organic carbon. The growth of algae in most lakes is limited by the availability of phosphorus. If phosphorus is in sufficient supply, nitrogen is usually the limiting nutrient insofar as the nitrogen is the nutrient that controls the rate of algal growth because the nutrient is not readily available.

When phosphorus and nitrogen are introduced into the lake, either naturally from storm runoff, or from a pollution source, the nutrients promote rapid growth of algae in the epilimnion. When the algae die, they drop to the lake bottom (the hypolimnion) and become a source of carbon for decomposing bacteria. Aerobic bacteria will use all available dissolved oxygen in the process of decomposing this material, and the dissolved oxygen may be depleted enough to cause the hypolimnion to become anaerobic. As more and more algae die, and more and more dissolved oxygen is used in their decomposition, the metalimnion may also become anaerobic. When this occurs, biological activity is restricted to the epilimnion (eutrophication).

Eutrophication is the continually occurring natural process of lake aging and occurs in three stages: (i) the oligotrophic stage, which is characterized by low levels of biological productivity and high levels of oxygen in the hypolimnion, (ii) the mesotrophic stage, which is characterized by moderate levels of biological productivity and the beginnings of declining oxygen levels following lake stratification, and (iii) the eutrophic stage, at which point the lake is very productive, with extensive algal blooms, and increasingly anaerobic conditions in the hypolimnion. Natural eutrophication may take thousands of years. If enough nutrients are introduced into a lake system, as may happen as a result of human activity, the eutrophication process may be shortened to as little as ten years.

Because phosphorus is usually the nutrient that limits algal growth in lakes, the addition of phosphorus, in particular, can speed eutrophication. If only phosphorus is introduced into a lake, it will cause some increase in algal growth, but nitrogen quickly becomes a limiting factor for most species of algae. One group of photosynthetic organisms, however, is uniquely adapted to take advantage of high phosphorus concentrations—the cyanobacteria, which are also referred to as blue green algae. Cyanobacteria are autotrophic bacteria that can store excess phosphorus inside their cells and use the excess phosphorus to support future cell growth and also have the ability to use dissolved nitrogen gas as a nitrogen source, which is rapidly replenished by atmospheric nitrogen. Most other aquatic autotrophs cannot use nitrogen as a nitrogen source and, as a result, cyanobacteria proliferate in ecosystems where nitrogen has become limiting to other algae, and can sustain their growth using cellular phosphorus for long periods of time.

3.2 Groundwater

If has been, and probably still is, believed that the water that moves through soil and mineral strata is purified naturally from the soil and/or the mineral strata in a relatively pure condition. Unfortunately, is not the case and groundwater pollution is becoming an increasing concern throughout on a worldwide basis. Although many soils and mineral strata do have the ability to remove certain types of pollutants (such as phosphorus, heavy metals, bacteria, and suspended solids) from water, the pollutants that dissolve in water (such as nitrate derivatives) may pass through the soil bed or mineral bed and remain in the groundwater. In agricultural regions, the nitrogen and other soluble chemicals in fertilizers or animal wastes can seep into the groundwater and show up in high (harmful) concentrations in water taken from wells in the region.

Other potential sources of groundwater pollution include leaking underground storage tanks, solid waste landfills, improperly stored hazardous waste, careless disposal of solvents and hazardous chemicals on ground surfaces, and road salts and deicing compounds.

3.3 Oceans

For millennia, even up to the early part of the 20th Century, the oceans were considered infinite sinks for the disposal of waste materials. Even some coastal areas sewage was discharged into the ocean through a large diameter pipeline from a municipality. But, the oceans are fragile water systems and an ocean is also a complex chemical solution that has remained relatively unchanged for millennia that has allowed the oceanic life to evolve. However the susceptibility of the oceans to pollutants is real and requires continued care. For example, the continental shelf, especially near major estuaries, receives the greatest load of pollutants. In fact, many estuaries have become so badly polluted that they are closed to commercial fishing.

Ocean disposal of untreated wastewater is severely restricted in many countries but many ocean-side major cities—where pollutants monitoring is, if practiced at all—still discharge untreated sewage into the ocean. Even though the sewage may be carried a considerable distance from shore by pipeline and discharged through diffusers to achieve maximum dilution, the practice remains highly controversial, and the long-term consequences are debatable and must be presumed to fall on the side of having an adverse effect on the oceans.

4 Effects of pollutants

Environmental pollution has existed for centuries but only started to be significant after the onset of the Industrial Revolution in the 19th Century. In the current context, pollution occurs when a natural water system cannot destroy a non-indigenous chemical or substance without creating harm or damage to itself. The chemicals involved are not produced by nature, and the destruction process can vary from days to years, even thousands of years such as is the case for radioactive pollutants). In other words, pollution takes place when nature does not know how to decompose an element that has been brought to it in an unnatural way.

Water pollution occurs when unwanted materials enter in to water, changes the quality of water and harmful to environment and human health (Pawari and Gawande, 2015; Alrumman et al., 2016. Water is an important natural resource used for drinking and other developmental purposes in our lives. Safe drinking water is necessary for human health all over the world. Being a universal solvent, water is a major source of infection. According to world health organization (WHO) 80% diseases are water borne. Drinking water in various countries does not meet WHO standards [4]. 3.1% deaths occur due to the unhygienic and poor quality of water (Pawari and Gawande, 2015).

Discharge of domestic and industrial effluent wastes, leakage from water tanks, marine dumping, radioactive waste and atmospheric deposition are major causes of water pollution. Heavy metals industrial waste can accumulate in lakes and river, proving harmful to humans and animals. Toxins in industrial waste are the major cause of immune suppression, reproductive failure and acute poisoning. Infectious diseases, such as cholera, typhoid fever, and other diseases gastroenteritis, diarrhea, vomiting, skin and kidney problem are spreading through polluted water. Human health is affected by the direct damage of plants and animal nutrition. Water pollutants are killing sea weeds, mollusks, marine birds, fishes, crustaceans and other sea organisms that serve as food for human. Insecticides like DDT concentration is increasing along the food chain (Khan and Ghouri, 2011).

Thus, pollution must be taken seriously, as it has a negative effect on natural water systems (and other ecosystems) that are an absolute need for life to exist on the Earth, such as water and air. Indeed, without it, or if they were present on different quantities, animals—including humans—and plants could not survive.

Maintaining environmental—in terms of maintaining healthy ecosystems that are free of pollutants—requires an understanding the influence of pollutants on the environment as well as understanding the effects of human-made hazards and insulation of human health and environmental systems from these hazards. This involves examining and evaluating the effects of chemicals made by humans to human health or wildlife and the manner by which the ecological systems impacts spread of illnesses. It can include everything from managing the use of pesticides to the quality of drywall used in construction.

There is a wide variety of environmental hazards that occur every day and can be produced from contaminated water. To better understand environmental health, the environmental hazards can be classified into four categories: (i) biological hazards, (ii) physical hazards, (iii) chemical hazards, and (iv) cultural hazards. In fact, one of the most common types of water system pollutants is the introduction of biodegradable organic material into the water system.

When a high-energy organic material such as raw sewage is discharged into a water system, a number of changes occur downstream from the point of discharge. As the organic components of the sewage are oxidized, oxygen is used at a rate greater than that upstream from the sewage discharge, and the dissolved oxygen in the water system decreases markedly. The rate of re-aeration, or solution of oxygen from the air, also increases, but is often not enough to prevent total depletion of oxygen in the water system. If the dissolved oxygen is totally depleted, the water system becomes anaerobic. Often, however, the dissolved oxygen does not drop to complete depletion and the water system can recover without a period of anaerobiosis (life in the absence of air or free oxygen).

However, when the rate of oxygen use overwhelms the rate of oxygen reaeration, the water system may become anaerobic which is identifiable by the presence of floating sludge, bubbling gas, and a foul smell. The gas is formed because oxygen is no longer available to act as the hydrogen acceptor, and ammonia, hydrogen sulfide, and other gases are formed. Some of the gases dissolve in water, but others can attach themselves as bubbles to sludge (solid black or dark benthic deposits) and buoy the sludge to the surface. In addition, the odor of hydrogen sulfide will indicate the anaerobic condition. Increased turbidity, settled solid matter, and low dissolved oxygen all contribute to a decrease in fish life.

Biological hazards emanate from environmental relations between organisms. Some examples of biological hazards include bacteria, viruses, fungi, spores, pathogenic micro-organisms, tuberculosis, and malaria. When these diseases and pathogens are transmitted between two or more organisms, it is an infectious disease. The real reason humans suffer from these pathogens and diseases is that they are being infested by other organisms, which is a natural process, but at the same time hazardous.

The water system will react much differently to inorganic contaminants, such as a waste effluent from a metal-plating plant. If the waste is toxic to aquatic life, the type and total number of organisms will decrease downstream from the outfall. The dissolved oxygen will not typically decrease and might even increase. There are many types of pollution, and a water system will react differently to each. When two or more wastes are involved, the situation is even more complicated.

Physical hazards rely upon physical processes which happen naturally in the environment, for example, natural disasters like volcanoes, earthquakes, droughts, landslides, blizzards, and tornadoes. Physical hazards are considered secret events, but not all are (for example, ultraviolet, UV) radiation are openly happening each day. Ultraviolet radiation is categorized as hazardous since too much exposure to it destroys the DNA and triggers health complications in humans such as cataracts and skin cancer.

Chemical hazards occur in ecological systems, such as in water systems, in two ways: (i) human-made, i.e. anthropogenic, or (ii) natural. Examples of naturally occurring chemical hazards include mercury and lead, which are considered heavy metals. Human-made chemical hazards encompass lots of synthetic chemicals human produce such as pesticides, plastics, and disinfectants. Some organisms even generate natural chemicals, which are hazardous to the environment, for instance, elements contained in peanuts and dairy that trigger allergic reactions to humans.

In the current context, guaranteeing that portable or enough drinking water will be readily obtainable to the local community is yet another main aspect of environmental health. Environmental health campaigners will look to find ways of developing water reserves that can be wholly cushioned from the possibility of contamination. These environmental advocates not only look towards drinking and cooking water, but making sure there is availability of enough water for crop irrigation. Another aspect of water control is to make sure that proper drainage systems are in place. Ensuring appropriate drainage is in place is due to health concerns that may be instigated by stagnant water and how it can contribute to existence of diseases carried along by mosquitoes that commonly breed in stagnant water.

Water pollution is a global issue and world community is facing worst results of polluted water (Briggs, 2003). Major sources of water pollution are discharge of domestic and agriculture wastes, population growth, excessive use of pesticides and fertilizers and urbanization. Bacterial, viral and parasitic diseases are spreading through polluted water and affecting human health. It is recommended that there should be proper waste disposal system and waste should be treated before entering in to river.

4.1 Human health

Water pollution can affect the quality of life for people and other living things around the globe. By affecting the water in the ocean, in land-based water sources like rivers, and in the groundwater, water contamination becomes a worldwide threat to human health. The effects of water pollution are wide-ranging (Table 5.2), but they can damage life in the ocean and freshwater, which continues up the global food chain as animals that feed on fish and other aquatic life take in pollution. Examples include mercury pollution, which comes from human industrial activity and puts toxic mercury in the ocean, where fish eat it. When humans then eat the fish, they are at risk for mercury poisoning.

Chemicals in water that affect human health—some of the chemicals affecting human health are the presence of heavy metals such as fluoride, arsenic, lead, cadmium, mercury, petrochemicals, chlorinated solvents, pesticides and nitrates.

In fact, in terms of heavy metal pollutants, the most anthropogenic sources of heavy metals found in wastewater are industrial, such as contamination from crude oil operations and from sewage disposal. Although, some heavy metals, such as zinc, copper and iron are described to be essential in aquatic environment because of their roles in several biochemical processes, when present in high concentrations, they become detrimental (Saeed and Shaker, 2008). The incorporation of heavy metals into food chains could lead to their in aquatic organisms to a level that affects their physiological state. Because most heavy metals are known to be toxic and carcinogenic, they represent serious threat to human health and the fauna and flora of receiving water bodies.

Table 5.2 Chemicals in water that have a negative effect on health

Chemical	Effects
Pesticides	Can damage the nervous system Carbonates and organophosphates can cause cancer Chlorides can cause reproductive Can cause endocrinal damage
Nitrates	Especially dangerous to babies that drink formula milk Restricts the amount of oxygen in the brain
Lead	Can accumulate in the body and damage the central nervous system
Arsenic	Causes liver damage, skin cancer Causes vascular diseases
Fluorides	In excessive amounts can make teeth yellow Can cause damage to the spinal cord
Petrochemicals	Even with very low exposure, can cause cancer

Arsenic is a natural component of the earth's crust and is widely distributed throughout the environment in the air, water and land. However, arsenic is a very toxic chemical that reaches the water naturally or from wastewater of tanneries, ceramic industry, chemical factories, and from insecticides such as lead arsenate, effluents from fertilizers factories and from fumes coming out from burning of coal and crude oil. Arsenic is highly dangerous for human health causing respiratory cancer, arsenic skin lesion from contaminated drinking water. Long exposure leads to bladder and lungs cancer. Lead is contaminated in the drinking water source from pipes, fitting, solder, household plumbing systems. In the human beings, it affects the blood, central nervous system and the kidneys. Child and pregnant women are mostly prone to lead exposure. Mercury is used in industries such as smelters, manufactures of batteries, thermometers, pesticides, and fungicides.

A number of microorganisms and thousands of synthetic chemicals have the potential to contaminate water. Drinking water containing bacteria and viruses can result in illnesses such as hepatitis, cholera, or giardiasis. Methemoglobinemia (blue baby syndrome) an illness affecting infants, can be caused by drinking water that is high in nitrates. Benzene, a component of gasoline, is a known human carcinogen. The properties are:

Property	Value
Physical state	Colorless liquid
Melting point	5.5 °C (43 °F)
Boiling point	80.1 °C (176 °F)
Density	0.88 g/cm^3 at 20 °C (68 °F)
Vapor pressure	13.3 kPa at 26.1 °C (79 °F)

The serious health effects of lead are well known—learning disabilities in children; nerve, kidney, and liver problems; and pregnancy risks. Concentrations in drinking water of these and other substances are regulated by federal and state laws. Hundreds of other chemicals, however, are not yet regulated, and many of their health effects are unknown or not well understood. Preventing contaminants from reaching the water is the best way to reduce the health risks associated with poor drinking water quality.

4.2 Water-borne disease

Most diseases in the world are related to water and sanitation and in order to break the cycle of disease, there must be improvements in the quality of water that people use. In all cases, sewage disposal systems should be located away from water systems destined for human use and consumption. Groundwater, which is another water source, can become contaminated through unclean irrigation water.

Waterborne diseases are caused by pathogenic microorganisms that most commonly are transmitted in contaminated fresh water. Infection commonly results during bathing, washing, drinking, in the preparation of food, or the consumption of food thus infected. Thus, waterborne pathogens, in the form of disease-causing bacteria and viruses from human and animal waste, are a major cause of illness from contaminated drinking water. Diseases spread by unsafe water include cholera, giardia, and typhoid. Even in wealthy nations, accidental or illegal releases from sewage treatment facilities, as well as runoff from farms and urban areas, contribute harmful pathogens to waterways.

Microorganisms play a major role in water quality and the microorganisms that are concerned with water borne diseases are Salmonella sp., Shigella sp., *Escherichia coli* and Vibrio cholera. All these cause typhoid fever, diarrhea, dysentery, gastroenteritis and cholera. The most dangerous form of water pollution occurs when feces enter the water supply. Many diseases are perpetuated by the fecal-oral route of transmission in which the pathogens are shed only in human feces. The presence of fecal coliforms of *E. coli* is used as an indicator for the presence of any of these water borne pathogens. In fact, water contamination is the leading worldwide cause of deaths and diseases, and that it accounts for the deaths of >14,000 people daily, and the majority of them being children under 5 years old. In recent years, the widespread reports of pollutants in water have increased public concern about the quality of water. Children are generally more vulnerable to intestinal pathogens and it has been reported that about 1.1 million children die every year due to diarrheal diseases.

Climatic systems display complex interactions of interconnected components, including the atmosphere, hydrosphere, cryosphere, biosphere and geosphere. Global climate change will interfere with these interactions and alter the hydrologic cycle not only by altering mean meteorological measures but by increasing the frequency of extreme events such as excessive precipitation, storm surges, floods and droughts. These extreme weather-related events can affect water availability, quality, or access, posing a threat to human populations. Water-borne pathogens often act in concert through two major exposure pathways: drinking water and recreational water use.

Theoretical, simulation, and empirical data corroborate that increased water vapor, due to higher mean temperatures, triggers more intense precipitation events, even if the precipitation quantity remains constant. Such events can result in run-off and loading of coastal waters with pathogens, nutrients and toxic chemicals that may adversely affect aquatic life and public health. This is particularly true in coastal watersheds where human development and population increases have led to urbanization of coastal areas. Storm surges greatly increase the risk and the amount of pollutants entering recreational coastal waters. Climate change events can negatively impact public health: exposure to southern.

Risk of gastroenteritis and respiratory infections due to recreational water use are much higher during the rainy season rather than the dry season. Precipitation is projected to increase in northern Europe, but no similar studies have been published for Europe so far. Conversely, extended periods of hot weather can increase the mean temperature of water bodies which can be favorable for microorganism reproduction cycles and algal blooms. Water-borne outbreaks have the potential to be rather large and of mixed etiology but the actual disease burden in Europe is difficult to approximate and most likely underestimated.

Erratic and extreme precipitation events can overwhelm water treatment plants and lead to *cryptosporidium* outbreaks due to oocysts infiltrating drinking-water reservoirs from springs and lakes and persisting in the water distribution system. Droughts or extended dry spells can reduce the volume of river flow possibly increasing the concentration of effluent pathogens posing a problem for the clearance capacity of treatment plants.

Aging water treatment and distribution systems are particularly susceptible to weather extremes posing a significant vulnerability of the drinking water supply. Environmental pollutants can synergistically interact with climatic conditions and exacerbate exposure of human populations. Infrastructure improvements and environmental protection can attenuate potential negative consequences of climate change from water-borne diseases.

4.3 Phytotoxicity

The destruction of plant life by contaminated water is often referred to as phytotoxicity which can occur when (i) a chemical is properly applied directly to the plant during adverse environmental conditions, (ii) a chemical material is applied improperly, (iii) a spray, dust, or vapor drifts from the target crop to a sensitive crop, (iv) a runoff carries a chemical to a sensitive crop, and (v) persistent residues accumulate in the soil or on the plant.

High concentrations of mineral salts in solution within the growing medium can have phytotoxic effects. Sources of excessive mineral salts include infiltration of seawater and excessive application of fertilizers. For example urea (H_2NCONH_2) is used in agriculture as a nitrogenous fertilizer but if too much is applied, phytotoxic effects can result, either by urea toxicity or by the ammonia produced through hydrolysis of urea by soil urease. Acid soils may contain high concentrations of aluminum (as Al^{3+}) and manganese (as Mn^{2+}) which can be phytotoxic.

Herbicides are designed to kill plants, and are used to control unwanted plants such as agricultural weeds. However herbicides can also cause phytotoxic effects in plants that are not within the area over which the herbicide is applied, for example as a result of wind-blown spray drift or from the use of herbicide-contaminated material (such as straw or manure) being applied to the soil.

When chemical pollutants build up in aquatic or terrestrial environments, plants can absorb these chemicals through their roots. Phytotoxicity occurs when toxic chemicals poison plants. The symptoms of phytotoxicity on plants include poor growth, dying seedlings and dead spots on leaves. For example, mercury poisoning which many people associate with fish can also affect aquatic plants, as mercury compounds build up in plant roots and bodies result in bioaccumulation. As animals feed on polluted food the increasing levels of mercury is built up through food chain.

Many of the gases from acid, aerosols and other acidic substances released into the atmosphere from industrial or domestic sources of combustion from fossil fuels finally fall down to ground and reach the water bodies along with run-off rainwater from polluted soil surfaces thereby causing acidification of water bodies by lowering its pH. In many countries chemical substances like sulfates derivatives, nitrate derivatives, and chloride derivatives have been reported to make water bodies such as lakes, river and ponds acidic. ii. Nutrient deficiency in aquatic ecosystem: Population of decomposing microorganisms like bacteria and fungi decline in acidified water which in turn reduces the rate of decomposition of organic matter affecting the nutrient cycling.

The critical pH for most of the aquatic species is 6.0. The diversity of species decline below this pH whereas the number and abundance of acid tolerant species increases. Proliferation of filamentous algae rapidly forms a thick mat at the initial phase of the acidification of water. Diatoms and green algae disappear below pH 5.8. Cladophora is highly acid tolerant species and is abundant in acidic freshwater bodies. Macrophytes are generally absent in acidic water as their roots are generally affected in such water resulting in poor plant growth. Potamogeton pectinalis is found in acidified water. It is observed that plants with deep roots and rhizomes are less affected while plants with short root systems are severely affected in acidic water.

Organic matter from dead and decaying materials of plants and animals is deposited directly from sewage discharges and washed along with rainwater into water bodies causing increase in decomposers/microbes such as aerobic and anaerobic bacteria. Rapid decomposition of organic matter increase nutrient availability in water favoring the luxuriant growth of planktonic green and blue-green algal bloom. In addition many of the macrophytes such as Salvinia, Azolla and Eicchhornia grow rapidly causing reduced penetration of light into deeper layer of water body with gradual decline of the submerged flora. This condition results in reducing the dissolved Oxygen and increase in the biological oxygen demand (BOD). The biological oxygen demand of unpolluted fresh water is usually below 1 mg per/l while that of organic matter polluted water is >400 mg per liter.

Thus, phytotoxicity shows itself in many ways. Damage may appear as yellowing, death or destruction of tissues, stunting, growth retardation, abnormal growth or defoliation. The use of new chemicals-or using a familiar product in a new way-brings uncertainty to a chemical application. Phytotoxicity can result from a variety of scenarios including: (i) adverse reaction due to sensitivity of a particular species or variety of plant to a particular chemical, (ii) use of excessive rates or too many applications at close intervals, (iii) poor application of the chemical, and (iv) environmental conditions at the time of application. Understanding the various reasons for plants' phytotoxic response to a chemical application can lead to safer, more effective use of chemicals in plant-management programs.

Perhaps the most frequently observed phytotoxic damage is due to the fact that not all plants get along with all chemicals (plant sensitivity). For example, Malathion, a commonly used insecticide, is phytotoxic to Hibiscus, Lantana, petunias, white pine

(*Pinus strobus*), maples and many species of fern. Sulfur is toxic to Viburnum spp. Horticultural oils can harm tender new growth and foliage of particularly sensitive species such as mountain ash, beech and birch. Atrazine, a commonly used herbicide, selectively controls weeds in St. Augustine grass but may damage ryegrass.

References

Alrumman, S.A., El-kott, A.F., Kehsk, M.A., 2016. Water pollution: Source and treatment. Am. J. Environ. Eng. 6 (3), 88–98.

Bibi, S., Khan, R.L., Nazir, R., 2016. Heavy metals in drinking water of Lakki Marwat District, KPK, Pakistan. World Appl. Sci. J. 34 (1), 15–19.

Boyd, C.E., 2015. Water Quality, second ed. Springer, New York.

Briggs, D., 2003. Environmental pollution and the global burden of disease. Br. Med. Bull. 68, 1–24.

Easterbrook, G., 1995. A Moment on the Earth: The Coming Age of Environmental Optimism. Viking Press, New York.

Eglinton, G. (Ed.), 1975. Environmental Chemistry. Vol. 1. Specialist Periodical Reports. The Chemical Society, London, England.

Khan, M.A., Ghouri, A.M., 2011. Environmental pollution: Its effects on life and its remedies. J. Arts Sci. Commerce 2 (2), 276–285.

Laws, E.A., 2017. Aquatic Pollution: An Introductory Text, fourth ed. John Wiley & Sons Inc., Hoboken, NJ.

Melchior, D.C., Bassett, R.L. (Eds.), 1990. Chemical Modeling of Aqueous Systems II. American Chemical Society, Washington, DC. Symposium Series No. 416.

Pawari, M.J., Gawande, S., 2015. Ground water pollution and its consequence. Int. J. Eng. Res. Gen. Sci. 3 (4), 773–776.

Radovanović, L., Speight, J.G., 2018. Global warming—Truth and myths. Petrol. Chem. Ind. Int. 1 (1). https://www.opastonline.com/wp-content/uploads/2018/10/Globalwarming-truthandmyths-pcii-18.pdf.

Saeed, S.M., Shaker, I.M., 2008. Assessment of heavy metals pollution in water and sediments and their effect on Oreochromisniloticus in the Northern Delta lakes, Egypt. In: Proceedings of the International Symposium on Tilapia in Aquaculture 8, Cairo. Page 475–489. https://pdfs.semanticscholar.org/ba97/6d52dfdf6922527a4e8e5599aed805cdff79.pdf.

Speight, J.G., Islam, M.R., 2016. Peak energy—Myth or reality. Scrivener Publishing, Beverly, MA.

Crude oil in water systems

Chapter outline

1 Introduction

Water pollution is the contamination of a water system and is the result of the introduction of contaminants into the water system. In the current context, the pollutants are crude oil and/or crude oil products. Furthermore, water pollution by crude oil and/or crude oil products can be grouped into pollution of (i) surface water systems, (ii) groundwater, and (iii) marine pollution, which is a sub-set of surface pollution.

Pollution by crude oil and crude oil products is one of the most predominant forms of ocean pollution causing severe damages to amenities, ecosystems, and resources. The hydrocarbons in crude oil and in crude oil products encompass a large number of chemicals, and these also find their way into the coastal zone from wastewater discharges. Waste oil (from spent oil from cars) is frequently discharged into municipal water works in particular in underdeveloped countries. This, together with oil washed out from roads and parking lots by rain, often finds its way to a water system. A third source is the occurrence of natural seepage in coastal environments.

The composition of the wasted oil and crude oil are variable. One of the main components is the family of saturated or aliphatic hydrocarbons. Their molecular size range from the one-carbon atom compound (methane) and other gaseous hydrocarbons to larger molecules that include liquid phase n-hydrocarbon components, waxes, and solid hydrocarbons. Other components include naphthene derivatives and aromatic derivatives, (Speight, 2014).

Crude oil is complex mixtures of natural products, with a wide range of molecular weights and structures. The low-boiling saturated hydrocarbons (naphtha range) at

Natural Water Remediation. https://doi.org/10.1016/B978-0-12-803810-9.00006-1

low concentrations produce anesthesia and narcosis and at greater concentration, cell damage, and death in a wide variety of soil invertebrates and lower vertebrates. The higher-boiling hydrocarbon derivatives (kerosene and lube oil range) also have varying degrees of toxicity and can interfere with nutrition and the reception of chemical clues that are necessary for communication between many animals. Aromatic hydrocarbon derivatives are abundant in crude oil and represent the most dangerous fraction and the low-boiling aromatic derivatives [benzene, C_6H_6, toluene, $C_6H_5CH_3$, and the xylene isomers, 1,2-, 1,3-,1,4-$C_6H_4(CH_3)_2$] can cause acute poisoning to many lung-breathing invertebrates and vertebrates, including humans.

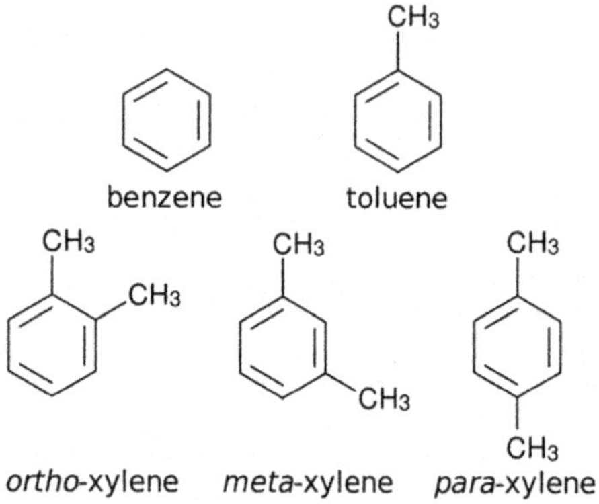

By way of clarification for the non-organic chemistry, 1,2-xylene is also known as ortho-xylene (or o-xylene), 1,3-xylene is also known as meta-xylene (or m-xylene), and 1,4-xylene is also known as para-xylene (or p-xylene) (Table 6.1).

Another important aspect of crude oil chemistry is the presence, in variable quantities, of aromatic hydrocarbons and in particular the polyaromatic hydrocarbons or polynuclear aromatic hydrocarbon derivatives (PNAs, also often referred to as polyaromatic hydrocarbon derivatives, PAHs). This is a very important family of compounds because they represent some of the most toxic components found in crude oil and crude oil products.

Most polynuclear aromatic hydrocarbon derivatives are insoluble in water, which limits their mobility in the environment, although polynuclear aromatic hydrocarbon derivatives adsorb onto to fine-grained sediments, such as clay minerals. The

Table 6.1 Properties of the BTX hydrocarbon derivatives at ambient conditions

	Benzene	Toluene	*o*-Xylene	*m*-Xylene	*p*-Xylene
Molecular formula	C_6H_6	C_7H_8	C_8H_{10}	C_8H_{10}	C_8H_{10}
Molecular weight	78.12	92.15	106.17	106.17	106.17
Boiling point, °C	80.1	110.6	144.4	139.1	138.4
Melting point, °C	5.5	–95.0	–25.2	–47.9	13.3

two-ringed polynuclear aromatic hydrocarbon derivatives, and to a lesser extent selected three-ring polynuclear aromatic hydrocarbon derivatives dissolve in water, making them more available for biological uptake and biodegradation. Furthermore, two-ringed to four-ringed polynuclear aromatic hydrocarbon derivatives are sufficiently volatile to appear in the atmosphere predominantly in gaseous form, although the physical state of four-ring polynuclear aromatic hydrocarbon derivatives can depend on temperature.

Two- and three-ringed polynuclear aromatic hydrocarbon derivatives can disperse widely while dissolved in water or as gases in the atmosphere, while polynuclear aromatic hydrocarbon derivatives with higher molecular weights can disperse locally or regionally adhered to particulate matter that is suspended in air or water until the particles land or settle out of the water system. Abiotic degradation on the top layers of a surface water system can produce nitrogenated, halogenated, hydroxylated, and oxygenated polynuclear aromatic hydrocarbon derivatives—some of these derivatives can be more toxic, water-soluble, and mobile than the parent polynuclear aromatic hydrocarbon derivatives. In contrast, compounds with five or more rings have low solubility in water and low volatility; they are therefore predominantly in solid state bound to sediments in a water system. In solid state, these compounds are less accessible for biological uptake or degradation, increasing their persistence in the water system.

Thus, the contamination of water systems by spilled crude oil and crude oil products is a persistent and widespread pollution problem which causes ecological disturbances and the associated health implications (Okoh, 2006; Salam et al., 2011). Once crude oil or a crude oil product is released and comes into contact with water, air, and the necessary salts, microorganisms present in the environment, the natural process of crude oil biodegradation begins (Antai, 1990; Davies et al., 2001; Antić et al., 2006.). However, some of the crude oil-related pollutants are carcinogenic and mutagenic (Miller and Miller, 1981; Obayori et al., 2009).

Remediation of oil spills is a serious issue because of its adverse effects on the biosphere. Oil spreads on the top surface of water and form a horizontal smooth and slippery surface (the oil slick). The oil forms thin coating on the feathers of birds which loses their insulating properties and results in freezing death of the bird. It will also reduce the amount of oxygen dissolving from air in water which is necessary for marine life. Oil spill has toxic impact on aquatic animals and damages their food resources and habitats. Therefore, proper remediation must be done after oil spillage. In order to understand the nature of an oil spill and an oil slick, it is necessary to understand the constituents of crude oil.

Crude oil and crude oil products are mixtures of differing molecular species hydrocarbon derivatives and the constituents of these molecular categories are present in varied proportions resulting in high variability in crude oil and crude oil products (Parkash, 2003; Gary et al., 2007; Speight, 2014, 2017; Hsu and Robinson, 2017). In terms of bulk fractions (Chapter 2), the resin constituents and the asphaltene constituents are of particular interest (or notoriety) because these constituents generally resist degradation.

The severity of the impact of an oil spill depends on the quantity of the crude oil and its chemical and physical properties. The physical and chemical properties of oil affect weathering/transformation processes (evaporation, spreading, emulsification, dissolution, sedimentation and photolysis). After a spill, the constituents of crude oil and crude oil products are subjected to physical and chemical processes such as evaporation or photochemical oxidation which produce changes in the composition of the spilled material (Taghvaei Ganjali et al., 2007; Speight, 2014). These processes collectively may lead towards the formation of numerous oxygenated products which make it difficult to recover the oil.

The recognized mechanical and chemical methods for remediation of a crude oil polluted ecosystem are often expensive, technologically complex and lack public acceptance (Speight, 1996, 2005; Speight and Lee, 2000; Vidali, 2001). Thus, bioremediation is often the method of choice for effective removal of hydrocarbon pollutants from a variety of ecosystems (Okoh and Trejo-Hernandez, 2006). In fact, crude oil and crude oil products are a rich, source of carbon and the reaction of the hydrocarbon derivatives contained therein with aerial oxygen (with the release of carbon-dioxide) is promoted by a variety of microorganisms (Odu, 1977; Atlas, 1981; Atlas and Bartha, 1992; Steffan et al., 1997).

However, the rate of microbial degradation of hydrocarbon derivatives in water is affected by several physicochemical and biological parameters including the number and species of microorganisms present, the conditions for microbial degradation activity (e.g. presence of nutrient, oxygen, pH and temperature) the quality, quantity and bioavailability or bioaccessibility of the contaminants; and the water characteristics such as particle size distribution (Chapter 1) (Atlas, 1991; Freijer et al., 1996; Margesin and Schinner, 1997a,b; Dandie et al., 2010).

Hydrocarbon degrading bacteria and fungi are mainly responsible for the mineralization (conversion of hydrocarbon derivatives to carbon dioxide and water) of crude oil-related pollutants and are distributed in diverse ecosystems (Leahy and Colwell, 1990; Song et al., 1990). Furthermore, the population of microorganisms found in a polluted environment will degrade crude oil-related constituents differently and at a different rate than microorganisms in a relatively clean environment (Obire, 1990, 1993; Obire and Okudo, 1997; Obire and Nwaubeta, 2001).

However, it is uncommon to find organisms that could effectively degrade both aliphatic constituents and aromatic constituents possibly due to differences in metabolic routes and pathways for the degradation of the two classes of hydrocarbon derivatives. There are indications of the existence of bacterial species with propensities for simultaneous degradation of aliphatic hydrocarbon derivatives and aromatic hydrocarbon derivatives (Amund et al., 1987; Obayori et al., 2009). This rare ability may be as a result of long exposure of the organisms to different hydrocarbon pollutants resulting in genetic alteration and *acquisition* of the appropriate degradative genes.

Moreover, it is essential to recognize that the environmental impact of crude oil spills is dependent on previous hydrocarbon exposures and the adaptive status of the local microbiota (Greenwood et al., 2009). The different structural and functional response of microbial sub-groups to different hydrocarbon derivatives confirms that the overall response of biota is sensitive to crude oil composition. This suggests that the preferred response to anticipated contaminants may be engineered by pre-exposure to representative substrates. The controlled adaption of microbes to a threatening contaminant is the basis of proactive bioremediation technology, including the augmentation of newly contaminated sites with locally remediated water in which the biota had already been adapted (bioaugmentation; Chapter 1).

Aquifer contamination can also take place because of the migration of the oil through the porous media and its subsequent adsorption on the rock surface. Many technologies have been proposed for the treatment of oil contaminated sites; these can be performed by two basic processes in-situ and ex-situ treatment using different cleaning technologies such as thermal treatment, biological treatment, chemical extraction, and aeration accumulation techniques (Speight, 1996; Speight and Lee, 2000).

Thus, establishing the chemical history of recently contaminated regions is an important aspect of environmental bioremediation. The premise being that microbial species adapted through a history of exposure to more bioavailable crude oil hydrocarbon derivatives is less severely impacted by a spill than microbes with no such pre-exposure or adaptation (Page et al., 1996; Peters et al., 2005). Indeed, the diversity of crude oil hydrocarbon degraders in most natural environments may be significant but, in the absence of a previous pollution history, the numbers of the microbes may be low due to lack of prior stimulus (Swannell et al., 1996).

Crude oil and crude oil products are mixtures of differing molecular species hydrocarbon derivatives and the constituents of these molecular categories are present in varied proportions resulting in high variability in crude oil and crude oil products (Speight, 2014). In terms of bulk fractions (Chapter 2), the resin constituents and the asphaltene constituents are of particular interest (or notoriety) because these constituents generally resist degradation. After a spill, the constituents of crude oil and crude oil products are subjected to physical and chemical processes such as evaporation or photochemical oxidation which produce changes in the composition of the spilled material (Taghvaei Ganjali et al., 2007; Speight, 2014).

2 Physical and chemical methods of oil spill remediation

Remediation is the removal of a toxic compound from a water system or the containment of a spill so that the area that is contaminated does not increase. Spill remediation refers to the response to an environmental spill that occurs in freshwater and marine water, and on land. The responses to the spill of crude oil or a crude oil product are varied and depend on a number of factors including the type of spill, and whether people or wildlife are immediately threatened by the spill. A spill may be handled at the site, or the contaminated soil and sand can be trucked to another site to be treated. Spill remediation can be a task that involves various experts including those familiar with the geology of the area, water flow patterns (hydrology), and the behavior of chemically diverse compounds.

The spill of crude oil or a crude oil product is the most important type of environmental disaster which usually occurs. It has impact on humans as wells as on plants and wild life, including birds, fish and mammals. Drilling and production accidents and everyday human activities are the main causes of oil spills. The oil spills are hazardous to environment and also affect human health. The adverse effects have been seen on soil, groundwater, plants and animals and immediate actions are required to eradicate the spill problem. The spillage of crude oil or crude oil products at any point may result into explosion and fire hazards. Effective attempts have been made for the remediation of the soil and cleaning the water resources on onshore and offshore. In this paper, some of the known and new methods of remediation have been discussed to solve the oil spill problem economically.

Oil spills that coat the rocks, soil, and sand beaches of coastal shorelines can be treated in several ways. Hot water sprayed onto the shore can help dislodge pollutants. However, the use of high pressure, steaming water can destroy microbial life in the area being treated. Most bacteria cannot tolerate the boiling water temperature that is used in spill cleanup. Although there are a few bacteria that are able to tolerate such temperatures, such as bacteria found naturally in hot springs, these extremely hardy bacteria typically do not degrade the pollutants, whereas the bacteria that are naturally present in the area of a spill may have developed the ability to use the pollutants as nutrients. The loss of the natural microbial population can reduce the efficiency of a spill remediation, since bacteria that are in the soil or sand may be able to use some of the pollutants as sources of food and degrade them. In an idea situation, the compounds that are left over following this degradation may not be as toxic and the original chemicals.

Physical remediation methods are mostly to control oil spills in a water environment. There are two main steps in controlling the oil spills and these (i) are containment and (ii) recovery. The methods are mainly used as a barrier (as soon as possible after the spill) to control the spreading oil spill without changing its physic al and chemical characteristics. Different equipment are used to control oil spills which are: (i) booms, (ii) skimmers, and (iii) adsorbent materials, as well as (iv) thermal methods (Chapter 8).

Briefly, *booms* are used for both containment and recovery and are floatation devices which act like physical barriers and which will not allow the oil to spread across the water so that oil could be recovered. *Skimmers* represent a variety of mechanical equipment used to physically remove floating spills from the water surface and are generally effective only in calm water, and suction skimmers are susceptible to clogging by floating debris. *Adsorbent materials* which are oleophilic and hydrophobic in nature come out as a good controller of oil spills. After skimming operation, adsorbent are used to clean the remaining oil. *Thermal remediation methods* involve burning the oil and by use of this method high rates of oil can be removed by using minimal specialized equipment, such as fire resistant boom or igniters.

The way in which an oil slick breaks up and dissipates depends largely on the chemical and physical persistence of the crude oil or the crude oil product. Low-boiling products such as kerosene tend to evaporate and dissipate quickly and naturally. These are called non-persistent oils. In contrast, persistent oils, such as many crude oils, break up and dissipate more slowly and usually require a clean-up response. Physical properties such as the density, viscosity and pour point of the oil all affect its behavior. However, dissipation does not occur immediately and the time this takes depends on a

Table 6.2 Some common factors affecting petroleum hydrocarbon biodegradation

Factor	Comment
Composition	Structure, amount, toxicity
Physical state	Spreading, adsorption
Weathered or not	Evaporation, photo-oxidation.
Temperature	Evaporation rate, degradation rate
Mineral nutrients	May be inhibitors
Reaction	low pH may be limiting
Micro-organisms	Inability to degrade polynuclear systems

series of factors, including the amount and type of oil spilled, the weather conditions and whether the oil remains on the water or is washed to the shoreline (Table 6.2).

Chemical remediation methods are among the best remediation techniques available for both on shore and offshore. They not only block the spreading of oil spill but also protect the sensitive marine habitat. They are usually used in addition with physical method in marine oil spill remediation. In this technique physical and chemical property of oil is being changed. Various chemicals are used to treat the oil spills as they have capabilities to change the physical and chemical properties of oil. The chemicals which control oil spills include dispersants and solidifiers. Dispersants have surface active agents included and have the capability to break down the slick of oil into smaller droplets and transfer it into the water column where it undergoes rapid dilution and can be easily degraded.

Dispersants weather the slick of oil into smaller droplets and transfer it into the water column where it undergoes to rapid dilution and can be easily degraded. Dispersants are applied through spraying the water with the chemical and confirming the mixing by wind or propeller of boat whereas solidifiers are those hydrophobic polymers which on reaction with oil convert it into solid rubber state which can be easily removed.

New techniques like soil vapor extraction, degradation, bioremediation are major methods of remediation. Physical, chemical, thermal methods are used for cleaning oil spills in marine environment. Many government guidelines and policies have also been discussed in this paper which is made for national and international level to control and prevent oil spills problem. Thus, effective attempts have been made in remediating the aqueous environment, especially the marine environment. The remedial actions that are being taken by these companies are as follows as: (i) chemical methods, (ii) physical methods, (iii) thermal methods.

3 Biodegradation

The biodegradability of any crude oil constituent is a measure of the ability of that constituent to be metabolized (or co-metabolized) by bacteria or other microorganisms through a series of biological process, which include ingestion by organisms as well as microbial degradation (Payne and McNabb Jr., 1984). The chemical characteristics of the contaminants influence biodegradability; in addition, the location and distribution of contamination by crude oil and/or crude oil products in the subsurface can significantly influence the likelihood of success for bioremediation.

The biodegradability of crude oil and crude oil products is inherently influenced by the composition of the substrate upon which the bacteria are acting (Chapter 1). For example, crude oil is quantitatively biodegradable and kerosene, which consists almost exclusively of medium chain-length alkanes, is completely biodegradable under suitable conditions but for heavy asphaltic crudes, approximately only 6–10% of the material oil may be biodegradable within a reasonable time period, even when the conditions are favorable for biodegradation (Bartha, 1986; Okoh et al., 2002; Okoh, 2003; Okoh, 2006). In addition, biodegradation of crude oil-related constituents can be enhanced by use of a consortium of different bacteria compared to the activity of single bacterium species (Ghazali et al., 2004; Milić et al., 2009).

Hydrocarbon degradation rates in the water ecosystems generally follow the order:

$$n - \text{Alkanes} > \text{branched alkanes} > \text{low molecular weight aromatics} > \text{cyclic alkanes}$$

The process is usually aerobic, requiring terminal or sub-terminal oxidation of the alkanes (Harayama et al., 1999), while aromatic hydrocarbon ring structures are broken through hydroxylation and carboxylation processes (Cerniglia, 1992). Hydrocarbon biodegradation in water is associated both with water-soluble or -miscible compounds, and with the oil-water interphases, mainly on droplets and thin oil films with high surface/volume ratio (Bartha and Atlas, 1987; Button et al., 1992; Floodgate, 1984). However, it is difficult to predict the extent and rates of hydrocarbon degradation processes in marine environment, due to the many factors involved in the process (Table 6.2) (Leahy and Colwell, 1990; Atlas and Bartha, 1992; Margesin and Schinner, 1999).

Biodegradation rates of polynuclear aromatic hydrocarbon derivatives (PAHs) in spilled oil stranded on tidal flats were evaluated using model reactors to clarify the effects of non-aqueous phase liquid (NAPL) on the biodegradation of polynuclear aromatic hydrocarbon derivatives in stranded oil on tidal flat with special emphasis on the relationship between dissolution rates of polynuclear aromatic hydrocarbon derivatives into water and viscosity of the non-aqueous phase liquid. Biodegradation of polynuclear aromatic hydrocarbon derivatives in non-aqueous phase liquids was limited by the dissolution rates of polynuclear aromatic hydrocarbon derivatives into water. Biodegradation rate of chrysene was smaller than that of acenaphthene and phenanthrene due to the smaller dissolution rates. Dissolution rates of polynuclear aromatic hydrocarbon derivatives in heavy fuel oil are typically lower than the dissolution rates of polynuclear aromatic hydrocarbon derivatives because of the higher viscosity of heavy fuel oil—the solubility of the fuel oil constituents (i.e., penetration of the solvent into the heavy fuel oil) is more than likely diffusion controlled. Hence, biodegradation rates of polynuclear aromatic hydrocarbon derivatives in heavy fuel oil are lower than the biodegradation rates of crude oil.

Biodegradation rates of polynuclear aromatic hydrocarbon derivatives in non-aqueous phase liquids with slow rate of decrease like fuel oil C was slower than those in non-aqueous phase liquids with rapid rate of decrease like crude oil. The smaller rate of decrease of fuel oil C than crude oil was due to the higher viscosity of fuel oil C. Therefore, not only the dissolution rate of polynuclear aromatic hydrocarbon derivatives but also the rates of decrease of non-aqueous phase liquids were important factors for the biodegradation of polynuclear aromatic hydrocarbon derivatives (Kose et al., 2003). Generally, the data show that the biodegradation of polynuclear aromatic hydrocarbon derivatives in non-aqueous phase liquids was limited by the dissolution rates of polynuclear aromatic hydrocarbon derivatives into water.

In addition to the composition of the crude oil-related substrate, crude oil and crude oil products introduced to the environment are immediately subject to a variety of changes caused by physical, chemical, and biological effects—usually (incorrectly) referred to collectively as *weathering*. Physical and chemical processes include (i) evaporation, (ii) dissolution of crude oil constituents in a water system (or aquifer), (iii) dispersion, (iv) photochemical oxidation, (v) formation of water-oil emulsions, and (vi) adsorption on to suspended particulate material. These processes (Table 6.3) are not sequential and typically occur simultaneously and cause important changes in the composition and properties of the original pollutant, which in turn may affect the rate or effectiveness of biodegradation.

Specifically, the biodegradation of crude oil typically: (i) raises the viscosity and decreases the API gravity, which adversely reduces the ability of the degraded product to flow, (ii) decreased the hydrocarbon content, thereby increasing the residuum content (iii) increases the concentration of certain metals, (iv), increases the sulfur content (v) increases oil acidity, and adds compounds such as carboxylic acids and phenols. All of these changes are seen in the product relative to the unchanged (non-biodegraded) crude oil.

The commercial practice of bioremediation focuses primarily on the cleaning up of crude oil hydrocarbon derivatives (Del'Arco and de França, 1999). Thus, successful application of bioremediation technology to a contaminated ecosystem requires knowledge of the characteristics of the site and the parameters that affect the microbial biodegradation of pollutants (Sabate et al., 2004). However, a number of limiting factors have been recognized to affect the biodegradation of crude oil constituents (Table 6.2).

Despite the difficulty of degrading certain fractions, some hydrocarbon derivatives are among the most easily biodegradable naturally occurring compounds. Biodegradation gradually destroys crude oil-related spills by the sequential metabolism of various classes of compounds present in the oil (Bence et al., 1996). When biodegradation occurs in an oil reservoir, the process dramatically affects the fluid properties and hence the value and producibility of an oil accumulation. Specifically, crude oil biodegradation typically raises viscosity of the residual material (which reduces oil producibility) and reduces the API gravity (which reduces the value of the produced oil). It increases the asphaltene content (relative to the saturated and aromatic hydrocarbon content and the starting material), the concentration of certain metals, the sulfur content and oil acidity.

There are indications that crude oil biodegradation involves more biological components than just the microorganisms that directly attack crude oil constituents (the

Table 6.3 Processes that occur after and oil spill into a water system (listed alphabetically)

Process	Comment
Advection	The movement of oil due to overlying winds or underlying currents, which increase the surface area of the oil, thereby increasing its exposure to air, sunlight and underlying water; effects are not uniform and do not affect the chemical composition of the oil.
Biodegradation	Occurs on the water surface, in the water, in the sediments, and at the shoreline; may be an induction period) after a spill and can continue as long as degradable hydrocarbon derivatives are present.
Dissolution	The transfer of constituents from the spilled material into the aqueous phase. Once dissolved these constituents are bioavailable and if exposed to marine life can cause environmental impacts and injuries.
Emulsification	The mixing of water droplets into oil spilled on water surface forming a water-in-oil emulsion.
Evaporation	Subject to atmospheric conditions such as temperature; the transfer of lower-boiling constituents of the crude oil from the liquid phase to the vapor phase; the remaining components have higher viscosity and specific gravity leading to the formation of tar balls.
Natural dispersion	The process of forming small oil droplets that become incorporated in the water in the form of a dilute oil-in-water suspension and occurs when breaking waves mix the oil in the water column.
Photo-oxidation	Occurs when sunlight in the presence of oxygen transforms hydrocarbon derivatives (by increasing the oxygen content of a hydrocarbon) into new by-products; results in changing in the interfacial properties of the oil, affecting spreading and emulsion formation; may result in transfer of toxic by-products into the water column due to the enhanced water solubility of the by-product.
Sedimentation	The incorporation of the constituents of crude oil and crude oil products within both bottom and suspended sediments.
Spreading	Dominates the initial stages of the spill and involves the whole oil and is the movement of the entire oil slick horizontally on the surface of water due to effects of gravity, inertia, friction, viscosity and surface tension—on calm water spreading occurs in a circular pattern outward from the center of the release point.

primary degraders) and shows that the primary degraders interact with these components (Head et al., 2006). In addition, primary degraders need to compete with other microorganisms for limiting nutrients and the non-crude oil-degrading microorganisms can be affected by metabolites and other compounds that are released by oil-degrading bacteria and vice versa.

The environment, having been exposed to crude oil inputs for thousands of years, can assimilate the hydrocarbon derivatives under the proper conditions. However, areas of particular concern are low energy environments common to estuarine systems. These environments, such as marshes, mud flats, and subtidal areas, are vital to marine fisheries and estuarine productivity and are especially sensitive to contaminant impacts. These systems are particularly vulnerable to impacts of crude oil where the crude oil can persist in these systems for years.

The removal processes for crude oil in wetlands are (i) evaporation, (ii) photo-oxidation, (iii) dissolution of specific constituents, (iv) microbial degradation, and (v) physical flushing. However, once incorporated into the sediment, biodegradation and dissolution are the primary removal mechanisms. Crude oil biodegradation in wetland environments can be limited by anoxia and nutrient availability. Consequently, estuarine wetlands are the most vulnerable of the low-energy intertidal areas to spills of crude oil and crude oil products.

Following the 1994 San Jacinto River flood and spill in southeast Texas, a crude oil-contaminated wetland was reserved for a long-term research program to evaluate bioremediation as a viable spill response tool (Mills et al., 2003). Sediment samples from six test plots were collected eleven times over an eleven-month period to assess the concentration of any remaining crude oil. The rapid degradation rates of the crude oil hydrocarbon derivatives are attributed to conditions favorable to biodegradation. It was suggested that elevated nutrient levels from the flood deposition and the unconsolidated nature of the freshly deposited sediment possibly provided a nutrient rich, oxic environment and that an active and capable microbial community was present due to prior exposure to crude oil. These factors provided an environment conducive for the rapid bioremediation of the crude oil in the contaminated wetland leading to 95% of the spilled material.

The conditions of the contaminated area plays a major role on whether bioremediation is the appropriate method of cleanup for the given oil spill. The success of bioremediation is dependent upon physical conditions and chemical conditions. Physical parameters include temperature, surface area of the oil, and the energy of the water. Chemical parameters include oxygen and nutrient content, pH, and the composition of the oil. Temperature affects bioremediation by changing the properties of the oil and also by influencing the crude oil-degrading microbes (Nedwell, 1999). When the temperature is lowered, the viscosity of the crude oil is increased which changes the toxicity and solubility of the oil, depending upon its composition (Zhu et al., 2001). Temperature also has an effect on the growth rate of the microorganisms, as well as the degradation rate of the hydrocarbon derivatives, depending upon their characteristics.

Biostimulation, the addition of nutrients, is practiced for cleanup of crude oil spills in water systems when there is an existing population of oil degrading microbes present. However, when different kinds of crude oil or crude oil products enter a water system, many physical, chemical and biological degradation processes start acting on the oil. These processes change the properties and behavior of the oil. Some processes (such as the introduction of oxygen function during oxidation) cause the oil to disappear from the surface of the water. The introduction of oxygen functions increases the density (the oil is heavier) and the hydrophilicity of the oil thereby making it (the oil) more amenable to the water. However, the fact that it is no longer visible on the water surface does not necessarily mean that it is gone or has been rendered environmentally harmless. It may even reappear on a distant shoreline or even po the water surface if the temperature of the water can cause it to do so. In fact, every time oil enters a water system, a number of factors will decide the physical, chemical and biological degradation of the oil, as well as the potential environmental damage of the spill or discharge, in that particular area (Table 6.2).

When an oil spill occurs, the result is a large increase in carbon and this also stimulates the growth of the indigenous crude oil-degrading microorganisms. However, these microorganisms are limited in the amount of growth and remediation that can occur by the amount of available nitrogen and phosphorus. By adding supplemental nutrients in the appropriate concentration (i.e. appropriate to the water system and the type of crude oil or crude oil product spilled) concentrations, the hydrocarbon degrading microbes are capable of achieving their maximum growth rate and hence the maximum rate of pollutant uptake. It has been found that when using nitrogen for the supplemental nutrient, a maximum growth rate is achieved by the oil degrading microorganisms (Boufadel et al., 2007). Biostimulation has been proven to be an effective way of achieving increased hydrocarbon degradation by the indigenous microbial population (Coulon et al., 2006).

Supplemental nutrients tended to move downward during rising or lowering of the water levels (Boufadel et al., 2007). This is very useful information in determining the proper timing to add nutrients in order to allow for the maximum residence time of the nutrients in the contaminated areas. The results of this experiment concluded that the nutrients should be applied during low tide at the high tide line, which results in maximum contact time of the nutrients with the oil and hydrocarbon degrading microorganisms.

The water environment encompassing the vast majority of surface of the Earth is a repertoire of a large number of microorganisms. The environmental roles of the biosurfactants produced by many such marine microorganisms have been reported earlier (Poremba et al., 1991; Schulz et al., 1991; Abraham et al., 1998; Das et al., 2008).

4 Bioremediation

Many water systems contain a range of micro-organisms or microbes that can partially or completely degrade crude oil and crude oil products to water soluble compounds and eventually to carbon dioxide and water. Many types of microbe exist and each tends to degrade a particular group of compounds in crude oil. However, some compounds in oil are very resistant to attack and may not degrade. The main factors affecting the efficiency of biodegradation, are the levels of nutrients (nitrogen and phosphorus) in the water, the temperature and the level of oxygen present. As biodegradation requires oxygen, this process can only take place at the oil-water interface since no oxygen is available within the oil itself. The creation of oil droplets, either by natural or chemical dispersion, increases the surface area of the oil or increases the area available for biodegradation to take place.

In fact, many species of aqueous micro-organisms or bacteria, fungi and yeasts feed on the compounds that make up oil. Hydrocarbon derivatives consumed by these micro-organisms can be partially metabolized or completely metabolized to carbon dioxide and water. The rate of biodegradation depends on several or all of the factors (Table 6.2) and a wide range of micro-organisms is required for a significant reduction of the oil. To sustain biodegradation, nutrients such as nitrogen and phosphorus are sometimes added to the water to encourage the micro-organisms to grow and reproduce. Biodegradation tends to work best in warm water environments.

Thus, bioremediation method is a very simple and cheap remediation technique. In this method, microorganisms degrade and metabolize any chemical substance and re-establish environmental quality. Microorganisms fasten the natural weakling process by assimilating organic molecules to cell biomass with carbon dioxide, water and heat as by products. Biostimulation and bioaugmentation are the two bioremediation methods. In biostimulation, nutrients are added to stimulate the growth of the microorganism while in case of bioaugmentation microorganisms are added to existing native oil degrading population. For different hydrocarbon derivatives there are different microorganisms that have been used. They work with different degradation mechanisms depending on the type of hydrocarbon present in oil. The best degrader in the entire microorganism is bacteria. Good concentration of nutrients like nitrogen and phosphorus are growth inhibitor of hydrocarbon-degrader. Because of environment friendly and economic properties, bioremediation has become an advantageous technique for remediation.

In a water ecosystem a spill of crude oil or a crude oil product represents the introduction of a non-aqueous phase liquid composed of a large amount of organic matrices in addition to other toxic constituent, such as polynuclear aromatic hydrocarbon derivatives as occur in resin constituents and asphaltene constituents.

Obviously, there are many different processes acting on a marine oil spill, modifying its physical and chemical characteristics, such as biodegradation, scattering, evaporation, dissolution, dispersion, sedimentation and photo-oxidation (Prince, 1993). These processes occur simultaneously but at different speeds, depending on the physical and chemical properties of the oil and the ambient conditions, like the environmental temperature and energy. On several occasions, however, these processes can be so slow that it is necessary to interfere in order to hasten degradation of the pollutants (Rosa and Triguis, 2007).

Oil type, weather, wind and wave conditions, as well as air and water temperature, all play important roles in ultimate fate of spilled oil in marine environment. After oil is discharged in the environment, a wide variety of physical, chemical and biological processes begin to transform the discharged oil. These chemical and physical processes are collectively called weathering and act to change the composition, behavior, route of exposure and toxicity of discharged oil. The weathering processes are described as follows: (i) spreading, (ii) advection, (iii) evaporation, (iv) dissolution, (v) natural dispersion, (vi) emulsification, (vii) photo-oxidation, (viii) sedimentation, (ix) shoreline stranding, and (x) biodegradation (Table 6.3).

Weathering processes occur simultaneously—one process does not stop before the other begins. The order in which these processes presented is instantaneous, the relative significance of these processes may change if the spill occurred below the water surface or in tropical or ice conditions. The spill chronology may also vary if the spill is near the shoreline in which case it can contaminate the groundwater even before the weathering or cleanup processes start. Also there are many onshore and offshore operations in a crude oil industry that can cause aquifer contamination.

4.1 Conditions

The composition and inherent biodegradability of the crude oil hydrocarbon pollutant is, perhaps, the first and most important consideration when the suitability of a cleanup approach is to be evaluated (Atlas, 1975). Heavier crude oil is generally much more difficult to biodegrade than lighter ones, just as heavier crude oils could be suitable for inducing increased selection pressure for the isolation of crude oil hydrocarbon degraders with enhanced efficiency. Also, the amount of heavy crude oil metabolized by some bacterial species increases with increasing concentration of the contaminant while degradation rates may appear to be more pronounced within a specific concentration range (Okoh et al., 2002; Rahman et al., 2002).

An important aspect of the conditions for biodegradation at a spill site is the activity of microorganisms is the ability of the organisms to produce enzymes to catalyze metabolic reactions, which governed by the genetic composition of the organism(s). Enzymes produced by microorganisms in the presence of carbon sources cause initial attack on the hydrocarbon constituents while other enzymes are utilized to complete the breakdown of the hydrocarbon. Thus, lack of an appropriate enzyme either prevents attack or is a barrier to complete hydrocarbon degradation.

Biodegradation of crude oil-related constituents by bacteria can occur under both aerobic (oxic) and anaerobic (anoxic) conditions (Zengler et al., 1999), usually by the action of different consortia of micro-organisms. In the subsurface, biodegradation occurs primarily under anaerobic conditions, mediated by sulfate reducing bacteria in cases where dissolved sulfate is present (Holba et al., 1996), or methanogenic bacteria in cases where dissolved sulfate is low (Bennett et al., 1993). Although subsurface oil biodegradation does not require oxygen, there is a requirement for the presence of essential nutrients (such as nitrogen, phosphorus, potassium), which can be provided by dissolution/alteration of minerals in the water layer. In the absence of nutrients, the potential for hydrocarbon degradation in anoxic sediments is markedly reduced (Dibble and Bartha, 1976).

In-situ groundwater can be an effective medium for biodegradation of crude oil hydrocarbon derivatives. While there is some notable exceptions (such as MTBE, which is not a hydrocarbon; see Chapter 11) the short-chain, low molecular weight, more water-soluble constituents are degraded more rapidly and to lower residual levels than are long-chain, high molecular weight, less soluble constituents. However, as with all bioremediation efforts, crude oil and crude oil products (such as residual fuel oil and asphalt) typically have a residual high-boiling) fraction composed of resin and asphaltene constituents, which is composed of complex, polynuclear aromatic systems (Speight, 2014).

Microbial utilization of hydrocarbon derivatives (being fully reduced substrates) requires an exogenous electron sink. In the initial attack, this electron sink has to be molecular oxygen. In the subsequent steps too, oxygen is the most common electron sink. In the absence of molecular oxygen, further biodegradation of partially oxygenated intermediates may be supported by nitrate or sulphate reduction.

Uptake and utilization of water insoluble substrates, such as crude oil alkanes, require specific physiological adaptations of the microorganisms. The synthesis of specific amphiphilic molecules (i.e., biosurfactants) is often surmised to be a prerequisite for either specific adhesion mechanisms to large oil drops or emulsification of oil, followed by uptake of submicron oil droplets. In fact, various species of bacteria have been observed to adopt the requisite strategy to deal with water insoluble substrates, such as hydrocarbon derivatives (Rosenberg, 1991). Hence, to facilitate hydrocarbon uptake through the hydrophilic outer membrane, many hydrocarbon-utilizing microorganisms produce cell wall-associated or extracellular surface-active agents (Haferburg et al., 1986). This includes such low molecular weight compounds such as fatty acids, triacyl-glycerol derivatives, and phospholipids, as well as the heavier glycolipids, examples of which include *emulsan* and *liposan* (Cirigliano and Carman, 1984).

Emulsan is the extracellular form of a polyanionic, cell-associated heteropolysaccharide produced by the oil-degrading bacterium *Acinetobacter calcoaceticus* RAG-1 (Rosenberg et al., 1979b; Zuckerberg et al., 1979). The biopolymer stabilizes emulsions of hydrocarbon derivatives in water and has optimal activity when a mixture of aromatic and aliphatic components is present, such as in crude oil (Rosenberg et al., 1979a). The activity of the amphipathic emulsifier is due primarily to its high affinity for the oil-water interface (Zosim et al., 1982) and its ability to orient itself at the interface to form a hydrophilic film around the oil droplets (Zosim et al., 1982, 1986; Shabtai et al., 1986).

Studies with bacteriophages, antibodies, and emulsan-deficient mutants have demonstrated that: (i) as the cells approach stationary phase, emulsan accumulates on the cell surface before release into the medium (Goldman et al., 1982), (ii) cell-bound emulsan serves as a specific receptor and acts as stabilizer for the oil-water interface (Pines and Gutnick, 1981, 1984a), (iii) this indicates that the cell-bound form of emulsan is required for growth on crude oil—species without cell-bound emulsan no longer grow well on crude oil-related (Rosenberg et al., 1983a,b; Pines and Gutnick, 1984b, 1986), (iv) the affinity of emulsan for the oil-water interface suggests that it might affect microbial degradation of emulsified oils (Gutnick and Minas, 1987.

In another work, crude oil was also treated with purified emulsan, the heteropolysaccharide bioemulsifier produced by *Acinetobacter calcoaceticus* RAG-1. A mixed bacterial population as well as nine different pure cultures isolated from various sources was tested for biodegradation of emulsan-treated and untreated crude oil. Biodegradation was measured both quantitatively and qualitatively. Biodegradation of linear alkanes and other saturated hydrocarbon derivatives, both by pure cultures and by the mixed population, was reduced after emulsan pretreatment. In addition, degradation of aromatic compounds by the mixed population was also reduced in emulsan-treated oil. In sharp contrast, aromatic biodegradation by pure cultures was either unaffected or slightly stimulated by emulsification of the oil (Foght et al., 1989).

4.2 Effects

These early stages of oil biodegradation (loss of n-paraffins followed by loss of acyclic isoprenoid derivatives) can be readily detected by gas chromatography (GC) analysis of the crude oil. However, in heavily biodegraded crude oils, gas chromatographic analysis alone cannot distinguish differences in biodegradation due to interference of the unresolved complex mixture that dominates the gas chromatographic traces of heavily degraded crude oils. Among such crude oils, differences in the extent of biodegradation can be assessed using gas chromatography–mass spectrometry (GC–MS) to quantify the concentrations of biomarkers with differing resistances to biodegradation.

During biodegradation, the properties of the crude oil fluid changes because different classes of compounds in crude oil have different susceptibilities to biodegradation (Goodwin et al., 1983). The early stages of biodegradation (in addition to any evaporation effects) are characterized by the loss of n-paraffins (n-alkanes or branched alkanes) followed by loss of acyclic isoprenoid derivatives (e.g., norpristane, pristane, and phytane). Compared with those compound groups, other compound classes (such as highly branched and cyclic saturated hydrocarbon derivatives as well as aromatic compounds) are more resistant to biodegradation. However, even the more-resistant compound classes are eventually destroyed as biodegradation proceeds.

The type of oil, its concentration and the types of hydrocarbon derivatives it contains influence the rate of biodegradation. The Gulf spill released light, sweet crude oil, which is more readily broken down than heavy, sour oil. *Mousse*, tar balls, or oil slicks that wash onshore are concentrated compared to dispersed oil, more protected from wind and wave action than oil in open water and have less surface area for microbes to access. Smaller droplets of oil are more biodegradable.

4.3 Effect of nutrients

Different types of nutrients (primarily nitrogen and phosphorus) have been applied to improve crude oil hydrocarbon degradation, including classic (water soluble) nutrients and oleophilic and slow-release fertilizers. In addition to carbon and oxygen, bacteria need nitrogen and phosphorus to survive. These nutrients are found naturally in the ocean environment. Nitrogen and phosphorus-based fertilizers from farms and gardens on land also enter marine waters through storm water runoff.

Bioavailability is one main factor that influences the extent of biodegradation of hydrocarbon derivatives. Generally, hydrocarbon derivatives have low-to-poor solubility in water and, as a result, are adsorbed on to clay or humus fractions, so they pass very slowly to the aqueous phase where they are metabolized by microorganisms. Cyclodextrins are natural compounds that form soluble inclusion complexes with hydrophobic molecules and increase degradation rate of hydrocarbon derivatives in vitro.

In the perspective of an in situ application, β-cyclodextrin does not increase eluviation (the lateral or downward movement of the suspended material during the percolation of water) of hydrocarbon derivatives through the soil and consequently does not increase the risk of groundwater pollution (Sivaraman et al., 2010). Furthermore, the combination of bioaugmentation and enhanced bioavailability due to β-cyclodextrin was effective for a full degradation (Bardi et al., 2003). Thus, in situ bioremediation of polynuclear aromatic hydrocarbon-polluted water can be improved by the augmentation of degrading microbial populations and by the increase of hydrocarbon bioavailability (Bardi et al., 2007).

Inadequate mineral nutrients, especially nitrogen, and phosphorus, often limit the growth of hydrocarbon utilizers in water. Nutrients are very important ingredients for successful biodegradation of hydrocarbon pollutants, especially nitrogen, phosphorus and in some cases iron. Depending on the nature of the impacted environment, some of these nutrients could become limiting thus affecting the biodegradation processes.

When a major oil spill occurs in freshwater and/or marine ecosystems, the supply of carbon is dramatically increased and the availability of nitrogen and phosphorus generally becomes the limiting factor for oil degradation. This is more pronounced in marine environments, due to the low background levels of nitrogen and phosphorus in water (Floodgate, 1984), unlike in freshwater systems that regularly fluctuate in nutrient status as result of perturbations and receipt of industrial and domestic effluents and agricultural runoff. Freshwater wetlands are typically considered to be nutrient limited, due to heavy demand for nutrients by the plants, which can be considered to be *nutrient traps* since a substantial amount of nutrients is often found in the indigenous biomass (Mitsch and Gosselink, 1993).

Generally, the additions of nutrients is necessary to enhance the biodegradation of crude oil-related pollutants (Choi et al., 2002; Kim et al., 2005). In fact, even in harsh sub-Arctic climates, it has been observed that the effectiveness of fertilizers for crude oil increases the chemical, microbial and toxicological parameters compared to the used of various fertilizers in a pristine environment (Pelletier et al., 2004).

However, excessive nutrient concentrations can inhibit the biodegradation activity, and there can be a negative effect on the biodegradation of hydrocarbon derivatives in the presence of high nitrogen-phosphorus-potassium levels (Oudot et al., 1998; Chaîneau et al., 2005)—this effect is more pronounced on the bioremediation of aromatic hydrocarbon derivatives (Carmichael and Pfaender, 1997). The biodegradation of various aromatic hydrocarbon derivatives is also sensitive to acidity or alkalinity and also to by-products of the biodegradation of the saturate fraction, which serves to explain the persistence of aromatic crude oil hydrocarbon derivatives in certain ecosystems.

On the other hand, in an investigation of the role of the nitrogen source in biodegradation of crude oil components by a defined bacterial consortium under cold, marine conditions (10 °C/50 °F), it was observed that nitrate did not affect the pH, whereas ammonium amendment led to progressive acidification, accompanied by an inhibition of the degradation of aromatic (particularly polynuclear aromatic) hydrocarbon derivatives (Foght et al., 1999). However, the aromatic systems were degraded or co-metabolized in the absence of nutrients where the pH remained almost unchanged. The best overall biodegradation was observed in the presence of nitrate without ammonium, plus high phosphate buffering—a disadvantage of nitrate is that significant emulsification of the crude oil occurs. Generally, it is worth bearing in mind that acidity/alkalinity (pH) is an important factor that requires consideration as it affects the solubility of both polynuclear aromatic hydrocarbon derivatives as well as the metabolism of the microorganisms, showing an optimal range for bacterial degradation between 5.5 and 7.8 (Bossert and Bartha, 1984; Wong et al., 2001).

The molar ratio of carbon, nitrogen and phosphorus (C/N/P) is very important for the metabolism of the microorganisms and, therefore, for degradation of polynuclear aromatic hydrocarbon derivatives (Bossert and Bartha, 1984; Alexander, 1994; Kwok and Loh, 2003). A molar ratio 100:10:1 is frequently considered optimal for contaminated water and soil (Bossert and Bartha, 1984; Alexander, 1994), whilst some authors have reported negative or no effects (Chaîneau et al., 2005). These contradictory results are due to the nutrient ratio required by bacteria that degrade polynuclear aromatic hydrocarbon derivatives, which depends on environmental conditions, type of bacteria and type of hydrocarbon (Leys et al., 2005).

Furthermore, it is not surprising that the chemical form of those nutrients is also important, the soluble forms being (i.e. iron or nitrogen in form of phosphate, nitrate and ammonium) the most frequent and efficient due to their higher availability for microorganisms. Depending on the microbial community and their abundance, another factor that may improve polynuclear aromatic hydrocarbon degradation is the addition of readily assimilated carbon sources, such as glucose (Zaidi and Imam, 1999).

4.4 Effect of temperature

Temperature plays an important role in the biodegradation of crude oil-related hydrocarbon derivatives not only because of the direct effect on the chemistry of the pollutants but also because of the effect on the physiology and diversity of the microbial surroundings (Atlas, 1975). In fact, water temperature is important because most of the physical, chemical and biological characteristics of a river are directly affected by temperature. All aquatic organisms have preferred temperature ranges in which they can survive and reproduce optimally. Temperature also has an important influence on water chemistry. The rates of chemical reactions generally increase with increasing temperature. Temperature is a regulator of the solubility of gases and minerals (solids)—or how much of these materials can be dissolved in water. The solubility of important gases, such as oxygen and carbon dioxide increases as temperature decreases. In addition certain pollutants become more toxic at increased temperatures. Also, plant growth increases with warmer temperatures. When plants die, they are decomposed by bacteria, which use up the oxygen. Increased plant growth means more oxygen being removed from the water during the decomposition process. The sensitivity of organisms is also affected by temperature and many organisms require a specific temperature range and changing that range may eliminate some organisms from the ecosystem. Under temperature extremes, organisms may become stressed, which makes them more vulnerable to toxic wastes, parasites and disease.

Typically, biodegradation of crude oil and crude oil products occurs at temperatures $<80\,^{\circ}C$ ($<176\,^{\circ}F$) (Conan, 1984; Barnard and Bastow, 1991)—at higher temperatures (unless the microbes are of a specific thermophilic type) many of the microorganisms involved in subsurface oil biodegradation cannot exist.

Thus, the ambient temperature of an environment affects both the properties of spilled crude oil or crude oil product (Speight, 2014) and the activity or population of microorganisms (Venosa and Zhu, 2003). At low temperatures, the viscosity of the oil increases, while the volatility of toxic low-molecular weight hydrocarbon derivatives is reduced, delaying the onset of biodegradation (Atlas, 1981). Temperature also variously affects the solubility of hydrocarbon derivatives (Foght et al., 1996).

Although hydrocarbon biodegradation can occur over a wide range of temperatures, the rate of biodegradation generally decreases with decreasing temperature. Highest degradation rates generally occur in the range of 30–40 °C (86–104 °F) in ecosystems, 20–30 °C (68–86 °F) in some freshwater environments, and 15–20 °C (59–68 °F) in marine environments (Bossert and Bartha, 1984). In fact, the biodegradability of crude oil is highly dependent not only on composition but also on microbial incubation temperature (Atlas, 1975)—at 20 °C (68 °F) conventional crude oil has higher abiotic losses and is more susceptible to biodegradation than heavy oil. As expected from crude oil chemistry and composition (Speight, 2014), the rate of mineralization for the heavy oil is significantly lower at 20 °C (68 °F) than for conventional oil.

During biodegradation, some preference is shown for removal of the paraffin constituents over the aromatic and asphaltic constituents, especially at low temperatures. Branched paraffins, such as pristane, are degraded at both 10 and 20 °C (50 and 68 °F). This was confirmed by showing that the residual material (after an incubation period of 42 days) had a lower relative percentage of paraffins and higher percentage of asphaltic constituents (usually resin and asphaltene constituents) than fresh or weathered oil (Atlas, 1975).

Finally, the relative resistance of conventional (light) crude oil and even crude oil distillate products to degradation at low temperatures should be considered in choosing shipping routes for these materials. Accidental spillages in the Arctic will most likely be of regional heavy type oil (such as the Prudhoe Bay crude oil), which will be subject to slow but constant microbial degradation (Atlas, 1975). As a quick note, the Prudhoe Bay oil field is a large oil field on North Slope of Alaska and contains an estimated 25 billion (25×10^9) barrels) of crude oil (ca. 21° API).

Water temperature affects the physical and chemical properties of oil and the rate of biodegradation. Colder temperatures slow the rate; warmer temperatures increase the rate.

The ubiquitous distribution of oil-degrading bacteria in the polar environment has already been reported (Leahy and Colwell, 1990). Before contamination, counts indicated that hydrocarbon-degrading microorganisms comprised <0.01% of the total number of bacteria but represented 10% of the cultivable heterotrophic microorganisms. These results are consistent with those that indicate that hydrocarbon-degrading microorganisms comprised 1–10% of the total number of saprophytic bacteria in marine bacterial communities (Wright et al., 1997).

Temperature can affect the physical and chemical characteristics of the discharged oils, hydrocarbon derivatives with high pour points could be expected to show more temperature-related biodegradation than contaminants with low pour points. However, globally low temperatures have been reported to play a significant role in controlling the nature and extent of microbial hydrocarbon metabolism (Nedwell, 1999; Gerdes et al., 2005).

Furthermore, heterotrophic and hydrocarbon-degrading bacteria counts are often similar in cold and warm climates (Aislabie, 1997; Eckford et al., 2002). These results seem in conflict with the typical temperature-related assumption predicting an increase of microbial metabolism with a temperature increase (Leahy and Colwell, 1990; Bossert and Bartha, 1984). According to the Arrhenius equation, any decrease in temperature should cause an exponential decrease of the reaction rates, the magnitude of which depending on the value of the activation energy (Margesin et al., 2007b).

Low temperatures may affect the utilization of substrates comprising a mixture of hydrocarbon derivatives (Baraniecki et al., 2002). Altered growth responses might result from beneficial changes to membrane fluidity. The production and/or activity of cold-active enzymes are usually significantly below the "optimal" growth temperature (as determined from growth rate) of the enzyme producer, which reflects the thermal characteristics of the secretion process (Margesin et al., 2007b).

In another investigation (Coulon et al., 2006), it was found that when crude oil was added to the test the presence of crude oil-degrading microbes increased by two orders of magnitude at 4 °C (39 °F), more than three orders of magnitude at 12 °C (54 °F), and more than four orders of magnitude at 20 °C (68 °F). The surface area of the oil is also a significant parameter in the success of bioremediation because the growth of oil degrading microorganisms occurs at the interface of the water and oil. The larger the surface area of the oil results in a larger area for growth and hence larger numbers of microbes. The energy of the water is important because rough waters will disperse and dilute essential nutrients for the microorganisms and also spread the oil, contaminating more areas.

A more recent investigation (Delille et al., 2009) provides further evidence of the high potential of indigenous Antarctic bacterial communities for bioremediation action even at low temperatures. Little difference in data obtained under three incubation temperatures and with two different concentrations of oil is clearly indicating that temperature had only a rather limited influence on crude oil degradation in Antarctic seawater. This point is important when considering bioremediation as an efficient means to clean up contaminated waters in in remote polar locations.

4.5 Effect of dispersants

The effect of dispersants on the fate of dispersed oil has often been the subject of conflicting reports—some workers have proposing that dispersants had (i) little effect on oil biodegradation, (ii) a positive effect, and (iii) a negative effect (Robichauz and Myrick, 1972; Mulkins-Phillips and Stewart, 1974; Traxler and Bhattacharya, 1979; Foght and Westlake, 1982; Litherathy et al., 1989). Biodegradation occurs more readily when oil droplets are dispersed. Chemical dispersants break oil into smaller droplets, which increases the surface area available for bacteria to access. The exact effects of dispersants such as Corexit on the rate of biodegradation are unknown, especially in deep water. The effectiveness of dispersants also depends on the type and consistency of oil and the oil-dispersant ratio.

On the other hand, and perhaps more specifically, it has been suggested that dispersants tend to increase oil biodegradation by increasing the surface area for microbial attack, and encouraging migration of the droplets through the water column making oxygen and nutrients more readily available (Mulyono et al., 1994). However, dispersants can have a detrimental (toxic) effect on microbial processes thereby retarding the rate of crude oil degradation (Mulyono et al., 1994; Varadaraj et al., 1995). It would appear that the dual capability of dispersants (increasing the surface area of dispersed oil and affecting the growth of hydrocarbon-degraders) is related to the chemistry of the dispersant which influences the effectiveness of dispersants for bioremediation (Varadaraj et al., 1995; Davies et al., 2001).

It is clear that the introduction of external nonionic surfactants (the main components of oil spill dispersants) will influence the alkane degradation rate (Bruheim and Eimhjelle, 1998; Rahman et al., 2003). There are indications that the use of surfactants in situations of crude oil-related contamination may have a stimulatory, inhibitory, or neutral effect on the bacterial degradation of the crude oil constituents (Liu et al., 1995). Thus, there is the need to accurately characterize the roles of chemical and biological surfactants in order that performance in biological systems may be predicted (Rocha and Infante, 1997; Lindstrom and Braddock, 2002).

However, in contrast to chemical dispersants, which caused ecological damage after application for abatement of spilled crude oil-related constituents in marine ecosystems (Smith, 1968), biosurfactants from freshwater microorganisms are less toxic and partially biodegradable (Poremba et al., 1991).

4.6 Rates of biodegradation

Different results for microbial activity measurements may be obtained in laboratory studies depending on pretreatment and size of the sample, even when the environmental conditions are mimicked. These differences may be related to, among other factors, differences in the bioavailability of the contaminant in different analyses.

Thus, modeling chemical reactions to determine the rate of the reaction is a common practice in chemistry. However, blending of chemical reactants, which is a part of modeling studies, can have beneficial or adverse effects on chemical reactions. Blend time predictions are usually based on empirical correlations and when a competitive side reaction is present, the final product distribution may be suspect. In biodegradation, the effect of the microbe in relation to the hydrocarbon stream on the reaction outcome is crucial. Also, the scale up of such reactions from the laboratory to the field may not be straightforward. Thus, there is a need for comprehensive, for optimistic caution when models of biodegradation chemistry are used to predict information such as incubation period, product, byproducts, and whether or not the microbes will survive the duration of the treatment. It is also equally important to recognize the potential for interference of one chemical species with another—a situation that is not often determined in the laboratory when simple model substrates (such as single chemical entities or mixtures of two-to-twelve compounds that are not truly representative of crude oil and crude oil products) are used for the experiments.

The most important critical stage of crude oil degradation during the first 48 h of a spill is usually evaporation, the process by which the lower molecular weight (lower-boiling constituents) constituents of crude oil volatilize into the atmosphere. Evaporation can be responsible for the loss of one- to two-thirds of the mass of spilled material (assuming that the spilled material is conventional crude oil or a distillate product) during this period, with the loss rate decreasing rapidly over time. The constituents of heavy oil, tar sand bitumen, and asphalt do not evaporate to the same extent—the lower boiling constituents being generally absent from these materials.

Evaporative loss is controlled by the composition and physical properties of the crude oil or crude oil constituents, the surface of the spill, wind velocity, and temperature (Payne and McNabb Jr., 1984). Derivation of a universal model for such a process may insert inaccuracies because of the complex and changing nature of crude oil and crude oil products. In addition, the material left behind is richer in metals (mainly nickel and vanadium), waxes, resin constituents, and asphaltene constituents than the original oil. With evaporation, the specific gravity and viscosity of the original oil also increase—after several days, spilled conventional (light) crude oil may begin to resemble the more viscous Bunker C oil (heavy fuel oil) in composition and properties (Mielke, 1990).

Although, there has been a reported case of lack of correlation between degradation rates, specific growth rates and concentration of the starter oil (Thouand et al., 1999), in such a case, it would appear that biomass was required only to a particular threshold enough to produce the appropriate enzyme system that carry through the degradation process even when biomass production had ceased (Pitter and Chudoba, 1990), where production of variance with the theory of microbial growth in batch cells is totally dependent on the consumed carbon source.

Many reports on the effect of sunlight irradiation so far published have focused on the physicochemical changes on intact crude oil other than to biodegraded crude oil (Jacquot et al., 1996; Nicodem et al., 1998). Recent studies have reported that photo-oxidation increases the biodegradability of crude oil hydrocarbon derivatives by increasing its bioavailability and thus enhancing microbial activities (Maki et al., 2005).

In fact, bioremediation is a multi-variable process and optimization through classical methods is subject to question. To overcome the disadvantages of the process, response surface methodology (RSM) has been advocated for analyzing the effects of several independent variables on the bioremediation process in order to assess the optimum conditions for the process (Nasrollahzadeh et al., 2007; Huang et al., 2008; Pathak et al., 2009; Vieira et al., 2009; Mohajeri et al., 2010; Zahed et al., 2010). The outcome is the suggestion that the rates of biodegradation can be increased by modifying selected physical and chemical conditions that control biodegradation in multiphase systems, namely, (i) bioavailability and (ii) terminal electron acceptor availability (Sandrin et al., 2006).

Possible approaches to increasing contaminant bioavailability will depend on the system in question. However, any approach that either increases the physical mixing of the non-aqueous phase liquids and aqueous phases or that increases the solubility of the contaminant in the aqueous phase will result in increased biodegradation. Such approaches include the addition of surfactants or co-solvents (to increase solubility), aeration, or hydraulic pulsing, to increase mixing of the two phases.

The complex array of factors that influence biodegradation of crude oil-related is not realistic to expect a simple rate model or kinetic model to provide precise and accurate descriptions of concentrations during different seasons and in different environments. Therefore, it is nearly impossible to predict the rates of biodegradation of such a process. To give a final answer on how much time remediation processes require and what the final mineral oil concentrations will be, experiments should be continued until the biodegradation processes have stopped completely. In future, it will be necessary to use complex models to yield a more exact assessment of remediation to the desired level (Maletić et al., 2009).

Finally, when predicting by modeling the behavior of an aged contaminant it is relevant to adapt the models in use to correspond to conditions relevant at the contaminated sites. As with all efforts at modeling the outcome of complex processes, the variable parameters used in the models must be based on (i) the properties of the material (in this case the crude oil-based contaminant), (ii) data retrieved about the conditions of the actual site, and on (iii) experiments performed using the original aged contaminant without any additions (model compounds or analytical *spikes*).

4.7 Effect of weathering

Weathered crude oil (i.e. crude oil and crude oil related products) that has been exposed to air and oxidized and to other influence such as evaporation offer a differ challenge to bioremediation efforts. Microbes adapt to gradual exposure to oil. The more oil a microbial community has been exposed to in the past, the greater its capacity and availability to biodegrade oil in the future. In one study, microbes from sediments previously contaminated with oil were able to metabolize oil 10 to 400 times faster than those from sediments that had never been contaminated. Once a species of bacteria is exposed to oil and metabolizes it, the next generations inherit that ability, a concept known as *genetic adaptation*.

Several areas at many sites, especially refinery sites with a long history of operation, have weathered oil floating on the surface of the groundwater. However, site conditions can be manipulated to enhance bioremediation and speed up the degradation rates of the contaminants. There are several techniques that can be applied to enhance the biological degradation of contaminants: (i) supplementation with suitable sources of nitrogen and phosphorus, (ii) manipulation of redox potential by the injection of air, oxygen, or nitrate to enhance aerobic biodegradation, (iii) addition of surfactants to make the contaminants bioavailable, (iv) site microbial inoculation, and (v) injection of co-substrates such as molasses, or lactate to enhance the biodegradation of chlorinated contaminants;

Oxygen is often the limiting factor in aerobic bioremediation at many sites. The degradation of crude oil hydrocarbon derivatives occurs much faster under aerobic conditions compared to anaerobic conditions. Therefore, the addition of oxygen can significantly increase the remediation rates. In open water, oil hydrocarbon derivatives undergo aerobic biodegradation by bacteria that use oxygen dissolved in the water. Scientists have monitored dissolved oxygen levels around the spill since it occurred. Early measurements of the deep plume showed a rate of 30% oxygen depletion, which demonstrated the presence of biodegrading microbes. Sediment on the ocean floor and along the coast is, for the most part, *anoxic*: it does not contain oxygen. Hydrocarbon derivatives that settle into sediments on the ocean floor and along coasts undergo anaerobic biodegradation, a much slower process. Therefore, onshore oil lingers longer than oil on the water and can become a chronic pollutant.

Oxygen addition is most frequently used to address dissolved phase contamination, such as total crude oil hydrocarbon derivatives and BTEX, as well as contamination in the capillary fringe zone. Oxygen can only be effective if the hydrocarbon derivatives are bioavailable and there is no nutrient limitation.

4.8 Effect of flowing water

Oil discharges generate a dynamic situation on the oil–water interphases, caused by a variety of physical and chemical processes, including wind, currents, oil weathering, film generation, and oil dispersion. Thus, immiscible oil surfaces are constantly washed with water, and new microbes are continuously contacting the oil film or droplet surfaces. Simulation of these dynamic conditions may be achieved rather in flow-through systems than by closed static experiments. Various controlled flow-through systems have been described which simulate biodegradation under marine conditions, including microcosms, and large tank-based mesocosms (Siron et al., 1993). However, since dispersed oil was used in most of these systems it was impossible to differentiate between HC processes in the oil and water phases.

In more recent studies with immobilized oil in flow-through systems may give valuable contributions to the determination and understanding of various processes after oil discharges to aquatic environments (Brakstad et al., 2004). For instance, hydrocarbon dissolution and biodegradation from oil films, and the potential impacts in the water column, may be predicted after an oil spill in combination with remote sensing of oil films thickness (Brown and Fingas, 2003). Further, if depletion from films and droplets are comparable, the system described here may be used for gaining new insights into dissolution, biodegradation and microbiological processes of dispersed oil. This may also have implications for studies of oil spill treatment with chemical dispersants.

4.9 Effect of deep-sea environments

The deep sea (or deep layer) is the lowest layer in the ocean. Little or no light penetrates this part of the ocean, and most of the organisms that live there rely for subsistence on falling organic matter. Furthermore, most deep ocean waters have sufficient oxygen to support life.

The oxygen in these waters does not appear to be depleted by the respiration of deep-water organisms, presumably because the overall density of these organisms is quite low and these organisms typically have low metabolic rates (exceptions to this generalization are some species that live in the vicinity of seamounts and deep-sea chemosynthetic communities). Thus, in the deep-sea environment, organisms (particular organisms that are capable of using crude oil constituents for sustenance, i.e. bioremediation) must be adapted to survive under extreme pressure, limited light, variable water movement, and other factors not present in surface waters.

Light is virtually absent in the deep ocean, which means that deep-sea organisms cannot rely on vision to find food and mates and to maintain various interspecific and intraspecific associations. However, the lack of light precludes photosynthesis and contributes to a general scarcity of food in the deep ocean. As a result, the overall density of organisms is low in many parts of the deep ocean, and many of these organisms are relatively small with low metabolic rates. Chemosynthetic communities in the vicinity of hydrothermal vents and cold seeps are an exception, since they are not dependent upon photosynthesis as a primary food source.

Temperature is relatively stable in the deep ocean, typically about 5 °C (41 °F) at 3500 ft depth, although water temperatures in the vicinity of hydrothermal vents can vary by nearly 400 °C (the water does not boil because of the high pressure) in the space of a few meters. The deep sea (as well as other habitats, such as deep groundwater, deep sediments or oilfields) is influenced by high pressure. Barophiles (piezophiles) are microorganisms that require high pressure for growth, or grow better at pressures higher than atmospheric pressure (Prieur and Marteinsson, 1998).

A pollutant with a density greater than that of marine waters (such as oxidized crude oil-water emulsions) will be expected to sink to a level commensurate with the density and even to the deep benthic zone, where the hydrostatic pressure is notably high. A combination of high pressure and low temperatures in the deep ocean results in low microbial activity (Alexander, 1999).

References

Abraham, W.R., Meyer, H., Yakimov, M., 1998. Novel Glycine containing glucose lipids from the alkane using bacterium *Alcanivorax borkumensis*. Biochim. Biophys. Acta 1393, 57–62.

Aislabie, J., 1997. Hydrocarbon-degrading Bacteria in oil-contaminated soils near Scott Base, Antarctica. In: Lyons, W.B., Howard-Williams, C., Hawes, I. (Eds.), Ecosystem Processes in Antarctica's Ice-Free Landscape. Balkema Publishers Ltd, Rotterdam, Netherlands, pp. 253–258.

Alexander, M., 1994. Biodegradation and Bioremediation. Academic Press Inc, New York.

Alexander, M., 1999. Biodegradation and Bioremediation, second ed. Academic Press, London, United Kingdom.

Amund, O.O., Adewale, A.A., Ugoji, E.O., 1987. Occurrence and characterization of hydrocarbon utilizing Bacteria in Nigerian soils contaminated with spent motor oil. Indian J. Microbiol. 27, 63–87.

Antai, S.P., 1990. Biodegradation of bonny light crude oil by Bacillus sp and Pseudomonas sp. Waste Manag. 10, 61–64.

Antić, M.P., Jovanćićević, B.S., Ilić, M., Vrvić, M.M., Schwarzbauer, J., 2006. Petroleum pollutant degradation by surface water microorganisms. Environ. Sci. Pollut. Res. 13 (5), 320–327.

Atlas, R.W., 1975. Effects of temperature and crude oil composition on petroleum biodegradation. Appl. Microbiol. 30 (3), 396–403.

Atlas, R.M., 1981. Microbial degradation of petroleum hydrocarbons: an environmental perspective. Microbiol. Rev. 45, 180–209.

Atlas, R.M., 1991. Microbial hydrocarbon degradation—bioremediation of oil spills. J. Chem. Technol. Biotechenol. 52, 149–156.

Atlas, R.M., Bartha, R., 1992. Hydrocarbon biodegradation and oil spill bioremediation. Adv. Microbial Ecol. 12, 287–338.

Baraniecki, C.A., Aislabie, J., Foght, J.M., 2002. Characterization of *Sphingomonas sp. Ant 17*, an aromatic hydrocarbon-degrading bacterium isolated from Antarctic soil. Microb. Ecol. 43, 44–54.

Bardi, L., Ricci, R., Mario Marzona, M., 2003. In situ bioremediation of a hydrocarbon polluted site with cyclodextrin as a coadjuvant to increase bioavailability. Water Air Soil Poll Focus Vol. 3, 15–23.

Bardi, L., Martini, C., Opsi, F., Bertolone, E., Belviso, S., Masoero, G., Marzona, M., Ajmone Marsan, F., 2007. Cyclodextrin-enhanced in situ bioremediation of Polyaromatic hydrocarbons-contaminated soils and plant uptake. J. Incl. Phenom. Macrocycl. Chem. 57, 439–444.

Barnard, P.C., Bastow, M.A., 1991. Hydrocarbon generation, migration, alteration, entrapment and mixing in the central and northern North Sea. In: England, W.A., Fleet, A.J. (Eds.), Petroleum Migration. Geological Society, London, United Kingdom, pp. 167–190. Special Publication.

Bartha, R., 1986. Microbial Ecology: Fundamentals and Applications. Addisson-Wesley Publishers, Reading, MA.

Bartha, R., Atlas, R.M., 1987. Transport and transformation of petroleum: biological processes. In: Long Term Environmental Effects of Offshore Oil and Gas Development. Applied Science Publishers, London, United Kingdom, pp.287–341.

Bence, A.E., Kvenvolden, K.A., Kennicutt, M.C., 1996. Organic geochemistry applied to environmental assessments of Prince William sound, Alaska, after the Exxon Valdez oil spill—A review. Org. Geochem. 24, 7–42.

Bennett, P.C., Siegel, D.E., Baedecker, M.J., Hult, M.F., 1993. Crude oil in a shallow sand and gravel aquifer 1. Hydrogeology and inorganic geochemistry. Appl. Geochem. 8, 529–549.

Bossert, I., Bartha, R., 1984. The fate of Petroleum in Soil Ecosystems. In: Atlas, R.M. (Ed.), Petroleum Microbiology. Macmillan, New York, pp. 453–473.

Boufadel, M.C., Suidan, M.T., Venosa, A.D., 2007. Tracer studies in a Laboratory Beach subjected to waves. J. Environ. Eng. 133 (7), 722.

Brakstad, O.G., Bonaunet, K., Nordtug, T., Johansen, Ø., 2004. Biotransformation and dissolution of petroleum hydrocarbons in natural flowing seawater at low temperature. Biodegradation 15, 337–346.

Brown, C.E., Fingas, M.F., 2003. Development of airborne oil thickness measurements. Mar. Pollut. Bull. 47, 485–492.

Bruheim, P., Eimhjelle, K., 1998. Chemically emulsified crude oil as substrate for bacterial oxidation: differences in species response. Can. J. Microbiol. 44 (2), 195–199.

Button, D.K., Robertson, B.B., McIntosh, D., Jüttner, F., 1992. Interactions between marine Bacteria and dissolved-phase and beached hydrocarbons after the Exxon Valdez oil spill. Appl. Environ. Microbiol. 58, 243–251.

Carmichael, L.M., Pfaender, F.K., 1997. The effect of inorganic and organic supplements on the microbial degradation of Phenanthrene and pyrene in soils. Biodegradation 8, 1–13.

Cerniglia, C.E., 1992. Biodegradation of polycyclic aromatic hydrocarbons. Biodegradation 3, 351–368.

Chaîneau, C.H., Rougeux, G., Yepremian, C., Oudot, J., 2005. Effects of nutrient concentration on the biodegradation of crude oil and associated microbial populations in the soil. Soil Biol. Biochem. 37, 1490–1497.

Choi, S.-C., Kwon, K.K., Sohn, J.H., Kim, S.-J., 2002. Evaluation of fertilizer additions to stimulate oil biodegradation in sand seashore Mescocosms. J. Microbiol. Biotechnol. 12, 431–436.

Cirigliano, M.C., Carman, G.M., 1984. Isolation of a bioemulsifier from *Candida lipolytica.* Appl. Environ. Microbiol. 48, 747–750.

Coulon, F., McKew, B.A., Osborn, A.M., McGenity, T.J., Timmis, K.N., 2006. Effects of temperature and biostimulation on oil-degrading microbial communities in temperate estuarine waters. Environ. Microbiol. 9 (1), 177–186.

Dandie, C.E., Weber, J., Aleer, S., Adetutu, E.M., Ball, A.S., Juhasz, A.L., 2010. Five bioaccessibility assays for predicting the efficacy of petroleum hydrocarbon biodegradation in aged contaminated soils. Chemosphere 81, 1061–1068.

Das, P., Mukherjee, S., Sen, R., 2008. Improved bioavailability and biodegradation of a model Polyaromatic hydrocarbon by a biosurfactant producing bacterium of marine origin. Chemosphere 72, 1229–1234.

Davies, L., Daniel, F., Swannell, R., and Braddock, J. 2001. Biodegradability of Chemically-Dispersed Oil. Report No. AEAT/ENV/R0421. AEA Technology Environment, Abingdon, Oxfordshire, United Kingdom.

Del'Arco, J.P., de França, F.P., 1999. Biodegradation of crude oil in a Sandy sediment. Int. Biodeter. Biodegr. 44, 87–92.

Delille, D., Pelletier, E., Rodriguez-Blanco, A., Ghiglione, J., 2009. Effects of nutrient and temperature on degradation of petroleum hydrocarbons in sub-Antarctic coastal seawater. Polar Biol. 32, 1521–1528.

Dibble, J.T., Bartha, R., 1976. The effect of Iron on the biodegradation of Petroleum in Seawater. Appl. Environ. Microbiol. 31, 544–550.

Eckford, R., Cook, F.D., Saul, D., Aislabie, J., Foght, J., 2002. Free-living nitrogen-fixing bacteria from antarctic soils. Appl. Environ. Microbiol. 68, 5181–5185.

Floodgate, G.D., 1984. The fate of petroleum in marine ecosystems. In: Atlas, R.M. (Ed.), Petroleum Microbiology. MacMillan, New York, pp. 355–397.

Foght, J.M., Westlake, D.W.S., 1982. Effect of the dispersant Corexit 9527 on the microbial degradation of Prudhoe Bay oil. Can. J. Microbiol. 28, 117–122.

Foght, J.M., Gutnick, D.L., Westlake, D.W.S., 1989. Effect of Emulsan on biodegradation of crude oil by pure and mixed bacterial cultures. Appl. Environ. Microbiol. 36–42. January.

Foght, J.M., Westlake, D., Johnson, W.M., Ridgway, H.F., 1996. Environmental gasoline-utilizing isolates and clinical isolates of *Pseudomonas aeruginosa* are taxonomically indistinguishable by chemotaxonomic and molecular techniques. Microbiology 142, 1333–1338.

Foght, J., Semple, K., Gauthier, C., Westlake, D.S., Blenkinsopp, S., Sergy, G., Wang, Z., Fingas, M., 1999. Effect of nitrogen source on biodegradation of crude oil by a defined bacterial consortium incubated under cold, marine conditions. Environ. Technol. 20, 839–849.

Freijer, J.I., de Jonge, H., Bouten, W., Verstraten, J.M., 1996. Assessing mineralization rates of petroleum hydrocarbons in soils in relation to environmental factors and experimental scale. Biodegradation 7, 487–500.

Gary, J.G., Handwerk, G.E., Kaiser, M.J., 2007. Petroleum Refining: Technology and Economics, fifth ed. CRC Press, Taylor & Francis Group, Boca Raton, FL.

Gerdes, B., Brinkmeyer, R., Dieckmannm, G., Helmke, E., 2005. Influence of crude oil on changes of bacterial communities in Arctic Sea-ice. FEMS Microbiol. Ecol. 53, 129–139.

Ghazali, F.M., Rahman, R.N.Z.A., Salleh, A.B., Basri, M., 2004. Biodegradation of hydrocarbons in soil by microbial consortium. Int. Biodeter. Biodegr. 54, 61–67.

Goldman, S., Shabtai, Y., Rubinovitz, C., Rosenberg, E., Gutnick, D.L., 1982. Emulsan in *Acinetobacter calc oaceticus* RAG-1: Distribution of cell-free and cell-associated cross-reacting material. Appl. Environ. Microbiol. 44, 165–170.

Goodwin, N.S., Park, P.J.D., Rawlinson, A.P., 1983. Crude oil biodegradation under simulated and natural conditions. In: Bjorøy, M. (Ed.), Advances in Organic Geochemistry 1981. John Wiley & Sons Inc, New York, pp. 650–658.

Greenwood, P.F., Wibrow, S., George, S.J., Tibbett, M., 2009. Hydrocarbon biodegradation and soil microbial community response to repeated oil exposure. Org. Geochem. 40, 293–300.

Gutnick, D.L., Minas, W., 1987. Perspectives on microbial surfactants. Biochem. Soc. Biochem. Soc. Trans. 15, 22S–35S.

Haferburg, D., Hommel, R., Claus, R., Kleber, H.P., 1986. Extracellular microbial lipids as biosurfactants. Adv. Biochem. Eng. Biotechnol. 33, 53–93.

Harayama, S., Kishira, H., Kasai, Y., Shutsubo, K., 1999. Petroleum biodegradation in marine environments. J. Mol. Microbiol. Biotechnol. 1 (1), 63–70.

Head, I.M., Jones, D.M., Röling, W.F.M., 2006. Marine microorganisms make a meal of oil. Nat. Rev. Microbiol. 4, 173–182.

Holba, A.G., Dzou, I.L., Hickey, J.J., Franks, S.G., May, S.J., Lenney, T., 1996. Reservoir geochemistry of south pass 61 field, Gulf of Mexico: Compositional heterogeneities reflecting filling history and biodegradation. Org. Geochem. 24, 1179–1198.

Hsu, C.S., Robinson, P.R. (Eds.), 2017. Handbook of Petroleum Technology. Springer International Publishing AG, Cham, Switzerland.

Huang, L., Ma, T., Li, D., Liang, F., Liu, R., Li, G., 2008. Optimization of nutrient component for diesel oil degradation by *Rhodococcus erythropolis*. Mar. Pollut. Bull. 56, 1714–1718.

Jacquot, F., Guiliano, M., Doumenq, P., Munoz, D., Mille, G., 1996. In vitro photo-oxidation of crude oil Maltenic fractions: evolution of fossil biomarkers and polycyclic aromatic hydrocarbons. Chemosphere 33, 671–681.

Kim, S., Choi, D.H., Sim, D.S., Oh, Y., 2005. Evaluation of bioremediation effectiveness on crude oil-contaminated sand. Chemosphere 59, 845–852.

Kose, T., Mukai, T., Takimoto, K., Okada, M., 2003. Effect of non-aqueous phase liquid on biodegradation of PAHs in spilled oil on tidal flat. Water Res. 37, 1729–1736.

Kwok, C.-K., Loh, K.-C., 2003. Effects of Singapore soil type on the bioavailability of nutrients in soil bioremediation. Adv. Environ. Res. 7, 889–900.

Leahy, J.G., Colwell, R.R., 1990. Microbial degradation of hydrocarbons in the environment. Microbiol. Rev. 54, 305–315.

Leys, M.N., Bastiaens, L., Verstraete, W., Springael, D., 2005. Influence of the carbon/nitrogen/phosphorus ratio on polycyclic aromatic hydrocarbons degradation by *Mycobacterium* and *Sphingomonas* in soil. Appl. Microbiol. Biotechnol. 66, 726–736.

Lindstrom, J.E., Braddock, J.F., 2002. Biodegradation of petroleum hydrocarbons at low temperature in the presence of the dispersant Corexit 9500. Mar. Pollut. Bull. 44, 739–747.

Litherathy, P., Haider, S., Samhan, O., Morel, G., 1989. Experimental studies on biological and chemical oxidation of dispersed oil in seawater. Water Sci. Technol. 21, 845–856.

Liu, Z., Jacobson, A.M., Luthy, R.G., 1995. Biodegradation of naphthalene in aqueous nonionic surfactant systems. Appl. Environ. Microbiol. 61, 45–151.

Maki, H., Sasaki, T., Haramaya, S., 2005. Photo-oxidation of biodegradable crude oil and toxicity of the photo-oxidized products. Chemosphere 44, 1145–1151.

Maletić, S., Dalmacija, B., Rončević, S., Agbaba, J., Petrović, O., 2009. Degradation kinetics of an aged hydrocarbon-contaminated soil. Water Air Soil Pollut. 202, 149–159.

Margesin, R., Schinner, F., 1997a. Efficiency of indigenous and inoculated cold-adapted soil microorganisms for biodegradation of diesel oil in alpine soils. Appl. Environ. Microbiol. 63, 2660–2664.

Margesin, R., Schinner, F., 1997b. Bioremediation of diesel-oil-contaminated alpine soils at low temperatures. Appl. Microbiol. Biotechnol. 47, 462–468.

Margesin, R., Schinner, F., 1999. Biological decontamination of oil spills in cold environments. J. Chem. Technol. Biotechnol. 74, 381–389.

Margesin, R., Neuner, G., Storey, K.B., 2007b. Cold-loving microbes, plants, and animals—fundamental and applied aspects. Naturwissenschaften 94, 77–99.

Mielke, J.E. 1990. Oil in the ocean: the short- and long-term impacts of a spill. Congressional research service, 90-356 SPR, July 24, p. 11.

Milić, J.S., Beškoski, V.P., Ilić, M.V., Ali, S.A.M., Gojgić-Cvijović, G.Đ., Vrvić, M.M., 2009. Bioremediation of soil heavily contaminated with crude oil and its products: Composition of the microbial consortium. J. Serb. Chem. Soc. 74 (4), 455–460.

Miller, E.C., Miller, J.A., 1981. Search for the ultimate chemical carcinogens and their reaction with cellular macromolecules. Cancer 47, 2327–2345.

Mills, M.A., Bonner, J.S., McDonald, T.J., Page, C.A., Autenrieth, R.L., 2003. Intrinsic bioremediation of a petroleum-impacted wetland. Mar. Pollut. Bull. 46, 887–899.

Mitsch, W.J., Gosselink, J.G., 1993. Wetlands, second ed. John Wiley & Sons Inc, New York.

Mohajeri, L., Aziz, H.A., Isa, M.H., Zahed, M.A., 2010. A statistical experiment design approach for optimizing biodegradation of weathered crude oil in coastal sediments. Bioresour. Technol. 101, 893–900.

Mulkins-Phillips, G.J., Stewart, J.E., 1974. Effect of environmental parameters of bacterial degradation on bunker C oil, crude oil and hydrocarbons. Appl. Environ. Microbiol. 28, 547–552.

Mulyono, M., Jasjfi, E., Maloringan, M., 1994. Oil dispersants: do they do any good? In: Proceedings of the 2nd International Conference on Health, Safety, and Environment in Oil and Gas Exploration and Production, Jakarta, vol. 1, pp. 539–549.

Nasrollahzadeh, H.S., Najafpour, G.D., Aghamohammadi, N., 2007. Biodegradation of Phenanthrene by mixed culture consortia in a batch bioreactor using central composite face-entered design. Int. J. Environ. Res. 1, 80–87.

Nedwell, D.B., 1999. Effect of low temperature on microbial growth: Lowered affinity for substrates limits growth at low temperature. FEMS Microbiol. Ecol. 30, 101–111.

Nicodem, D.E., Guedes, C.L.B., Correa, R.J., 1998. Photochemistry of petroleum. I. Systematic Study of a Brazilian Intermediate Crude Oil. Mar. Chem. 63, 93–104.

Obayori, O.S., Ilori, M.O., Adebusoye, S.A., Oyetibo, G.O., Omotayo, A.E., Amund, O.O., 2009. Degradation of hydrocarbons and biosurfactant production by *Pseudomonas sp.* strain LP1. World J. Microbiol. Biotechnol. 25, 1615–1623.

Obire, O., 1990. Bacterial degradation of three different crude oils in Nigeria. Nig. J. Bot. 1, 81–90.

Obire, O., 1993. The suitability of various Nigerian petroleum fractions as substrate for bacterial growth. Discov. Innov. 9, 25–32.

Obire, O., Nwaubeta, O., 2001. Biodegradation of refined petroleum hydrocarbons in soil. J. Appl. Sci. Environ. Mgt. 5 (1), 43–46.

Obire, O., Okudo, I.V., 1997. Effects of crude oil on a freshwater stream in Nigeria. Discov. Innov. 9, 25–32.

Odu, C.T.I., 1977. Pollution and the environment. Bull. Sci. Assoc. Nigeria 3 (2), 284–285.

Okoh, A.I., 2003. Biodegradation of bonny light crude oil in soil microcosm by some bacterial strains isolated from crude oil flow stations saver pits in Nigeria. Afr. J. Biotechnol. 2 (5), 104–108.

Okoh, A.I., 2006. Biodegradation alternative in the cleanup of petroleum hydrocarbon pollutants. Biotechnol. Mol. Biol. Rev. 1 (2), 38–50.

Okoh, A.I., Trejo-Hernandez, M.R., 2006. Remediation of petroleum hydrocarbon polluted systems: exploiting the bioremediation strategies. Afr. J. Biotechnol. 5, 2520–2525.

Okoh, A.I., Ajisebutu, S., Babalola, G.O., Trejo-Hernandez, M.R., 2002. Biodegradation of Mexican heavy crude oil (Maya) by *Pseudomonas aeruginosa*. J. Trop. Biosci. 2 (1), 12–24.

Oudot, J., Merlin, F.X., Pinvidic, P., 1998. Weathering rates of oil components in a bioremediation experiment in estuarine sediments. Mar. Environ. Res. 45, 113–125.

Page, D.S., Boehm, P.D., Douglas, G.S., Bence, A.E., Burns, W.A., Mankiewic, P.J., 1996. The natural petroleum hydrocarbon background in subtidal sediments of Prince William Sound Alaska, USA. Environ. Toxicol. Chem. 15, 1266–1281.

Parkash, S., 2003. Refining Processes Handbook. Gulf Professional Publishing, Elsevier, Amsterdam, Netherlands.

Pathak, H., Kantharia, D., Malpani, A., Madamwar, D., 2009. Naphthalene biodegradation using Pseudomonas Sp. HOB1: In vitro studies and assessment of naphthalene degradation efficiency in simulated microcosms. J. Hazard. Mater. 166, 1466–1473.

Payne, J.R., McNabb Jr., G.D., 1984. Weathering of Petroleum in the Marine Environment. Mar. Technol. Soc. J. 18 (3), 24.

Pelletier, E., Delille, D., Delille, B., 2004. Crude oil bioremediation in sub-Antarctic intertidal sediments: chemistry and toxicity of oil residues. Mar. Environ. Res. 57, 311–327.

Peters, K.E., Walters, C.C., Moldowan, J.M., 2005. The Biomarker Guide—Edition II. Cambridge University Press, Cambridge, United Kingdom.

Pines, O., Gutnick, D.L., 1981. Relationship between phage resistance and Emulsan production. Interaction of phages with the cell-surface of *Acinctobacter alcoaceticus* RAG-1. Arch. Microbiol. 130, 129–133.

Pines, O., Gutnick, D.L., 1984a. Specific binding of a bacteriophage at a hydrocarbon-water Interface. Appl. Environ. Microbiol. 157, 179–183.

Pines, O., Gutnick, D.L., 1984b. Alternate hydrophobic sites on the cell surface of *Acinetobacter calcoaceticuls* RAG-1. FEMS Microbiol. Lett. 22, 307–311.

Pines, O., Gutnick, D.L., 1986. Role for Emulsan in growth of *Acinetobacter calcoaceticus* RAG-1 on crude oil. Appl. Environ. Microbiol. 51, 661–663.

Pitter, P., Chudoba, J., 1990. Biodegradability of Organic Substances in the Aquatic Environment. CRC Press, Boca Raton, FL7–83.

Poremba, K., Gunkel, W., Lang, S., Wagner, F., 1991. Marine biosurfactants. III. Toxicity testing with marine microorganisms and comparison with synthetic detergents. Zeitsch. Natureforsch. 46c, 210–216.

Prieur, D., Marteinsson, V.T., 1998. Prokaryotes living under elevated hydrostatic pressure. In: Antranikian, G. (Ed.), Biotechnology of Extremophiles. Advances in Biochemical Engineering/BiotechnologyVol. 61. Springer, New York, pp. 23–35.

Prince, R., 1993. Petroleum spill bioremediation in marine environments. Crit. Rev. Microbiol. 19, 217–242.

Rahman, K.S.M., Thahira-Rahman, J., Lakshmanaperumalsamy, P., Banat, I.M., 2002. Towards efficient crude oil degradation by a mixed bacterial consortium. Bioresour. Technol. 85, 257–261.

Rahman, K.S.M., Thahira-Rahman, T., Kourkoutas, Y., Petsas, I., Marchant, R., Banat, I.M., 2003. Enhanced bioremediation of n-alkanes in petroleum sludge using bacterial consortium amended with Rhamnolipid and micronutrients. Bioresour. Technol. 90 (2), 159–168.

Robichauz, T.J., Myrick, H.N., 1972. Chemical enhancement of the biodegradation of crude oil pollutants. J. Petrol. Tech. 24, 16–20.

Rocha, C., Infante, C., 1997. Enhanced oily sludge biodegradation by a tensio-active agent isolated from *Pseudomonas aeruginosa* USBCS1. Appl. Microbiol. Biotechnol. 47, 615–619.

Rosa, A.P., Triguis, J.A., 2007. Bioremediation process on the Brazil shoreline. Environ. Sci. Pollut. Res. 14 (7), 470–476.

Rosenberg, E., 1991. The hydrocarbon-oxidizing bacteria. In: Bolows, A., Truger, H.G., Dworkin, M., Schleifer, K.H., Harder, W. (Eds.), The Prokaryotes. second ed. vol. 1. Springer-Verlag, New York, pp. 446–458.

Rosenberg, E., Perry, A., Gibson, D.T., Gutnick, D.L., 1979a. Emulsifier of Arthrobacter RAG-1: Specificity of hydrocarbon substrate. Appl. Environ. Microbiol. 37, 409–413.

Rosenberg, E., Zuckerberg, A., Rubinovitz, C., Gutnick, D.L., 1979b. Emulsifier of Arthrobacter RAG-1: isolation and emulsifying properties. Appl. Environ. Microbiol. 37, 402–408.

Rosenberg, E., Kaplan, N., Pines, O., Rosenberg, M., Gutnick, D.L., 1983a. Capsular polysaccharides interfere with adherence of *Acinetobacter calcoaceticus* to hydrocarbon. FEMS Microbiol. Lett. 17, 157–160.

Rosenberg, E., Gottlieb, A., Rosenberg, M., 1983b. Inhibition of bacterial adherence to hydrocarbons and epithelial cells by Emulsan. Infect. Immun. 39, 1024–1028.

Sabate, J., Vinas, M., Solanas, A.M., 2004. Laboratory-scale bioremediation experiments on hydrocarbon-contaminated soils. Int. Biodeter. Biodegr. 54, 19–25.

Salam, L.B., Obayori, O.S., Akashoro, O.S., Okogie, G.O., 2011. Biodegradation of bonny light crude oil by Bacteria isolated from contaminated soil. Int. J. Agric. Biol. 13 (2), 245–250.

Sandrin, T.R., Kight, W.B., Maier, W.J., Maier, R.M., 2006. Influence of a non-aqueous phase liquid (NAPL) on biodegradation of Phenanthrene. Biodegradation 17, 423–435.

Schulz, D., Passeri, A., Schmidt, M., Lang, S., Wagner, F., Wray, V., Gunkel, W., 1991. Marine biosurfactants, I. Screening for biosurfactants among crude oil degrading marine microorganisms from the North Sea. Z. Naturforsch. 46c, 197–203.

Shabtai, Y., Pines, O., Gutnick, D.L., 1986. Emulsan: a case study of microbial capsules as industrial products. Dev. Ind. Microbiol. 26, 291–307.

Siron, R., Pelletier, E., Delille, D., Roy, S., 1993. Fate and effects of dispersed crude oil under icy conditions simulated in Mesocosms. Mar. Environ. Res. 35, 273–302.

Sivaraman, C., Ganguly, A., Mutnuri, S., 2010. Biodegradation of hydrocarbons in the presence of Cyclodextrins. World J. Microbiol. Biotechnol. 26, 227–232.

Smith, J., 1968. Torrey Canyon—Pollution and Marine Life. In: Report by the Plymouth Laboratory of the Marine Biological Association of the United Kingdom, London. Cambridge University Press, London, United Kingdom, pp. 1906.

Song, H., Wang, X., Bartha, R., 1990. Appl. Environ. Microbiol. 56, 652.

Speight, J.G., 1996. Environmental Technology Handbook. Taylor & Francis, Washington, DC.

Speight, J.G., 2005. Environmental Analysis and Technology for the Refining Industry. John Wiley & Sons Inc, Hoboken, New Jersey.

Speight, J.G., 2014. The Chemistry and Technology of Petroleum, fifth ed. CRC Press, Taylor & Francis Group, Boca Raton, FL.

Speight, J.G., 2017. Handbook of Petroleum Refining. CRC Press, Taylor & Francis Group, Boca Raton, FL.

Speight, J.G., Lee, S., 2000. Environmental Technology Handbook, second ed. Taylor & Francis, New York.

Steffan, R.J., McCloy, K., Vainberg, S., Condee, C.W., Zhang, D., 1997. Biodegradation of the gasoline oxygenates methyl tert-butyl ether, ethyl tert-butyl ether and tert—Amyl methyl ether by propane-oxidizing Bacteria. Appl. Environ. Microbiol. 63 (11), 4216–4222.

Swannell, R.P.J., Lee, K., McDonagh, M., 1996. Field evaluation of marine oil spill bioremediation. Microbiol. Rev. 342–365. June.

Taghvaei Ganjali, S., Nahri Niknafs, B., Khosravi, M., et al., 2007. Photo-oxidation of crude petroleum Maltenic fraction I natural simulated conditions and structural elucidation of photoproducts. Iran J. Environ. Health Sci. Eng. 4 (1), 37–42.

Thouand, G., Bauda, P., Oudot, J., Kirsch, G., Sutton, C., Vidalie, J.F., 1999. Laboratory evaluation of crude oil biodegradation with commercial or natural microbial Inocula. Can. J. Microbiol. 45 (2), 106–115.

Traxler, R.W., Bhattacharya, L.S., 1979. Effect of a chemical dispersant on microbial utilization of petroleum hydrocarbons. In: McCarthy, L.T., Lindblom, G.P., Walter, H.F. (Eds.), Chemical Dispersants for the Control of Oil Spills. American Society for Testing and Materials, West Conshohocken, Pennsylvania.

Varadaraj, R., Robbins, M.L., Bock, J., Pace, S., MacDonald, D., 1995. Dispersion and biodegradation of oil spills on water. In: Proceedings of the 1995 Oil Spill Conference. American Petroleum Institute, Washington, DC, pp. 101–106.

Venosa, A.D., Zhu, X., 2003. Biodegradation of crude oil contaminating marine shorelines and freshwater wetlands. Spill Sci. Tech. Bull. 8 (2), 163–178.

Vidali, M., 2001. Bioremediation: an overview. Pure Appl. Chem. 73, 1163–1172.

Vieira, P.A., Faria, S.R., Vieira, B., De Franca, F.P., Cardoso, V.L., 2009. Statistical analysis and optimization of nitrogen, phosphorus, and inoculum concentrations for the biodegradation of petroleum hydrocarbons by response surface methodology. J. Microbiol. Biotechnol. 25, 427–438.

Wong, J.W.C., Lai, K.M., Wan, C.K., Ma, K.K., Fang, M., 2001. Isolation and optimization of PAHs-degradative Bacteria from contaminated soil for PAHs bioremediation. Water Air Soil Pollut. 13, 1–13.

Wright, A.L., Weaver, R.W., Webb, J.W., 1997. Oil bioremediation in salt marsh Mesocosms as influenced by N and P fertilization, flooding and season. Water Air Soil Pollut. 95, 179–191.

Zahed, M.A., Aziz, H.A., Isa, M.H., Mohajeri, L., 2010. Enhancement of biodegradation of n-alkanes from crude oil contaminated seawater. Int. J. Environ. Res. 4 (4), 655–664.

Zaidi, B.R., Imam, S.H., 1999. Factors affecting microbial degradation of polycyclic aromatic hydrocarbon Phenanthrene in Caribbean coastal water. Mar. Pollut. Bull. 38, 738–749.

Zengler, K., Richnow, H.H., Rossello-Mora, R., Michaelis, W., Widdel, F., 1999. Methane formation from long-chain alkanes by anaerobic microorganisms. Nature 401, 266–269.

Zhu, X., Venosa, A.D., Suidan, M.T., Lee, K., 2001. Guidelines for the Bioremediation of Marine Shorelines and Freshwater Wetlands. US EPA Report, (September).

Zosim, Z., Gutnick, D.L., Rosenberg, E., 1982. Properties of hydrocarbon-in-water emulsions stabilized by *Acinetobacter* RAG-1 Emulsan. Biotechnol. Bioeng. 24, 281–292.

Zosim, Z., Rosenberg, E., Gutnick, D.L., 1986. Changes in hydrocarbon emulsification specificity of the polymeric bioemulsifier Emulsan: Effects of alkanols. Colloid Polym. Sci. 264, 218–223.

Zuckerberg, A., Diver, A., Peeri, Z., Gutnick, D.L., Rosenberg, E., 1979. Emulsifier of Arthrobacter RAG-1: Chemical and physical properties. Appl. Environ. Microbiol. 37, 414–420.

Further reading

Aislabie, J.M., McLeod, M., Fraser, R., 1998. Potential for biodegradation of hydrocarbons in soil from the Ross dependency, Antarctica. Appl. Microbiol. Biotechnol. 49, 210–214.

Aislabie, J.M., Foght, J., Saul, D., 2000. Aromatic hydrocarbon-degrading Bacteria from soil near Scott base, Antarctica. Polar Biol. 23, 183–188.

Aislabie, J.M., Balks, M.R., Foght, J.M., Waterhouse, E.J., 2004. Hydrocarbon spills on Antarctic soils: Effects and management. Environ. Sci. Technol. 38 (5), 1265–1274.

Atlas, R.M., 1995. Petroleum biodegradation and oil spill bioremediation. Mar. Pollut. Bull. 31, 178–182.

Braddock, J.F., Ruth, M.L., Walworth, J.L., McCarthy, K.A., 1997. Enhancement and inhibition of microbial activity in hydrocarbon-contaminated Arctic soils: Implications for Utrientamended bioremediation. Environ. Sci. Technol. 31, 2078–2084.
Delille, D., 2000. Response of Antarctic soil assemblages to contamination by diesel fuel and crude oil. Microb. Ecol. 40, 159–168.
Delille, D., Delille, B., 2000. Field observations on the variability of crude oil impact in indigenous hydrocarbon-degrading Bacteria from sub-Antarctic intertidal sediments. Mar. Environ. Res. 49, 403–417.
Delille, D., Vaillant, N., 1990. The influence of crude oil on the growth of sub-Antarctic marine Bacteria. Antarct. Sci. 2, 655–662.
Delille, D., Coulon, F., Pelletier, E., 2004. Effects of temperature warming during a bioremediation study of natural and nutrient-amended hydrocarbon-contaminated sub-Antarctic soils. Cold Reg. Sci. Technol. 40, 61–70.
Kennicutt, M.C., 1990. Oil spill in Antarctica. Environ. Sci. Technol. 24, 620–624.
Margesin, R., 2000. Potential of cold-adapted microorganisms for bioremediation of oil-polluted alpine soils. Int. Biodeter. Biodegr. 46, 3–10.
Margesin, R., Schinner, F. (Eds.), 1999a. Cold-Adapted Organisms. Springer, New York.
Margesin, R., Schinner, F., 2001a. Biodegradation and bioremediation of hydrocarbons in extreme environments. Appl. Environ. Microbiol. 56, 650–663.
Margesin, R., Schinner, F., 2001b. Potential of halotolerant and halophilic microorganisms for biotechnology. Extremophiles 5, 73–83.
Margesin, R., Hämmerle, M., Tscherko, D., 2007a. Microbial activity and community composition during bioremediation of diesel-oil contaminated soil: Effects of hydrocarbon concentration, fertilizers and incubation time. Microb. Ecol. 53, 259–269.
Means, J.C., 1995. Influence of salinity upon sediment-water partitioning of aromatic hydrocarbons. Mar. Chem. 51, 3–16.
Michaud, L., Giudice, A.L., Saitta, M., De Domenico, M., Bruni, V., 2004. The biodegradation efficiency on diesel oil by two Psychrotrophic Antarctic marine Bacteria during a two-month-long experiment. Mar. Pollut. Bull. 49, 405–409.
Michelsen, T.C., Petito Boyce, C., 1993. Cleanup standards for petroleum hydrocarbons. Part 1. Review of methods and recent developments. J. Soil Contamin. 2 (2), 1–16.
Okoh, A.I., Ajisebutu, S., Babalola, G.O., Trejo-Hernandez, M.R., 2001. Potentials of *Burkholderia cepacia* strain RQ1 in the biodegradation of heavy crude oil. Int. Microbiol. 4, 83–87.
Peters, K.E., Scheuerman, G.L., Lee, C.Y., Moldovan, J.M., Reynolds, R.N., 1992. Effects of refinery processes on biological markers. Energy Fuel 6, 560–577.
Reynolds, C.M., Bhunia, P., Koenen, B.A., 1997. Soil Remediation Demonstration Project: Biodegradation of Heavy Fuel Oils. Special Report 97-20, Cold Regions Research & Engineering Laboratory. US Army Corps of Engineers, Fort Belvoir, Virginia.
Siron, R., Pelletier, E., Brochu, C., 1995. Environmental factors influencing the biodegradation of petroleum hydrocarbons in cold seawater. Arch. Environ. Contam. Toxicol. 28, 406–416.
Soli, G., Bens, E.M., 1972. Bacteria which attack petroleum hydrocarbons in a saline medium. Biotechnol. Bioeng. 14, 319–330.
Swannell, R.P.J., Lepo, J.E., Lee, K., Pritchard, P.H., Jones, D.M., 1995. Bioremediation of oil-contaminated fine-grained sediments in laboratory microcosms. In: Proceedings of the 2nd International Oil Spill Research and Development Forum, 23–26 May 1995. International Maritime Organisation, London, United Kingdom, pp. 45–55.
Venosa, A.D., Suidan, M.T., Wrenn, B.A., Strohmeier, K.L., Haines, J.R., Eberhart, B.L., King, D., Holder, E., 1996. Bioremediation of an experimental oil spill on the shoreline of Delaware Bay. Environ. Sci. Technol. 30 (5), 1764.

Wagner-Döbler, I., Bennasar, A., Vancanneyt, M., Strömpl, C., Brümmer, I., Eichner, C., Grammel, I., Moore, E.R.B., 1998. Microcosm enrichment of biphenyl-degrading microbial communities from soils and sediments. Appl. Environ. Microbiol. 64, 3014–3022.

Zhu, X., Venosa, A.D., Suidan, M.T., 2004. Literature Review on the Use of Commercial Bioremediation Agents for Cleanup of Oil-Contaminated Environments. Report No. EPA/600/R-04/075, National Risk Management Research Laboratory, United States Environmental Protection Agency, Cincinnati, OH. July.

Water and hydraulic fracturing

7

Chapter outline

1 Introduction

The environmental impact of the hydraulic fracturing process is related to land use and water consumption, air emissions, including methane emissions, brine and fracturing fluid leakage, water contamination, noise pollution, and health. Water and air pollution are the biggest risks to human health from hydraulic fracturing. Adherence to regulations and safety procedures is required to avoid negative impacts of the process on the environment, more specifically to the current context, on water systems. But, first and foremost and to allay any confusion, hydraulic fracturing (informally known as hydrofracking, fracking, fracing, or hydrofracturing) is not a process for drilling or constructing a well.

Hydraulic fracturing is a process that typically involves injecting water, sand, and (or) chemicals under high pressure into a bedrock formation via a well. This process is intended to create new fractures in the rock as well as increase the size, extent, and connectivity of existing fractures. Hydraulic fracturing is a well-stimulation technique used commonly in low-permeability rocks like tight sandstone, shale, and some coal beds to increase oil and/or gas flow to a well from crude oil-bearing rock formations. A form of hydraulic fracturing is also used in low permeability sediments and other tight subsurface formations to increase the efficiency of soil vapor extraction and other technologies used in remediating contaminated sites. To fracture a formation, energy must be generated by injecting a fluid down a well and into the formation.

Hydraulic fracturing and other well stimulation methods have led to a rapid expansion of the development of crude oil and natural gas in unconventional reservoirs (tight reservoirs and shale reservoirs) on a worldwide basis. This expansion of efforts has brought crude oil and natural gas development to the point where there is the potential

Natural Water Remediation. https://doi.org/10.1016/B978-0-12-803810-9.00007-3

for human exposure to new contaminants and environmental damage with public and scientific emphasis on the effects of hydraulic fracturing in particular.

In the present context in terms of the effect of the process on water availability and water quality, there has been a great deal of emotional statements expressed and used to criticize hydraulic fracturing while investigation of many of the issues have been hampered by the lack of meaningful scientific data on the potential cumulative risks posed by the combined emissions from a dense network of wells and associated infrastructure such as pipelines, compressor stations, and roads (Holloway and Rudd, 2013; Spellman, 2013; Uddameri et al., 2016). Furthermore, in many cases, State regulations are (in many cases) minimal and enforcement of any meaningful often cannot keep up with the rapid expansion of the unconventional crude oil and natural gas industry, resulting in insufficient protection from pollutants.

Briefly, and by way of explanation, hydraulic fracturing involves the pressurized injection of hydraulic fracturing fluids into geologic formations, causing fractures in the formations and enabling the release of crude oil and/or natural gas (Holloway and Rudd, 2013; Spellman, 2013; Uddameri et al., 2016). This technique has been employed in the United States since 1947 but as new technology for drilling horizontal wells has evolved and been deployed over the past two-to-three decades, the technique has been used to produce large quantities of natural gas from shale formations. It has been suggested that crude oil and natural gas from such sources is moving the United States to becoming a net energy exporter thereby providing the country with a desirable and measurable degree of self-sufficiency in energy production. As a result, hydraulic fracturing thus will play a central role in the future domestic energy policy of the United States and has removed the country from the immediate threat of petro-politics in which the country is subject to the stability (but, predominantly, the instability) of the governments of many oil-producing nations (Speight, 2011).

Hydraulic fracturing technology is responsible for the current success in the production of crude oil and natural gas from shale plays and from tight formations. The technology concept is to enhance natural fractures or create induced fractures by injecting fluids at pressures greater than the strength of the rock (Speight, 2014a, 2016). Addition of sand or other material (proppants) to the fluid is needed to keep the induced fractures open once the fluid has been removed and the pressure has subsided. Once the formation is fractured, the pressure exerted by the hydraulic fracturing fluid is reduced, which reverses the direction of the fluid flow in the well back toward the ground surface. Thus, the hydraulic fracturing fluid and any naturally-occurring substances released from the formation are allowed to flow to the surface (*flowback*).

In this process, most of the wells are in a horizontal configuration with one or more horizontal fractures extending into the target section of the formation—these fractures may extend more than a mile from the surface location of the well. Horizontal wells are more expensive to drill and develop but have better performance in terms of the production volumes of crude oil and natural gas because more of the reservoir is accessible thereby leading to increased production of the crude oil and natural gas.

In terms of the fracturing fluid, water has been the fluid of choice but a primary concern is the large amount of fresh water needed for the drilling—the water is either (i) produced from water sources in the area or (ii) brought to the drilling/production site by truck from local reservoirs. Preferentially, saltwater is not used as the fracturing fluid saltwater significantly increases the potential for corrosion and scale deposition in the formation, tubing, casing, and surface equipment, therefore inhibiting the production of crude oil and natural gas. Saltwater also significantly increases the potential for corrosion on the tubing, casing, and surface equipment, potentially shortening the life of a well (Speight, 2014b). In addition, chemicals needed to perform efficiently and effectively in a hydraulic fracturing project re not as effective or efficient in saltwater as in fresh water.

The hydraulic fracturing fluids also may contain chemicals that have become the subject of some public concern with respect to potential contamination of underground sources of drinking water. While there is a push from some members of the community to make the chemical composition of hydraulic fracturing fluids a matter of public record, many of the mixtures are considered to be proprietary and current law related to proprietary materials supports the maintenance of confidentiality with respect to the composition of those fluids. It should be noted that regardless of the chemical composition of a particular hydraulic fracturing fluid, the application of proper well design, completion, operations and monitoring according to rules and regulations that already exist in most states and provinces will ensure that fracturing operations do not negatively impact either the subsurface or surface environment.

Thus, as part of the hydraulic fracturing process, it is important to ensure that the induced fractures are contained within the target formation but this must be balanced against the frequent need to fracture as much of the reservoir as possible. Production of crude oil and natural gas is dependent upon the area of the reservoir that is fractured and the length of the fractures that are created. If the fractures penetrate a water-saturated formation, excess water is drawn into the hale formation that contains the crude oil or the natural gas shale which can potentially doom a well. The solution to undesirable penetrating fractures is to carefully plan the project using a multi-disciplinary teams of scientist and engineers—the plan should represent carefully-sized fracture project that is sufficiently focused on the locale and specify the use horizontal wells and multiple, simultaneous, and in the same well bore. However, of all the issues related to hydraulic fracturing, the possible effects on groundwater are without a doubt the most contentious. Numerous allegations have been made related to hydraulic fracturing, with particular emphasis on the impacts on water sources.

Finally, deductions made about the environmental effects of hydraulic fracturing must be made on the basis of scientific facts and not on emotion. In addition, critical evaluations of shale gas hydraulic fracturing and the potential impacts on the environment must be based on peer-reviewed, scientific analyses of quantitative data and not on the preferences of the funding organization. Agencies responsible for regulating or monitoring the environmental impacts of shale gas development need to be well-informed and the design of any national regulatory framework to protect the environment from hydraulic fracturing operations should start realistic directives and recommendations from factual data.

It is not the purpose of this chapter to present concerns that relate to policy, economics, and social areas that are outside the scope of this text but to present information to those persons who are interested in hydraulic fracturing (for whatever reason) in order to consider and answer some of the issues—either contentious issues or non-contentious issues. Neither it is the purpose of this chapter to wither extoll the virtues or condemn the hydraulic fracturing process but to present the potential advantages and disadvantages of the process in order that hydraulic fracturing may become a safe and environmentally-benign process as possible. Thus, the issues that may be associated with tight oil exploration and production are described below. Also described below are approaches that are either currently being employed or may be employed to mitigate the environmental issues associated with tight oil production.

It is the purpose of this chapter to present the effects of the process on water systems. Thus, the issues that may be associated with tight oil exploration and production are described below. Also described below are approaches that are either currently being employed or may be employed to mitigate the environmental issues associated with tight oil production.

2 Formation evaluation

The geologic characteristics of the formations in which crude oil and natural gas are found vary widely, along with the characteristics of the crude oil and the gas hydrocarbon derivatives themselves (Speight, 2014a). For example, in the context of this book, crude oil and natural gas formations are generally much less permeable than other formations causing the crude oil and natural gas to be much less free flowing which, along with a cap rock and a basement rock, that prevents spillage or leakage from the formation. In addition, heavy oil, due to its higher viscosity, has much less ability to flow freely through a formation compared to lighter oil. Tar sand bitumen, which is immobile in the deposit, has no ability to flow freely through the deposit (Speight, 2009, 2014a).

Some of the key geological issues with relevance to the potential environmental impacts of hydraulic fracturing are: (i) the potentially lack of understanding of rock fracture patterns and processes that occur in shale formations and in tight formations, (ii) the ability to predict and quantify permeable fracture networks in the subsurface before drilling, and (iii) the accuracy and precision with which the geometry (length, breadth, thickness, and position) of shale formations, tight formations, and aquifers in the subsurface can be determined, especially in areas with complex geological histories; this also include the relative position of aquifers to the aforementioned formations. Furthermore, the pore space in the various rock formations is made up of a variety of void spaces in the solid rock matrix which also and includes natural stress-cracks fractures. Thus, the aim of hydraulic fracturing is to improve permeability (fluid flow in the rock) by reopening natural fractures as well as creating new fractures to form a locally dense network of open and connected fractures without the potential damage to the environment (Holloway and Rudd, 2013; Spellman, 2013; Speight, 2014a, 2016; Uddameri et al., 2016).

In the present context, wells typically a mile or more below ground surface, often passing through groundwater aquifers to reach the crude oil and natural gas formations after which horizontal wells are then drilled into the formation to access the crude oil and natural gas. The well casings and cemented areas are designed to prevent contamination of any groundwater aquifers through which the well may be drilled. If the well casing remains intact the potential for the hydraulic fracturing process to pose any risk to underground water aquifers is minimized (even negated).

However, hydraulic fracturing fluid flow into fractures can be significant and, in addition, when flowback water is returned to the surface, hydraulic fracturing fluids and extracted naturally-occurring substances could potentially find a pathway into other geological features through natural faults or artificial penetrations, such as other wells, underground mine workings). Groundwater resources can be protected through careful well design, construction, operation and maintenance which are regulated by the states. In fact, many states require periodic well testing for well integrity and the regulations are (or should be) compiled to ensure that wells are constructed to prevent the migration of fracturing fluids into underground sources of drinking water as well as to ensure wells are built to prevent blowouts and prevent escape of the fracturing fluids.

3 The fracturing process

Hydraulic fracturing was developed in the 1940s as a process for the efficient extraction of crude oil and natural gas from small reservoirs. In the early 2000s, innovations in horizontal drilling through deep underground, fractured shale formations made fracking more economical, resulting in rapid growth of the technique throughout the world. The primary purpose of hydraulic fracturing is to increase the rate at which natural substances such as crude oil and natural gas can be recovered from subterranean low-permeability reservoirs.

The process is used to stimulate groundwater wells and also to measure stresses in the earth. Many companies also use hydraulic fracturing in dealing with wastewater such as hydrocarbon wastes by injecting them deep underground into the earth. Some other more unconventional uses of hydraulic fracturing may include electricity generation, induction of rock cave-ins for mining, and geologic sequestration of carbon dioxide. More specifically, the process allows access to narrow rock formations by drilling underground a mile or more vertically and then several thousand feet horizontally. The process requires the use of millions of gallons of water mixed with sand and chemicals that are then pumped into the drilling region, filling small fissures and releasing crude oil and/or the natural gas into the well borehole. Furthermore, hydraulics fracturing has brought about a significant increase in the production of crude oil and natural gas.

The hydraulic fracturing process involves pumping fluid through the perforations. The fracturing fluid itself exerts pressure against the rock, creating tiny cracks, or fractures, in the reservoir deep underground. The fluid is predominantly water, proppant (grains of sand or ceramic particles) and a small fraction of chemical additives. Once fluid injection stops, pressure begins to dissipate, and the fractures previously held open by the fluid pressure begin to close. Proppants then act as tiny wedges to hold open these narrow fractures, creating pathways for oil, natural gas and fracturing fluids to flow more easily to the well. A plug is set inside the casing to isolate the stimulated section of the well. The entire perforate-inject-plug cycle is then repeated at regular intervals along the targeted section of the reservoir. Finally, the plugs are drilled out, allowing the oil, natural gas and fluids to flow into the well casing and up to the surface.

However, before hydraulic fracturing of the reservoir is put into practice (or becomes a rarity), consideration of the toxicity level of to-be-injected chemicals (to-be-injected-additives) additives used in the hydraulic fracturing phase is necessity. In fact, defining the toxicity of such chemicals should be a relatively simple and quantifiable scientific task using the relevant MSDS (material safety data sheets) information.

As with many industrial operations and mining, the hydraulic fracturing process utilizes a range of potentially hazardous chemicals that must be carefully managed to prevent leaks and spills. Of additional concern is the management of the liquid, post-hydraulic fracturing fracking waste product. The hydraulic fracturing fluids may include hydrocarbon derivatives, heavy metal either as compounds or as the metal, and salts, in addition to indigenous earth contaminants, such as naturally occurring radioactive materials. Disposal of the wastewater by re-injection into the fissures or collection and secure storage of the wastewater (either onsite or by transport to a treatment facility) are common practices. The wastewater can also be recycled and used at the next well. While there is some concern about the possibility of leaks of the wastewater from storage sites, the same risk exists with any industrial waste production. With proper containment, storage of wastewater and other wastes from hydraulic fracturing projects should be no more of a problem than storage of other hazardous wastes.

3.1 Methods

Initial drilling is the same as for a conventional reservoir. A borehole is drilled vertically, then a casing is placed before cement and mud is pumped to place a barrier between the borehole and adjacent formation. Drilling of the well is now continued, to an adequate depth within the producing reservoir (the *kick-off point*) then the well bore is deviated gradually until it curves horizontally and drilled a distance of typically 1000 to more than 5000 ft (Arthur et al., 2009).

The hydraulic fracturing treatment follows the actual drilling and completion of the well (Arthur et al., 2009). The process is applied after well completion to facilitate movement of the reservoir fluids to the well and thence to the surface. The composition of these fluids varies according to the formation and ranges from a simple mixture of water and sand to more complex mixtures with a multitude of chemical additives (Speight, 2016). The sequence of fracturing a particular formation typically consists of: (i) an acid stage, (ii) a pad stage, (iii) a prop sequence stage, and (iv) a flushing stage.

The *acid stage* consists of several thousand gallons of water mixed with a dilute acid, such as hydrochloric or muriatic acid, which serves to clear cement debris in the wellbore and provide an open conduit for other fracturing fluids, by dissolving carbonate minerals and opening fractures near the wellbore. The *pad stage* consists of approximately 100,000 gal of slickwater without proppant material; the slickwater pad stage fills the wellbore with the slickwater solution (described below), opens the formation and helps to facilitate the flow and placement of proppant material. The *prop sequence stage* may consist of several sub-stages of water combined with proppant material, which consists of a fine mesh sand or ceramic material, intended to keep open (prop) the fractures created and/or enhanced during the fracing operation after the pressure is reduced. This stage may collectively use several hundred thousand gallons of water. The proppant material may vary from a finer particle size to a coarser particle size throughout this sequence. The *flushing stage* consists of a volume of fresh water sufficient to flush the excess proppant from the wellbore.

Once the well has been completed, it is ready for hydraulic fracturing process which involves pumping fluid through the perforations. The fracturing fluid itself exerts pressure against the rock, creating tiny cracks, or fractures, in the reservoir deep underground. The fluid is predominantly water, proppant (grains of sand or ceramic particles) and a small fraction of chemical additives. Once fluid injection stops, pressure begins to dissipate, and the fractures previously held open by the fluid pressure begin to close. Proppants then act as wedges to hold open these narrow fractures, creating pathways for oil, natural gas and fracturing fluids to flow more easily to the well. A plug is set inside the casing to isolate the stimulated section of the well. The entire perforate-inject-plug cycle is then repeated at regular intervals along the targeted section of the reservoir. Finally, the plugs are drilled out, allowing the crude oil, natural gas, and fluids to flow into the well casing and up to the surface.

The predominant fluid, or the predominant constituent of the fracturing fluid, is water—chemicals (typically 1% *v*/v) the fracturing fluid are added to keep the pipes cool by reducing friction and to prevent scale build-up and bacterial growth. The formulas for the fracturing fluid will vary, depending on the characteristics and structure of gas field or oil field. However, some of the chemical additives can be hazardous if not handled carefully. Safe handling of all water and fluids on site, including chemicals used for hydraulic fracturing, must be a high priority and compliance with all regulations regarding containment, transport and spill handling is essential.

When it comes to disposal of the fracturing fluid, there are options. The fluid, when environmentally possible, can be reused for additional wells in a single field, which reduces the overall use of fresh water and reduces the amount of recovered water and chemicals that must be sent for disposal. In addition, tanks (or lined storage pits) for the storage of recovered water are also a necessity until the water can be sent for disposal in a permitted disposal well or taken to a treatment plant for processing.

Thus, fractures from both horizontal and vertical wells can propagate vertically out of the intended zone thereby (i) reducing stimulation effectiveness, (ii) wasting proppant and fluids, and potentially connecting up with other hydraulic fracturing stages or unwanted water or gas intervals which can also lead to a variety of environmental issues. The direction of lateral propagation is largely dictated by the horizontal stress regime, but in areas where there is low horizontal stress anisotropy or in reservoirs that are naturally fractured, fracture growth is not always easy to predict. In shallow zones, horizontal hydraulic fractures can develop because the vertical stress component—the weight of the overburden—is smallest. A horizontal hydraulic fracture reduces the effectiveness of the stimulation treatment because it most likely forms along horizontal planes of weakness—such as the planes between formation strata—and is aligned preferentially to formation vertical permeability, which is typically much lower than horizontal permeability.

The fracturing fluid is pumped through the perforated intervals at high pressures in order to create fractures in the surrounding formation (pay zone). Hard particles (proppants) are added to the fracturing fluid and pumped into the formation after the fractures have been created. The proppant size and concentration is increased in stages over the entire course of one treatment. The propping agents hold open the newly created fractures, to facilitate hydrocarbon recovery. The design of fracture treatment is a complex task, which involves analysis, planning, experience and rigorous observation of different stages in the entire process (Cipolla and Wright, 2002).

Once fluid injection stops, pressure begins to dissipate, and the fractures previously held open by the fluid pressure begin to close. Proppants then act as tiny wedges to hold open these narrow fractures, creating pathways for oil, natural gas and fracturing fluids to flow more easily to the well. A plug is set inside the casing to isolate the stimulated section of the well. The entire perforate-inject-plug cycle is then repeated at regular intervals along the targeted section of the reservoir. Finally, the plugs are drilled out, allowing the oil, natural gas and fluids to flow into the well casing and up to the surface.

The crude oil/natural gas/fracturing fluid mixture is separated at the surface, and the fracturing fluid (also known as flowback water) is captured in tanks or lined pits. The fracturing fluids are then disposed of according to environmental regulations and approved methods. Hydraulic fracturing operations generally occur over a three-to-five day period. The entire well construction process (including hydraulic fracturing) takes only two-to-three months, compared to the 20-to-30 year productive life of a typical well.

The extent of a hydraulic fracture is a complex relationship between the strength of the rock and the pressure difference between the rock and the fracturing pressure. The extent is defined by the fracture dimensions—height, depth of penetration (wing length or fracture length), and aperture (width or opening). One measure of the strength of the rock is the Poisson Ratio. Thus, when a material is compressed in one direction, it usually tends to expand in the other two directions perpendicular to the direction of compression.

3.2 Well types

Success or failure of a hydraulic fracture treatment often depends on the quality of the candidate well selected for the treatment. Choosing an excellent candidate for stimulation often ensures success, while choosing a poor candidate normally results in economic failure. Horizontal drilling makes it possible for a well to be drilled vertically several thousand feet or meters and then curved to extend at an angle parallel to the surface of the Earth, threading the well through the horizontal gas formation to capture more pockets of gas. On the other hand, in some geological settings, it is more appropriate to directionally drill *s-shaped wells* from a single pad to minimize surface disturbance. These types of wells are drilled vertically several thousand feet and then extend in arc-shapes beneath the surface of the Earth. Whatever the type of well, multiple wells can be drilled from a central location to proceed in different directions within the reservoir.

All injection wells must be designed to meet the regulations set by the relevant regulatory agency to protect groundwater. In addition, production zones should have that have multiple confining layers above the zone to keep the injected fluids within the target formation. In addition, multiple layers of well casing and cement (similar to production wells) should be used with periodic mechanical integrity tests to verify that the casing and cement are holding the liquids. The amount and pressure of the injected fluid (specified in each well permit) should be monitored to maintain the fluids in the target zone and the pressure in the injection well and the spaces between the casing layers (also called the annuluses) should also be monitored check and verify the integrity of the injection well.

3.3 Fracturing fluids

Hydraulic fracturing is carried out using two broad classes of fracturing materials: fracturing fluid and proppants. The term *fracturing fluid* is a generic term that includes both the base fluid and additives (Tables 7.1 and 7.2). The additives are a wide range of chemicals that are used to influence the overall properties of the fracturing fluid. Since the default position of the fractures is the closed position, *propping agents* (*proppants*) are used to stop the fracture from closing after the fracture treatment to enable recovery of crude oil and natural gas—the most common proppant is fine sand.

In the process, the fracturing fluids are injected into the subsurface at a rate and pressure that are too high for the targeted formation to accommodate and, as the resistance to the injected fluids increases, the pressure in the injecting well increases to a level that exceeds the breakdown pressure of the rocks in the targeted formation. In this way, the hydraulic fracturing process fractures the targeted formation and, on occasion, other geologic strata within or around the targeted formation. This process sometimes does create new fractures, most often the process enlarges existing fractures thereby increasing the connections of the natural fracture networks in the targeted formation. The pressure-induced fracturing serves to connect the network of fractures in the formation to the hydraulic fracturing well (which subsequently will serve as the crude oil and/or natural gas production well). The fracturing fluids pumped into the subsurface under high pressure also deliver and emplace the proppant which, under pressure, is forced into the natural and/or enlarged fractures and acts to prop open the fractures even after the fracturing pressure is reduced. The increased permeability due to fracturing and proppant emplacement facilitates the flow and extraction of crude oil and natural gas from the fractured formation.

During hydraulic fracturing, the fluids that re-injected into the well and thence into the formation contain any one (usually several) of a variety of chemicals and once deep underground, the migration of the injected chemicals is not entirely predictable no matter what theoretical models indicate. Well failure, such as the use of insufficient well casing, could lead to the release of the chemicals at shallower depths that release the chemicals dangerously close to drinking water supplies. In addition, while some

Table 7.1 Examples of chemicals used in hydraulic fracturing fluids[a]

Chemical	Use
Acetic Acid	pH buffer
Acrylic copolymer	Lubricant
Ammonium persulfate	Breaker used to reduce viscosity
Boric Acid	Cross-linking agent to increase viscosity
Boric Oxide	Cross-linking agent to increase viscosity
2-Butoxyethanol	Reduction of surface tension to aid gas flow
Carbonic acid	Cross-linking agent to increase viscosity
Carboxy-Methyl Hydroxy-Propyl Guar	Gelling agent (thickens fluid)
Crystalline silica (cristobalite)	Proppant (holds open fractures)
Crystalline silica (quartz)	Proppant (holds open fractures)
Citric Acid	Iron control or for cleaning well bores
Diammonium Peroxidisulfate	Breaker used to reduce viscosity
Disodium Octaborate Tetrahydrate	Gelling agent to increase viscosity
Gas oils (crude oil), hydrotreated	Guar liquefier
Fumaric acid	pH buffer
Gelatin	Corrosion inhibitor or gelling agent
Guar Gum	Gelling agent
Hemicellulase Enzyme	Breaker used to reduce viscosity
Hydrochloric Acid	Cleaning of the wellbore prior to fraccing
Hydroxy-Ethyl Cellulose	Gelling agent
Hydroxy-Propyl Guar	Gelling agent
Magnesium silicate hydrate	Gelling agent
Methanol	Gelling agent
Mono ethanol amine	Reduction of surface tension to aid gas flow
Ethylene Glycol Monobutyl Ether	Gelling agent
Muriatic Acid	Mutual solvent
Non-crystalline silica	Proppant
Poly (oxy-1,2-ethanediyl)	Proppant
Polydimethyldiallylammonium chloride	Clay control
Potassium Carbonate	pH buffer
Potassium Chloride	Clay inhibitor
1-Propanol	Complexing agent
Quaternary Polyamines	Clay control
Sodium acetate	pH buffer
Sodium borate	pH buffer
Sodium Bicarbonate	pH buffer
Sodium Carbonate (Soda Ash)	pH buffer
Sodium Chloride	Viscosity reducer
Sodium Hypochlorite	Bactericide
Sodium Persulfate	Viscosity reducer
Terpenes	Reduction of surface tension to aid gas flow
Tetramethyl ammonium chloride	Clay control
Zirconium complex	Cross-linking agent to increase viscosity

See: Veil (2010) and Waxman et al. (2011) for more comprehensive lists of chemicals used in hydraulic fracturing projects.

[a] Listed alphabetically and not in order of preference. These additives are a relatively-common components of a water-based fracturing solution but are not necessarily used in *every* hydrofracturing operation. The exact blend and proportions of additives will vary based on the site-specific depth, thickness and other characteristics of the target formation.

Table 7.2 Types of additives used in fracturing fluids

Type	Compound	Comment
Acid	Hydrochloric acid (*muriatic acid*)	Used to clean cement from casing perforations and drilling mud clogging natural formation porosity, if any prior to fracturing fluid injection (dilute acids concentrations are typically on the order of 15% v/v acid)
Biocide	Glutaraldehyde	Fracture fluids typically contain gels which are organic and can therefore provide a medium for bacterial growth. Bacteria can break down the gelling agent reducing its viscosity and ability to carry proppant.
Breaker	Sodium Chloride	Chemicals that are typically introduced toward the later sequences of a fracturing project to break down the viscosity of the gelling agent to better release the proppant from the fluid as well as enhance the recovery or flowback of the fracturing fluid,
Corrosion inhibitor	*N,N*-dimethyl formamide	Used in fracture fluids that contain acids; inhibits the corrosion of steel tubing, well casings, tools, and tanks.
Cross-linking agent	Borate Salts	There are two basic types of gels that are used in fracturing fluids; linear and cross-linked gels. Cross-linked gels have the advantage of higher viscosity that do not break down quickly.
Friction Reducer	Crude oil distillate (*mineral oil*)	Minimizes friction allowing fracture fluids to be injected at optimum rates and pressures
Gel	Guar gum	Gels are used in fracturing fluids to increase fluid viscosity allowing it to carry more proppant than a straight water solution. In general, gelling agents are biodegradable.

of the fracturing fluid is removed from the well at the end of the fracturing process (flowback), a substantial amount of the fluid can remain underground.

The chemicals that are added to hydraulic fracturing fluids to facilitate the fracturing process vary depending on the location of the well and the geologic conditions of the sub-surface formations and cover a wide range of chemical types. Many of these chemicals (Tables 7.1 and 7.2), if not disposed of safely or are not prevented from allowed leaching into the drinking water supply (Table 7.3), could pose a serious environmental risk to the flora and fauna of the area. Thus, the primary environmental impacts associated with hydraulic fracturing (fracturing) result from the use of toxic chemicals during the fracturing process and the subsequent release of additional toxic chemicals and radioactive materials during well production.

Fracturing fluid flowback not only contains the chemical additives used in the drilling process volatile organic compounds (VOCs) and hazardous air pollutants (HAPs) such as benzene, toluene, ethylbenzene and xylene (BTEX) but can also contain

Table 7.3 The hydraulic fracturing water cycle and potential impacts on drinking water resources

Stage 1: Water Acquisition
• Large volumes of water are withdrawn from groundwater and surface water resources.
• Potential impacts on drinking water resources.
◦ Change in the quantity of water available for drinking
◦ Change in drinking water quality
Stage 2: Chemical Mixing
• The acquired water is combined with chemical additives and proppant.
• Potential impacts on drinking water resources.
◦ Release to surface and groundwater through on-site spills and/or leaks.
Stage 3: Well Injection
• Pressurized hydraulic fracturing fluid is injected into the well, creating cracks in the geological formation that allow oil or gas to escape through the well to be collected at the surface.
• Potential impacts on drinking water resources
◦ Release of hydraulic fracturing fluids to ground water due to inadequate well construction or operation.
◦ Movement of hydraulic fracturing fluids from the target formation to drinking water aquifers through man-made or natural features.
◦ Movement into drinking water aquifers of natural substances found underground, such as metals or radioactive materials which are mobilized during hydraulic fracturing activities
Stage 4: Flowback and Produced Water(Hydraulic Fracturing Wastewater)
• When pressure in the well is released, hydraulic fracturing fluid, formation water, and natural gas begin to flow back up the well. This combination of fluids, containing hydraulic fracturing chemical additives and naturally occurring substances, must be stored on-site In tanks or pits) before treatment, recycling, or disposal.
• Potential impacts on drinking water resources.
◦ Release to surface or ground water through spills or leakage from on-site storage
Stage 5: Wastewater Treatment and Waste Disposal
• Wastewater is dealt with in one of several ways, including but not limited to: disposal by underground injection, treatment followed by disposal to surface water bodies, or recycling (with or without treatment) for use in future hydraulic fracturing operations.
• Potential impacts on drinking water resources.
◦ Contaminants reaching drinking water due to surface water discharge and inadequate treatment of wastewater.
◦ Byproducts formed at drinking water treatment facilities by reaction of hydraulic fracturing contaminants with disinfectants

chemicals extracted from the underground formations, such as heavy metals and, on occasion, radioactive materials. Moreover, numerous pathways (such as older fractures that are opened and new fractures) are created by the hydraulic fracturing process for the release of these toxic and radioactive materials. As a result, the necessary protocols for handling of the toxic and radioactive materials must be in place as part of the original design plan as consultation of these protocols is essential throughout the life cycle of a well and the ensuing water cycle which involves five stages: (i) water acquisition,

(ii) chemical mixing, (iii) well injection, (iv) flowback and produced water which is waste water from the hydraulic fracturing process, and (v) wastewater treatment and waste disposal; each stage is subject to causing effects on drinking water resources (Table 7.3) (Holloway and Rudd, 2013; Spellman, 2013; Uddameri et al., 2016).

Not surprisingly, considering the variety of chemicals that can be added to the fracturing fluid (Tables 7.1 and 7.2), concerns also have been raised about the ultimate outcome of chemicals that are recovered and disposed of as wastewater, which is usually stored in tanks or pits at the well site, where spills are possible. For final disposal, the fluids must either be (i) recycled the fluids for use in future fracturing projects or (ii) injected it into underground storage wells (which, unlike the fracturing process itself, are subject to the Safe Drinking Water Act), discharge it to nearby surface water, or transport it to wastewater treatment facilities (Veil, 2010).

Thus, while, hydraulic fracturing has allowed access to domestic reserves of crude oil and natural gas that could provide an important stepping stone to a clean energy future as well as a measure of energy independence, questions about the safety of hydraulic fracturing persist, especially since many of the chemicals are known to be possible human carcinogens and/or regulated under the Safe Drinking Water Act or listed as hazardous air pollutants (Waxman et al., 2011).

In summary, the fracturing fluid is varied to meet the specific needs of each location; however, evaluating the widely reported percentage volumes of the fracturing fluid components reveals the relatively small volume of additives that are present. The overall, the concentration of additives in most fracturing fluids is a relatively consistent 0.5–2% *v*/v with water and proppants making up the remaining 98–99.5% v.v. However, a typical fracturing project uses upwards of five million gallons of fracturing fluid so a small percentage amount may actually result in a great deal of chemical usage, no matter how diluted it may be. The overall composition of fracturing fluids varies among companies and the drilling location. However, as a pretty good baseline, fracturing fluids typically contain: (i) approximately 90% water, (ii) approximately 9.5% proppant materials, (iii) approximately 0.5% chemicals—this percentage varies but is typically between 0.5% and 1.0% *w*/w of total fluid.

3.4 Water quality

A large proportion of the water used in hydraulic fracturing ends up becoming wastewater. Some is put away into man-made ponds, while some gets injected back into the underground, a process that is a danger itself. Still, other waste-water is treated in facilities and eventually put back into rivers. This is not good for communities that rely directly on flowing river water for drinking water. Even for communities that don't, most drinking water systems are somehow connected to rivers. In addition, most water treatment facilities are not equipped to remove all of the toxic compounds that come from the hydraulic fracturing fluid and into wastewater. Some of these compounds are carcinogenic in nature and can pose a threat to public health. In addition, pressure from the injection of wastewater can cause further cracking in rock layers beneath the surface of the Earth, which can then accelerate the migration of such dangerous water-borne constituents into aquifers. The pollution of water is too risky, and water contamination is ultimately inevitable with hydraulic fracturing. The contamination of groundwater, which is the main source for spring and well drinking waters, is also an effect of the hydraulic fracturing process. The danger that the hydraulic fracturing poses on drinking water is very serious and can even be life threatening.

Generally, the quality of produced water from hydraulic fracturing is somewhat less than pristine and cannot be readily used for another purpose without prior treatment. In fact, it is more correct to look upon the quality of produced water from any hydraulic fracturing project, although variable, as generally being poor and, in most situations, cannot be readily used for other purposes without prior treatment. Produced water may contain a wide range of contaminants in varying amounts, some of which occur naturally in the produced water but others are added through the process of hydraulic fracturing. The range of contaminants found in produced water can include: (i) salts, which include chloride derivatives, bromide derivatives, and sulfide derivatives of calcium, magnesium, and sodium, (ii) metals, which include barium, manganese, iron, and strontium, among others, (iii) oil, grease, and dissolved organics, which include benzene and toluene, among others, and (iv) production chemicals, which may include friction reducers to help with water flow, biocides to prevent/ growth of microorganisms, and additives to prevent corrosion, as well as a variety of other chemicals (Tables 8.1 and 8.3).

The quality of water generated by a specific well, however, can vary widely according to the same three factors that impact the volume of water produced from the well: (i) the type of crude oil or natural gas being produced, (ii) the geographic location of the well, and (iii) the method of oil or gas production. Typically, the type of hydrocarbon is a key driver of produced water quality, due to differences in geology across the formations in which the hydrocarbon derivatives are found. Specifically, the depth at which the hydrocarbon derivatives are found influences the salt and mineral content of produced water, and, in general, the deeper the formation is, the higher the salt and mineral content will be. Additionally, the amount of crude oil or natural gas that is mixed in with the produced water brought to the surface can also vary considerably. Also, the quality of the produced water can vary widely and, after treatment, some of the produced water is disposed of or reused by producers in other ways, such as: (i) discharging it to surface water, (ii) storing it in surface impoundments or ponds so that it can evaporate, (iii) irrigating crops, and (iv) reusing it for further hydraulic fracturing projects.

Managing produced water in these ways can require more advanced treatment methods, such as distillation but the manner in which produced water is managed and treated is primarily a decision that must be made within the bounds of federal and state regulations. By far the most serious local environmental concern, and probably the most contentious, is that of groundwater contamination. The potential risk to groundwater comes from two sources: the injected fluid (water plus chemical additives) and the released natural gas. However, the major issue that is not often determined scientifically is the exact site of this contamination, either: (i) percolation or diffusion from the hydraulically fractured formation at depth, or (ii) leakage from a defective well bore closer to the land surface.

However, the flow of hydraulic fracturing fluid into fractures can be significant and when flowback water is returned to the surface, fracturing fluids and naturally occurring substances (that have been extracted from the oil-containing or gas-containing formation) could potentially find a pathway into other geological features through natural faults or artificial penetrations (such as other non-related wells and underground mines). Groundwater resources can be protected through proper well design, construction, operation and maintenance—many states require periodic well integrity testing. However, in some geologic settings, methane can naturally originate from gas-producing rock layers below and close to the aquifer and be unrelated to the deeper fractured zone. Analysis of the crude oil and natural gas can be used to identify the origin of the crude oil and natural gas occurring in groundwater (Warner et al., 2012; Darrah et al., 2014).

Potential pathways for the fracturing fluids to contaminate water include events such as (i) surface spills prior to injection, (ii) migration of the injected fluid, (iii) surface spills of flowback water, and (iv) surface spills of produced water. Because the fracturing fluids are injected into the subsurface under high pressure, and because some of the fluids remain underground, there is concern that this mixture could move through the wellbore or fractures created in the reservoir rock by hydraulic pressure, and ultimately migrate up and enter shallow formations that are sources of freshwater (aquifers) (Cooley and Donnelly, 2012). There is also the possibility that geologic faults, previously existing fractures (which have not been identified due to an inadequate geological survey), and poorly plugged, abandoned wells could provide fluids with accessibility to aquifers (Osborn et al., 2011; Cooley and Donnelly, 2012; King, 2012; Molofsky et al., 2013; Vidic et al., 2013).

Thus, one challenge is to establish a baseline of the water quality before the fracturing operation commences. This will identify and distinguish the type and level of natural contaminants in the groundwater from those contaminants that are not indigenous to the groundwater as well as the amounts of indigenous contaminants that are in the groundwater. It is then possible, as an after the fact series of tests to determine any new contaminants or new levels of in-place contaminants that are the result of crude oil and natural gas development. Unfortunately, there often are no water quality analyses prior to hydraulic fracturing that can be used to provide a baseline comparison (Vidic et al., 2013).

Baseline water-quality testing, carried out prior to drilling for crude oil and natural gas, helps to document the quality of local natural groundwater and may identify natural or pre-existing contamination, or lack thereof, before crude oil and natural gas activity begins. Without such baseline testing, it is difficult to know if contamination existed before drilling, occurred naturally, or was the result of crude oil and natural gas activity. Many natural constituents, including methane, elevated chlorides, and trace elements occur naturally in shallow groundwater in oil-producing and gas-producing areas and are unrelated to drilling activities. The quality of water in private wells is not regulated at the state or federal level, and many owners do not have their well water tested for contaminants. States handle contamination issues differently.

4 Effects on water sources

Unconventional crude oil and natural gas development (through hydraulic fracturing) has the potential to impact water quality at many stages of the process. Water from beneath the ground has been exploited for domestic use, livestock use, and irrigation since the earliest times. Although the precise nature of its occurrence was not necessarily understood successful methods of bringing the water to the surface have been developed. In fact, if water for domestic use is obtained from a private well, there should be an acute awareness of the vat ours issues related to water quality. Private well water quality is not regulated in the United States, and well owners are responsible for performing their own monitoring.

Drilling for crude oil and natural gas, if properly conducted, does not necessarily contaminate groundwater. While properly drilled gas wells should keep contaminants from seeping into underground aquifers, some wells fail to do so, allowing methane and other chemicals to reach drinking water supplies. Unpredictable chemical releases from poorly managed drill sites, leaky wastewater pits, accidental spills, and truck accidents that occur above ground can also affect the quality of the well water.

After the fracturing process, a percentage of the water returns fairly quickly to the surface as wastewater (flowback). The briny water that has long been underground and comes up during continued operation of the well, called "produced water can contain naturally occurring contaminants like the radioactive element radium, along with other heavy metals and salts. All of this wastewater is toxic and must be collected and stored; it then must be treated or discharged—or reinjected into a deep disposal well.

The wastewater is often pumped into holding ponds where it can leak and settle into surrounding groundwater, and impact wildlife. The contamination of groundwater is of major concern for those who live near drilling operations and rely on drinking water wells. In addition, the contamination of watersheds that provide drinking water for millions of people in cities hundreds of miles away from any natural gas drilling sites poses a significant threat as well.

Additionally, some of the fluids pumped into the ground during the gas extraction process flow back to the surface. This wastewater is called "flowback" and can be contaminated with industrial and naturally occurring toxic substances. Some of the contaminants can alter the taste, odor, or clarity of well water while others are difficult to detect. Municipal water supplies may also be at risk from hydraulic fracturing due to surface water discharges, insufficient treatment of contaminated wastewater, and byproducts formed at drinking water treatment facilities by the reaction between hydraulic fracturing contaminants and disinfectants.

Reservoir management is a multi-faceted operation (Fig. 7.1) and while on-site environmental management is a typical practice associated with hydraulic fracturing, it should be considered as being site-specific. In addition, hydraulic fracturing base fluids (themselves variable in composition), most commonly water, are typically stored in tanks at the well site while additives may be stored on a flatbed truck or van enclosure that holds a number of containers. At the commencement of operations, the fracturing fluid and any chemical additives are sent to a blender for mixing after which the fluid is transferred to the wellhead for injection. It is during the time that the fluids and additives are being are transferred and moved around the well site and through various pieces of equipment that faulty equipment or human error can cause spills of the various components of fracturing fluids.

The type and amount of fluids and chemicals stored on-site is largely determined by the geological characteristics of the formation to be fractured as well as by

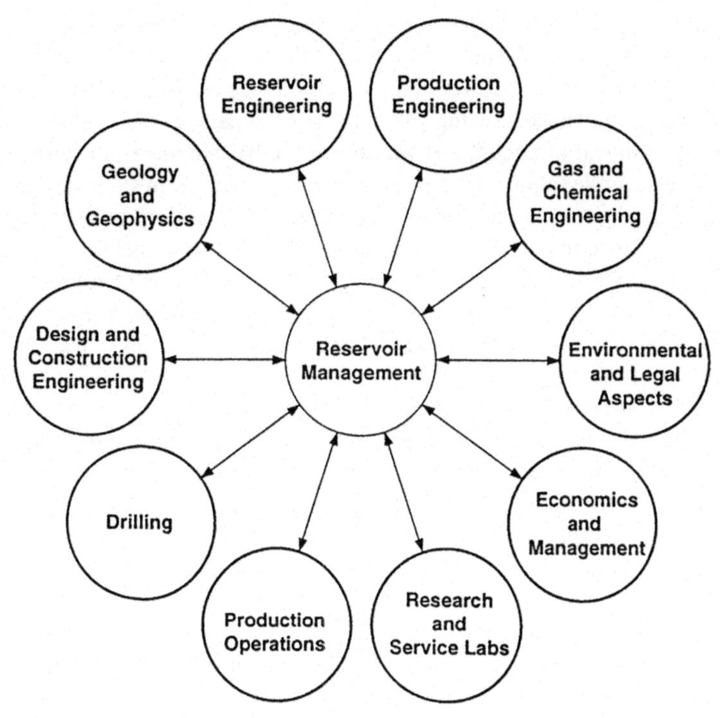

Fig. 7.1 Various aspects of reservoir management.

production goals and the chemical additives. Approximately 1–2% *v*/v or less of the volume of water-based hydraulic fracturing fluid is composed of chemical additives which indicates that approximately 500–260,000 gal or less of chemical additives may be brought on-site for hydraulic fracturing (US EPA, 2015). Chemical additives can be composed of one or more chemicals and can be used in hydraulic fracturing fluids as acids, friction reducers, surfactants, scale inhibitors, iron control agents, corrosion inhibitors, and biocides (Table 7.2) (Arthur et al., 2009; Gregory et al., 2011; US EPA, 2015). Moreover, for any analytical procedure or suite of analytical procedures, the following practices must be part of a strict testing protocol of sampling and identification of any chemicals by use of standards test methods that are not subject to intense criticism and can stand up to scrutiny in a court of law.

In order to withstand such scrutiny, the analytical records, as for any analytical process for materials such as crude oil, crude oil products, and natural gas (Speight, 2015), must be complete and should include but not necessarily restricted to the following information: (i) the precise (geographic or other) location from which the sample was obtained, (ii) the identification of the location by name, (iii) the character of the bulk material (solid, liquid, or gas) at the time of sampling, (iv) the means by which the sample was obtained, (v) the means and protocols that were used to obtain the sample, (vi) the date, the amount of sample that was originally placed into storage, (vii) any chemical analyses (elemental analyses, fractionation by adsorbents or by liquids, functional type analyses) that have been determined to date, (viii) any physical analyses that have been determined to date, (ix) the date of any such analyses, (x) the methods used for analyses that were employed, (xi) the analysts who carried out the work, and (xii) a log sheet showing the names of the persons (with the date and the reason for the removal of an aliquot) who removed the samples from storage and the amount of each sample (aliquot) that was removed for testing. In summary, there must be a means to accurately to track and identify the sample history so that each sample is tracked and defined in terms of source, activity, and the personnel involved in any of the above stages. Thus, the accuracy of the data from any subsequent procedures and tests for which the sample is used will be placed beyond a *reasonable doubt* and would stand the test of time in court should legal issues arise pertaining to the pollution of water-bearing formations.

Many shale formations contain quantities of potentially harmful chemical elements and compounds that could be dissolved into the hydraulic fracturing fluid and then return toward the surface during flowback. These include trace elements such as mercury, arsenic and lead; naturally occurring radioactive material (radium, thorium, uranium); and volatile organic compounds (VOCs). Thus, careful chemical monitoring of hydraulic fracturing fluids, including the flowback fluid and produced water, is required to mitigate the risks of contamination of the water sources.

Generally, the quality of produced water from hydraulic fracturing is somewhat less than pristine and cannot be readily used for another purpose without prior treatment. In fact, it is more correct to look upon the quality of produced water from any hydraulic fracturing project, although variable, as generally being poor and, in most situations, cannot be readily used for other purposes without prior treatment. Produced water may contain a wide range of contaminants in varying amounts, some of which occur naturally in the produced water but others are added through the process of hydraulic fracturing. The range of contaminants found in produced water can include: (i) salts, which include chlorides, bromides, and sulfides of calcium, magnesium, and sodium, (ii) metals, which include barium, manganese, iron, and strontium, among others, (iii) oil, grease, and dissolved organics, which include benzene and toluene, among others, and (iv) production chemicals, which may include friction reducers to help with water flow, biocides to prevent growth of microorganisms, and additives to prevent corrosion, as well as a variety of other chemicals (Tables 7.1 and 7.2).

The specific quality of water produced by a given project, however, can vary widely and, after treatment, some of the produced water is disposed of or reused by producers in other ways, such as (i) discharging it to surface water, (ii) storing it in surface impoundments or ponds so that it can evaporate, (iii) irrigating crops, and (iv) reusing it for further hydraulic fracturing projects. Managing produced water in these ways can require more advanced treatment methods, such as distillation but the manner in which produced water is managed and treated is primarily a decision that must be made within the bounds of federal and state regulations.

In fact, the quality of water generated by a given well, however, can vary widely according to the same three factors that impact the volume of water produced from the well: the hydrocarbon being produced, the geographic location of the well, and method of production used. Typically, the type of hydrocarbon is a key driver of produced water quality, due to differences in geology across the formations in which the hydrocarbon derivatives are found. Specifically, the depth at which the hydrocarbon derivatives are found influences the salt and mineral content of produced water, and, in general, the deeper the formation is, the higher the salt and mineral content will be. Additionally, the amount of crude oil or natural gas that is mixed with the produced water brought to the surface can also vary considerably.

By far the most serious local environmental concern, and probably the most contentious, is that of groundwater contamination. The potential risk to groundwater comes from two sources: the injected fluid (water plus chemical additives) and the released natural gas. However, the major issue that is not often determined scientifically is the exact site of this contamination, either: (i) percolation or diffusion from the hydraulically fractured formation at depth, or (ii) leakage from a defective well bore that is closer to the land surface.

The nature of the hydraulic fracturing process dictates that a significant amount of water is produced as a byproduct from hydraulic fracturing. Furthermore, concern has been raised over the increasing quantities of water for hydraulic fracturing in areas that experience water stress. Use of water for hydraulic fracturing can divert water from stream flow, water supplies for municipalities and industries such as power generation industries as well as recreation and aquatic life. The large volumes of water required for most common hydraulic fracturing methods have also raised concerns for arid regions, which may require water overland piping from distant sources (Nicot and Scanlon, 2012).

The hydraulic fracturing fluids is not only water but also a mixture of water, proppant, and chemical additives—the precise mix of additives depends on the formation to be fractured which dictates the process operations and the composition of the fluids and the proppants (Speight, 2016). Additives typically include gels to carry the proppant into the fractures, surfactants to reduce friction, hydrochloric acid to help dissolve minerals and initiate cracks, inhibitors against pipe corrosion and scale development, and biocides to limit bacterial growth (Table 7.1, Table 7.2). Chemical additives typically make up about 0.5% by volume of well fracturing fluids, but may be up to 2%. Some potential additives are harmful to human health, even at very low concentrations.

Production wells typically extend one mile or more below the ground surface, often passing through groundwater aquifers to reach crude oil-rich and natural gas-rich formations after which horizontal wells are drilled into the formation. Groundwater aquifers are typically separated from the shale formations by several thousand feet of rock, limiting the potential for any unreturned fracturing fluids to impact drinking water supplies. In addition, well casings and cement are designed to prevent contamination of any groundwater aquifers through which the well may be drilled.

There is the potential to over-pressure a well during the operation which may result in overlying formations becoming fractured, possibly serving as conduits for leakage of formation fluids and fracturing fluids into overlying formations, including aquifers. There is also the potential that over-pressured hydraulic fracturing operations could result in rapid upward leakage through the borehole into overlying formations, including aquifers and possibly even to the surface. The application of correct well design, completion, operations, and monitoring according to rules and regulations that already exist in most states and provinces will ensure that hydraulic fracturing operations do not negatively impact either the subsurface or surface environment. The result of mitigation efforts was that there was no quantifiable impact to local groundwater resources.

In addition, overweight drilling mud can cause a well bore to fail by fracture. The density (weight) of the drilling mud controls the fluid pressure exerted along the walls of the well bore. If the pressure of the mud exceeds the fracture pressure (the local minimum principal stress plus the fracture strength of the rock), a fracture can form and the drilling fluid can escape. However, pressures exceeding the rock fracture strength generated by overweight drilling muds are only likely at great depths (several thousand feet but within the depths of some shale formations), far beyond the extent of any groundwater aquifer and the risk of contamination from incorrect drilling mud composition may be limited.

To protect groundwater, proper well design, construction, and monitoring are essential. During well construction, multiple layers of telescoping pipe (or casing) are installed and cemented in place, with the intent to create impermeable barriers between the inside of the well and the surrounding rock. It is also common practice to pressure test the cement seal between the casing and rock or otherwise examine the integrity of wells. Wells that extend through a rock formation that contains high-pressure gas require special care in stabilizing the well bore and stabilizing the cement or its integrity can be damaged. Furthermore, differences in the type and sizes of well integrity datasets add to the challenge of generalizing well integrity failure rates (Davies et al., 2014, 2015).

However, the flow of hydraulic fracturing fluid into fractures can be significant and when flowback water is returned to the surface, fracturing fluids and naturally occurring substances (that have been extracted from the oil-containing or gas-containing formation) could potentially find a pathway into other geological features through natural faults or artificial penetrations (such as other non-related wells and underground mines). Groundwater resources can be protected through proper well design, construction, operation and maintenance—many states require periodic well integrity testing. However, in some geologic settings, methane can naturally originate from gas-producing rock layers below and close to the aquifer and be unrelated to the deeper fractured zone. Analysis of the crude oil and natural gas can be used to identify the origin of the crude oil and natural gas occurring in groundwater (Warner et al., 2012; Darrah et al., 2014).

Potential pathways for the fracturing fluids to contaminate water include events such as (i) surface spills prior to injection, (ii) migration of the injected fluid, (iii) surface spills of flowback water, and (iv) surface spills of produced water. Because the fracturing fluids are injected into the subsurface under high pressure, and because some of the fluids remain underground, there is concern that this mixture could move through the wellbore or fractures created in the reservoir rock by hydraulic pressure, and ultimately migrate up and enter shallow formations that are sources of freshwater (aquifers) (Cooley and Donnelly, 2012). There is also the possibility that geologic faults, previously existing fractures (which have not been identified due to an inadequate geological survey), and poorly plugged, abandoned wells could provide fluids with accessibility to aquifers (Osborn et al., 2011; Cooley and Donnelly, 2012; Molofsky et al., 2013; Vidic et al., 2013).

Thus, one challenge is to establish a baseline of the water quality before the fracturing operation commences. This will identify and distinguish the type and level of natural contaminants in the groundwater from those contaminants that are not indigenous to the groundwater as well as the amounts of indigenous contaminants that are in the groundwater. It is then possible, as an after the fact series of tests to determine any new contaminants or new levels of in-place contaminants that are the result of crude oil and natural gas development. Unfortunately, there often are no water quality analyses prior to hydraulic fracturing that can be used to provide a baseline comparison (Vidic et al., 2013).

Baseline water-quality testing, carried out prior to drilling for crude oil and natural gas, helps to document the quality of local natural groundwater and may identify natural or pre-existing contamination, or lack thereof, before drilling for crude oil and natural gas recovery begins. Without such baseline testing, it is difficult to know if contamination existed before drilling, occurred naturally, or was the result of crude oil and natural gas activity. Many natural constituents, including methane, elevated chlorides, and trace elements occur naturally in shallow groundwater in oil-producing and gas-producing areas and are unrelated to drilling activities. The quality of water in private wells is not regulated at the state or federal level, and many owners do not have their well water tested for contaminants. States handle contamination issues differently.

The current opinion is that all scientifically documented cases of groundwater contamination associated with hydraulic fracturing are related to poor well casings and their cements, or from leakages of fluid at the surface rather than from the hydraulic fracturing process itself. The absence of evidence implicating leakage from a fracture network could arise from the relatively short time span available for monitoring the signs of contamination, and potentially lower flow rates from a formation fracked at significant depth, although hydraulic fracturing has been performed in some areas for decades.

Surface spills related to the hydraulic fracturing—with the potential for seepage of the spilled chemical into a water system—occur mainly because of equipment failure or engineering or human misjudgments. Volatile chemicals held in waste water evaporation ponds can to evaporate into the atmosphere, or overflow. The runoff can also end up in groundwater systems. Groundwater may become contaminated by trucks carrying hydraulic fracturing chemicals and wastewater if they are involved in accidents on the way to hydraulic fracturing sites or disposal destinations. In addition, large quantities of chemicals must be stored at drilling sites and the volumes of liquid and solid waste that are produced and significant care must be taken that these materials do not contaminate surface water and soil during their transport, storage, and disposal.

Fluids used for slickwater hydraulic fracturing are typically more than 98% fresh water and sand by volume, with the remainder made up of chemicals that improve the effectiveness of the treatment, such as thickeners and friction reducers, and protect the production casing, such as corrosion inhibitors and biocides. These fluids are designed by service companies that tailor fracturing treatments to suit the needs of a particular project.

Because the fluids in each fracturing treatment would contain a different subset of these chemicals, and because these chemicals could be hazardous in sufficient concentrations, baseline water testing is necessary to enable regulatory agencies to conduct and respond appropriately should contamination or exposure occur. The use of more environmentally benign fracturing fluids would also help limit the environmental and health risks posed by fracturing fluids in the case of contamination. Chemicals to be used in fracturing fluids are generally stored at drilling sites in tanks before they are mixed with water in preparation for a fracturing project.

In fact, the most important requirement for an assessment of the impact of hydraulic fractions projects on the flora and fauna is complete testing of air and water prior to drilling and at regular intervals after drilling has commenced. This includes chemicals used in the drilling muds, fracturing fluids and any process wastewater (the latter contains heavy metals and radioactive compounds normally found in shale formations. Currently, the extent of testing (particularly for organic compounds) is frequently inadequate and limited by lack of information on what substances were used during the drilling process due to inadequate testing (Bamberger and Oswald, 2012).

After each fracturing stage, the fracturing fluid, along with any water originally present in the shale formation, is flowed back through the wellbore to the surface. The flowback period typically lasts for periods of hours to weeks, although some injected water can continue to be produced along with gas several months after production has started. Recycling water minimizes both the overall amount of water used for fracturing and the amount that must be disposed of. Many water treatment processes are currently being investigated that could be potentially be used at large scale and have a significant impact on this problem (Fig. 7.2).

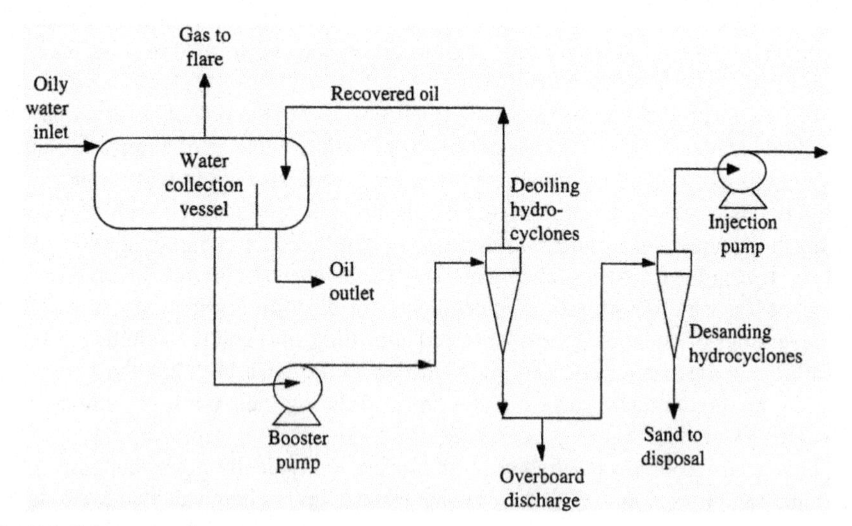

Fig. 7.2 Schematic of a water treatment process.

Flowback and water produced during the lifetime of a well can contain naturally occurring formation water that is millions of years old and therefore can display high concentrations of salts, naturally occurring radioactive material (NORM), and other contaminants including arsenic, benzene, and mercury. As a result, the water produced during hydraulic fracturing must be properly managed and disposed. Finally, one of the problematic aspects of handling flowback water is the temporary storage and transport of such fluids prior to treatment or disposal. In many cases, fluids may be stored in lined or even unlined open evaporation pits. Even if the produced water does not seep directly into the soil, a heavy rain can cause a pit to overflow and create contaminated runoff. Storing produced water in enclosed steel tanks, a practice already used in some wells, would reduce the risk of contamination while improving water retention for subsequent reuse. In addition, equipment used to move fluids between storage tanks or pits and the wellhead must be monitored and tested regularly to prevent spills, and precautions must be taken while transporting produced water to injection or treatment sites, whether via pipeline or truck.

Once production operations are in place and wells begin producing the fluids—produced water and oil—are often stored on site in large tanks while awaiting transport off-site. Safeguards put in place to protect the environment and public from tank releases include consistent measurements by pumpers, high level shut down sensors, continued equipment observations and maintenance, and secondary containments in place around the tanks to contain any fluids that may release from tank. Secondary containments may be constructed of properly packed and integrity tested earthen materials or up to specifically designed and manufactured metal containments with plastic liners. No matter the materials used in construction, secondary containments must be sufficiently large enough to contain all the fluids that could possibly escape the tanks.

Even with attempts to minimize the amount of on-site storage, some chemical and product storage is unavoidable and there are very valid concerns—including potential spills, leaks, tank or container overfill, and even the chance of traffic accidents on location or roadways leading to releases of chemicals and/or products. Release events could range from relatively small amounts from equipment leaks to possibly hundreds of barrels from tank release.

5 The future

Horizontal drilling in conjunction with hydraulic fracturing has made development of the shale gas resource an economic viable venture. As additional wells are drilled and more information is gathered on reservoir characteristics, additional advances may be realized in the fields of horizontal drilling and hydraulic fracturing which could further enhance development of other shale gas resources. However, development of the resources from tight formation and from shale formations includes many unique challenges, including water availability and water disposal.

To be successful, innovative solutions to developmental challenges associated with supplying the necessary volumes of water for hydraulic fracturing must be created. As more wells are drilled and development increases, it is anticipated that companies will continue designing leading edge technological solutions to water availability and disposal including expanding the volume of water that is re-used.

Hydraulic fracturing has been the primary focus of controversy because of surface and subsurface spills and releases, gas migration and groundwater contamination due to faulty well construction, blowouts, and leaks and spills of waste water and chemicals stored on pad sites. However, of all the issues related to hydraulic fracturing, the possible effects on groundwater are without a doubt the most contentious. Numerous allegations have been made related to hydraulic fracturing, with particular emphasis on the impacts on water sources. As additional regulatory measures, there will be continued focus on hydraulic fracturing. While questions relating to hydraulic fracturing wastewater disposal and the potential effects of hydraulic fracturing on water supply resources are reviewed, limits will continue to be placed on the drilling of new wells.

Thus, in order to establishing best environmental practice there is a need for assiduous monitoring and assessment in which national or local environmental agencies charged with monitoring the potential impacts of hydraulic fracturing in the exploration and production of shale gas should be fully funded and equipped to carry out the necessary tasks. In particular, if an agency is to approve or license the use of a specific chemical additive, it must have the means in place to detect and monitor the presence and movement of this chemical in local water supplies. Above all, baseline monitoring studies of groundwater are needed before any drilling activity begins.

Furthermore, while the mechanical hydraulic fracturing process itself arguably does not pose a significant environmental risk, there are potential risks to groundwater from poor well design or construction, especially in relation to the casings and the cements. Agencies need the resources and legal basis to investigate, analyze, approve or challenge the well designs and implementations used in the exploitation of shale gas. There must also be thorough testing of well casing and cement prior to injection of hydraulic fracturing of fluids. Moreover, active and regulated management of waste water from the hydraulic fracturing process is critical, as this fluid poses one of the greatest tangible risks to the environment.

There is also the need for a better geological understanding (in particular, geomechanical understanding) of the fracture networks produced by hydraulic fracturing operations is required, especially in more complex shale gas. Many shale gas formations in North America have a relatively simple sub-horizontal structure, but those in other parts of the world (especially Europe) are often folded and faulted on a variety of scales (Jackson and Mulholland, 1993). The more complex geometry of the shale gas formations in Europe, especially those of Carboniferous age, is due to an extended history of geological deformation spanning 300 million years. The generation and interaction of newly formed hydraulic fractures with much older pre-existing fault and fracture zones, and tilted bedding planes, is very poorly understood in terms of the mechanics and the hydrogeology, and new research programs are required to address these topics.

In terms of the importance of comprehensive hydraulic fracturing disclosure, one essential function of disclosure rules is to give nearby parties the information they need to fully understand the risks to their air and water and any impacts that may occur. With advance disclosure and proper notice, those who live or own property near a well can document pre-fracturing conditions, including air and water quality in the area, in case of pollution or spills. In particular, nearby water sources can be tested to determine baseline levels of the substances that will be used in the fracturing fluid in order to document whether water contamination was a result of hydraulic fracturing. To ensure that baseline testing can measure pre-hydraulic fracturing levels of all potential contaminants, disclosure of the chemicals must be made in advance, allowing sufficient time for testing to be arranged and performed before hydraulic fracturing begins.

In addition, parties must be aware that hydraulic fracturing is about to occur. In order to ensure that nearby parties are aware of upcoming hydraulic fracturing, disclosure rules should require advance notice to nearby landowners, those who own water wells, and non-owner residents. Prior disclosure and notification may also facilitate a conversation between local stakeholders, regulators and companies which can encourage the use of safer chemicals and practices, when they are available.

Disclosure rules should provide the public with information concerning the hydraulic fracturing process, and also on practices and materials employed throughout the lifecycle of a crude oil and natural gas well. Disclosure of the chemicals used in hydraulic fracturing, the waste generated and its management, and the details of how and where fracturing was completed, is essential. However, disclosure does not, by itself, make hydraulic fracturing safer. An adequate regulatory regime must also include, among other things, standards requiring best practices in well siting and construction, spill and leak reduction and containment, pollution capture, waste disposal, and the minimization of impacts from well pads, roads, and pipelines. Nevertheless, comprehensive disclosure rule is one important component of a full suite of hydraulic fracturing safeguards and is essential to investigate contamination that occurs when proper safeguards are not in place or accidents occur.

In addition, hydraulic fracturing is only one step in the crude oil and natural gas exploration and development process. Each phase of crude oil and natural gas development poses risks to the environment and public health. Among these are air, water, and soil pollution, and the use of dangerous chemicals during other phases of development. The public should be provided with accurate information on all hazards posed by the crude oil and natural gas industry.

Projections of the need for new wells to be drilled over the coming decades may lead to the construction of thousands of new well pads and thousands of miles of new access roads. The projected increase in roads has led some observers to be concerned about adverse effects of roads on ecosystems in areas of tight oil development. One way of mitigating this potential environmental issue is the increased use of *ecopads*, which allows for the drilling of multiple wells from a much smaller surface area, thereby reducing the number of well-drilling pads and the associated access roads that would need to be constructed during the development of these plays thereby leading to a reduction in the environmental footprint of the drilling operations.

Assuming that environmentally sustainable production methods are developed, this constraint could have a large and immediate impact on the start-up of the industry but decreasing as production expands. It is expected that environmental opposition will remain high despite industry proof that it can operate in an environmentally sustainable fashion. Sustainable production encompasses extraction methods that minimize the overall environment footprint including outright land disturbance, impacts on air, water, wildlife and the local population. The operations that employ processes with zero release of contaminants to the air, the water, and the land surface will be the ones that set the standard for environmental sustainability.

In summary, hydraulic fracturing remains a highly contentious public policy issue because of concerns about the environmental and health effects of its use, such as (i) the environmental risk, (ii) health risks from the chemicals injected into the ground, (iii) the occurrence of earthquakes, (iv) whether or not expansion of this technology for fossil fuels mean a decreased commitment to renewable energy technology, and (v) whether or not the environmental and health hazards are well understood and managed (Bamberger and Oswald, 2012; Jackson et al., 2013; Penning et al., 2014; Stern, 2014; Speight, 2016). In many cases it is unclear whether concerns raised relate specifically to hydraulic fracturing, or more generally to the development of unconventional crude oil resources, or to other aspects related to the development of all crude oil and natural gas resources.

Protecting the environment is an essential part of any project that focuses on the development of natural resources be they fossil fuel resources of other mineral resources. In the current context of hydraulic fracturing and protection of water resources, the project team should not be composed only of crude oil and natural gas engineers and drilling engineers, but should also include professionals (listed alphabetically) skilled in chemistry, environmental chemistry, environmental engineering, geochemistry, geology, hydrology, mineralogy, petrophysics, and petrology. The final team makeup will depend upon the project focus and may even be site specific.

Thus, in order to establishing best environmental practice for the protection of water systems (and other ecosystems), there is a need for assiduous monitoring and assessment in which national or local environmental agencies charged with monitoring the potential impacts of hydraulic fracturing in the exploration and production of shale gas should be fully funded and equipped to carry out the necessary tasks. In particular, if an agency is to approve or license the use of a specific chemical additive, it must have the means in place to detect and monitor the presence and movement of this chemical in local water supplies. Above all, baseline monitoring studies of groundwater are needed before any drilling activity begins.

Furthermore, while the mechanical hydraulic fracturing process itself arguably does not pose a significant environmental risk, there are potential risks to groundwater from poor well design or construction, especially in relation to the casings and the cements. Agencies need the resources and legal basis to investigate, analyze, approve or challenge the well designs and implementations used in the exploitation of shale gas. There must also be thorough testing of well casing and cement prior to injection of hydraulic fracturing of fluids to mitigate the potential of leakage of contaminated water and chemicals into a water system. Moreover, active and regulated management of waste water from the hydraulic fracturing process is critical, as this fluid poses one of the greatest tangible risks to the environment.

For the future, research and development should continue into the viability of removing all toxic additives from hydraulic fracturing fluids. The possibility of additive free hydraulic fracturing fluids (i.e. just water and sand) should be explored from a research perspective and industry sponsored testing. In fact, further research is needed into the treatment of flowback fluid, in particular a clearer understanding of those processes that work and those that do not and should include the quantification of risks and costs associated with the various options.

There is also the need for a better geological understanding (in particular, geomechanical understanding) of the fracture networks produced by hydraulic fracturing operations is required, especially in more complex shale gas. Many shale gas formations in North America have a relatively simple sub-horizontal structure, but those in other parts of the world (especially Europe) are often folded and faulted on a variety of scales (Jackson and Mulholland, 1993). The more complex geometry of the shale gas formations, especially those of Carboniferous age, is due to an extended history of geological deformation spanning 300 million years. The generation and interaction of newly formed hydraulic fractures with much older pre-existing fault and fracture zones, and tilted bedding planes, is very poorly understood in terms of the mechanics and the hydrogeology, and new research programmes are required to address these topics.

In terms of the importance of comprehensive hydraulic fracturing disclosure, one essential function of disclosure rules is to give nearby parties the information they need to fully understand the risks to their air and water and any impacts that may occur. With advance disclosure and proper notice, those who live or own property near a well can document pre-fracturing conditions, including air and water quality in the area, in case of pollution or spills. In particular, nearby water sources can be tested to determine baseline levels of the substances that will be used in the fracturing fluid in order to document whether water contamination was a result of hydraulic fracturing. To ensure that baseline testing can measure pre-hydraulic fracturing levels of all potential contaminants, disclosure of the chemicals must be made in advance, allowing sufficient time for testing to be arranged and performed before hydraulic fracturing begins.

In addition, parties must be aware that hydraulic fracturing is about to occur. In order to ensure that nearby parties are aware of upcoming hydraulic fracturing, disclosure rules should require advance notice to nearby landowners, those who own water wells, and non-owner residents. Prior disclosure and notification may also facilitate a conversation between local stakeholders, regulators and companies which can encourage the use of safer chemicals and practices, when they are available.

Disclosure rules should provide the public with information concerning the hydraulic fracturing process, and also on practices and materials employed throughout the lifecycle of a crude oil and natural gas well. Disclosure of the chemicals used in hydraulic fracturing, the waste generated and its management, and the details of how and where fracturing was completed, is essential for the following reasons:

- Adequate pre-hydraulic fracturing disclosure allows owners and users of nearby water sources to conduct baseline testing to establish the quality of their water prior to hydraulic fracturing, including the presence or absence of identified chemical constituents of frack fluids.
- Chemical disclosure is crucial to aid in determining the source of any subsequent groundwater contamination.

- First responders need the information to appropriately respond to accidents and emergencies.
- Chemical disclosure allows the public to fully assess the risks that chemical use, transport, and storage pose to their communities.
- Disclosure of water use provides the public information about the impacts of hydraulic fracturing on state supplies of fresh water.
- Disclosure of information regarding waste creation and disposition provides an accounting of the waste created, its contents, and the societal costs of its disposal.

However, disclosure does not, by itself, make hydraulic fracturing safer. An adequate regulatory regime must also include, among other things, standards requiring best practices in well siting and construction, spill and leak reduction and containment, pollution capture, waste disposal, and the minimization of impacts from well pads, roads, and pipelines. Nevertheless, comprehensive disclosure rule is one important component of a full suite of hydraulic fracturing safeguards and is essential to investigate contamination that occurs when proper safeguards are not in place or accidents occur.

In addition, hydraulic fracturing is only one step in the crude oil and natural gas exploration and development process. Each phase of crude oil and natural gas development poses risks to the environment and public health. Among these are air, water, and soil pollution, and the use of dangerous chemicals during other phases of development. The public should be provided with accurate information on all hazards posed by the crude oil and natural gas industry.

Projections of the need for new wells to be drilled over the coming decades may lead to the construction of thousands of new well pads and thousands of miles of new access roads. The projected increase in roads has led some observers to be concerned about adverse effects of roads on ecosystems in areas of tight oil development. One way of mitigating this potential environmental issue is the increased use of *ecopads*, which allows for the drilling of multiple wells from a much smaller surface area, thereby reducing the number of well-drilling pads and the associated access roads that would need to be constructed during the development of these plays thereby leading to a reduction in the environmental footprint of the drilling operations (National Petroleum Council, 2011).

As unconventional crude oil and natural gas resources in tight sandstone formations and in tight shale formations continue to be developed, the need for protection of the environment becomes even more necessary. The development of this industry affords an opportunity to employ production techniques that work within the regulatory constraints and serve to demonstrate environmentally responsible production to both the proponents of the industry and the detractors.

Assuming that environmentally sustainable production methods are developed, this constraint could have a large and immediate impact on the start-up of the industry but decreasing as production expands. It is expected that environmental opposition will remain high despite industry proof that it can operate in an environmentally sustainable fashion. Sustainable production encompasses extraction methods that minimize the overall environment footprint including outright land disturbance, impacts on air, water, wildlife and the local population. The operations that employ processes with zero release of contaminants to the air, the water, and the land surface will be the ones that set the standard for environmental sustainability.

References

Arthur, J., Bohm, B., Coughlin, B.J., Layne, M., Cornue, D., 2009. Evaluating the environmental implications of hydraulic fracturing in shale gas reservoirs. In: Proceedings of the SPE Americas Environmental and Safety Conference, San Antonio, Texas. March 23–25. Society of Petroleum Engineers, Richardson, Texas.

Bamberger, M., Oswald, R.E., 2012. Impacts of gas drilling on human and animal health. New Solut. 22 (1), 51–77.

Cipolla, C., Wright, C.A., 2002. Diagnostic techniques to understand hydraulic fracturing: What? Why and how? In: Proceedings of the 2002 SPE/CERI Gas technology symposium, Calgary, Canada. April 3–5. Society of Petroleum Engineers, Richardson, Texas.

Cooley, H., Donnelly, K., 2012. Hydraulic Fracturing and Water Resources: Separating the Frack from the Fiction. Pacific Institute, Oakland, CA.

Darrah, T.H., Vengosh, A., Jackson, R.B., Warner, N.R., Poreda, R.J., 2014. Noble gases identify the mechanisms of fugitive gas contamination in drinking-water Wells overlying the Marcellus and Barnett shales. Proc. Natl. Acad. Sci. 111 (39), 14076–14081.

Davies, R.J., Almond, S., Ward, R., Jackson, R.B., Adams, C., Worrall, F., Herringshaw, L.G., Gluyas, J.G., Whitehead, M.A., 2014. Oil and gas Wells and their integrity: Implications for shale and unconventional resource exploitation. Mar. Pet. Geol. 56, 239–254.

Davies, R.J., Almond, S., Ward, R., Jackson, R.B., Adams, C., Worrall, F., Herringshaw, L.G., Gluyas, J.G., 2015. Oil and gas Wells and their integrity: Implications for shale and unconventional resource exploitation. Mar. Pet. Geol. 59, 674–675.

Gregory, K.B., Vidic, R.D., Dzombak, D.A., 2011. Water management challenges associated with the production of shale gas by hydraulic fracturing. Elements 7, 181–186.

Holloway, M.D., Rudd, O., 2013. Fracturing: The Operations and Environmental Consequences of Hydraulic Fracturing. John Wiley & Sons Inc, Hoboken, NJ.

Jackson, D.I., Mulholland, P., 1993. Tectonic and stratigraphic aspects of the east Irish Sea basin and adjacent areas: contrasts in their post-carboniferous structural styles. In: Parker, J.R. (Ed.), Petroleum Geology of Northwest Europe. Proceedings of the 4th Geology Conference. The Geological Society, London, United Kingdom, pp. 791–808.

Jackson, R.B., Vengosh, A., Darrah, T.H., Warner, N.R., Down, A., Poreda, R.J., Osborn, S.G., Zhao, K., Karr, J.D., 2013. Increased stray gas abundance in a subset of drinking water Wells near Marcellus shale gas extraction. PNAS 110 (28), 11250–11255.

King, G.E. 2012. Hydraulic Fracturing 101: What Every Representative, Environmentalist, Regulator, Reporter, Investor, University Researcher, Neighbor and Engineer Should Know about Estimating Frac Risk and Improving Frac Performance in Unconventional Gas and Oil Wells. Paper No. 152596. Society of Petroleum Engineers, Richardson, Texas.

Molofsky, L., Conner, J.A., Wylie, A.S., Wagner, T., Farhat, S., 2013. Evaluation of methane sources in groundwater in Northeast Pennsylvania. Groundwater 51 (3), 333–349.

National Petroleum Council. 2011. Unconventional Oil. Paper No. 1–6. Working Document of the NPC North American Resource Development Study. Prepared by the Unconventional Oil Subgroup of the Resource & Supply Task Group. (September 15).

Nicot, J.-P., Scanlon, B.R., 2012. Water use for shale-gas production in Texas, US. Environ. Sci. Technol. 46, 3580–3586.

Osborn, S.G., Vengosh, A., Warner, N.R., Jackson, R.B., 2011. Methane contamination of drinking water accompanying gas-well drilling and hydraulic fracturing. Proc. Natl. Acad. Sci. 108 (20), 8172–8176.

Penning, T.M., Breysse, P., Gray, K., Howard, M., and Yan, B. 2014, Environmental Health Research Recommendations from the Inter-Environmental Health Sciences Core Center

Working Group on Unconventional Natural Gas Drilling Operations. Environmental Health Perspectives. National Institute of Environmental Health Sciences, National Institutes of Health, US Department of Health and Human Services, Research Triangle Park, North Carolina.

Speight, J.G., 2009. Enhanced Recovery Methods for Heavy Oil and Tar Sands. Gulf Publishing Company, Houston, TX.

Speight, J.G., 2011. An Introduction to Petroleum Technology, Economics, and Politics. In: Scrivener Publishing. Salem, MA, Massachusetts.

Speight, J.G., 2014a. The Chemistry and Technology of Petroleum, fifth ed. CRC Press, Taylor & Francis Group, Boca Raton, FL.

Speight, J.G. 2014b. Oil and Gas Corrosion Prevention. Gulf Professional Publishing, Elsevier, Oxford, United Kingdom.

Speight, J.G., 2015. Handbook of Petroleum Product Analysis, second ed. John Wiley & Sons Inc, Hoboken, NJ.

Speight, J.G., 2016. Handbook of Hydraulic Fracturing. John Wiley & Sons Inc, Hoboken, NJ.

Spellman, F.R., 2013. Environmental Impacts of Hydraulic Fracturing. CRC Press, Taylor & Francis Group, Boca Raton, FL.

Stern, P.C., 2014. Risks and Risk Governance in Shale Gas Development: Summary of Two Workshops. National Research Council, National Academies Press, Washington, DC. http://www.nap.edu/catalog.php?record_id=18953. (accessed April 14, 2015).

Uddameri, V., Morse, A., Tindle, K.J., 2016. Hydraulic Fracturing—Impacts and Technologies. CRC Press, Taylor & Francis Group, Boca Raton, FL.

US EPA, 2015. Review of State and Industry Spill Data: Characterization of Hydraulic Fracturing-Related Spills. Report No. EPA/601/R-14/001. Office of Research and Development, US Environmental Protection Agency, Washington, DC (May).

Veil, J.A., 2010. Water Management Technologies Used by Marcellus Shale Gas Producers. Argonne National Laboratory, Argonne, Illinois. United States Department of Energy, Washington, DC (July).

Vidic, R.D., Brantley, S.L., Venderbossche, J.K., Yoxtheimer, D., Abad, J.D., 2013. Impact of shale gas development on regional water quality. Science 340 (6134), 1–10.

Warner, N.R., Jackson, R.B., Darrah, T.H., Osborn, S.G., Down, A., Zhao, K., White, A., Vengosh, A., 2012. Geochemical evidence for possible natural migration of Marcellus formation brine to shallow aquifers in Pennsylvanian. Proc. Natl. Acad. Sci. 109 (30), 11961–11966.

Waxman, H.A., Markey, E.J., Degette, D., 2011. Chemicals Used in Hydraulic Fracturing. Committee on Energy and Commerce. United States House of Representatives, Washington, DC (April).

Further reading

Flewelling, S.A., Tymchak, M.P., Warpinski, N., 2013. Hydraulic fracture height limits and fault interactions in tight oil and gas formations. Geophys. Res. Lett. 40, 3602–3606.

Jones, J.R., Britt, L.K., 2009. Design and Appraisal of Hydraulic Fractures. Society of Petroleum Engineers, Richardson, TX.

Wright, C.A., Davis, E.J., Wang, G., Weijers, L., 1999. Downhole Tiltmeter fracture mapping: A new tool for direct measurement of hydraulic fracturing growth. In: Amadei, B., Krantz, R.L., Scott, G.A., Smeallie, P.H. (Eds.), Rock Mechanics for Industry. Balkema Publishers, Rotterdam, Netherlands.

Remediation technologies

Chapter outline

1 Introduction

Currently, an increased level of water consumption and correspondingly high levels of pollution have generated a prominent need for managing the water quality by maintaining safe levels for the water to be used in specific applications. In this respect, water remediation methods have taken a forward thrust in order to increase the water quality of potable water as well as that of industrial grade water in order to prevent contamination of natural water resources due to the discharge of industrial effluents. Several methods of water remediation such as physical, chemical, electrochemical and biological are discussed and worked out worldwide.

Environmental remediation deals with the removal of pollutants or contaminants from environmental systems such as, in the current context, groundwater, lakes, rivers, and oceans. Water remediation is the process of removing contaminants from water. Surface water in lakes, streams, and rivers can be directly contaminated by pollutants released directly into the water or by runoff from the ground. Groundwater can become polluted by contaminants leaching through the soil and sediment above it or as the result of industrial practices such as mining or drilling for natural gas and oil.

In any case, whatever the water system, immediate action should be taken as this can impact negatively on human health and the environment. However, remedial action is generally subject to a series of regulatory requirements, and also can be based on assessments of human health and ecological risks where no legislated standards exist or where standards are advisory.

A water system is a valuable source and remediation of contaminated water is critical in order to protect human health and the environment. It is of the utmost importance

Natural Water Remediation. https://doi.org/10.1016/B978-0-12-803810-9.00008-5

to properly characterize the site, and such a characterization includes defining the site in terms of (i) the geology, (ii) the hydrology, (iii) the contamination, and (iv) the potential releases to the environment, and locations and demographics of nearby populations. Once the site has been characterized, a risk assessment of hazards at the site is performed and a suitable remedial action may be selected (Reddy, 2008).

In order to perform these different, it is important that the entire remedial planning, from initial site characterization efforts until the completion of site cleanup, follows a rational strategy (Khan et al., 2004; Sharma and Reddy, 2004). If contamination has been detected and risk posed by the contamination is unacceptable, an appropriate remedial technology must be selected and properly implemented. By possessing knowledge of the available technologies, the remediation efforts will be better equipped to utilize proper judgment for the decisions regarding the remediation of a contaminated site.

Several technologies exist for the remediation of contaminated groundwater and many of these technologies are used in combination or other innovative technologies are being developed. Remediation technology for a particular site is selected based on the site specific hydrogeologic and contaminant conditions and the desired level of cleanup.

Thus, water remediation is the process that is used to treat polluted water by removing the pollutants or converting them into harmless products. The many and diverse activities of humans produce innumerable types of waste materials and by-products. Historically, the disposal of such waste have not been subject to many regulatory controls. Consequently, waste materials have often been disposed of or stored on land surfaces where they percolate into the underlying groundwater. As a result, the contaminated groundwater is unsuitable for use.

Current practices can still impact water, such as the over-application of fertilizers or pesticides, spills from industrial operations, infiltration from urban runoff, and leaking from landfills. Using contaminated groundwater causes hazards to public health through poisoning or the spread of disease, and the practice of groundwater remediation has been developed to address these issues. Contaminants found in groundwater cover a broad range of physical, inorganic chemical, organic chemical, bacteriological, and radioactive parameters. Pollutants and contaminants can be removed from groundwater by applying various remediation techniques, thereby bringing the water to a standard that is commensurate with various intended uses.

This chapter presents a review of the importance of water remediation, various methods that are employed to remediate and some automated techniques.

2 Sampling and analysis

The sources of water pollution/contamination might be the consequence of activities carried out in the past. Analogously, current waste disposal activities may affect water quality, and possibly human health, far into the future. This is a natural consequence of the fact that aquifers are open systems with exchange of water, substances and energy across the system boundaries. The main inputs of substances into the groundwater occur from the land surface through the soil zone by infiltration. A contaminant plume due to an industrial source can spread with time (often many years) with the natural groundwater flow over large distances of many kilometers and thus poses a long-term danger to water supply system based on the same aquifer. The water contamination is very high in most of the industrialized countries. Thus, the first step in water remediation is to identify potential contaminant sources and, it is at this point that sampling methods and analytical techniques need to be applied.

After site characterization (Table 8.1), water chemistry analyses are carried out to identify and quantify the chemical components and properties of water samples (Table 8.2) (Richardson, 2007). The type and sensitivity of the analysis depends on the purpose of the analysis and the anticipated use of the water. Chemical water analysis is carried out on water used in industrial processes, on waste-water stream, on rivers and stream, on rainfall and on the sea. In all cases the results of the analysis provides information that can be used to make decisions or to provide re-assurance that conditions are as expected. The analytical parameters selected are chosen to be appropriate for the decision-making process or to establish acceptable normality. Analytical methods routinely used can detect and measure all the natural elements and their inorganic compounds and a very wide range of organic chemical species using methods such as gas chromatograph and mass spectrometry (ASTM, 2019).

The quantitative measurements of pollutants are obviously necessary before water pollution can be controlled. In addition, many pollutants may be present at low concentrations, and very accurate methods of detection are required. However, before such test methods are applied to determine the extent of contamination and the types of contaminants, the water from must be samples so that the sample can be assessed as a representative sample.

Very often the choice of the methods of analysis will depend on local factors, e.g. the equipment already available or the cost of certain consumables. In addition,

Table 8.1 Site characterization: geologic and hydraulic properties of the water system (aquifer)

Aquifer	Comment
Extent of the aquifer	Depth Width Location
Physical properties	Pore size Sediment or rock type Hydraulic diffusivity Flow rate of water through the aquifer
Chemical properties	Chemical properties of the rock/sediment within the aquifer Purity of the aquifer upstream of the source of contamination Gases are dissolved in the aquifer
Impact	Fate of the contaminant in the aquifer Spread of the contaminant Change in the chemical composition due to natural effects

Table 8.2 Information needed about the contaminant

Contaminant information	Comments
Types	Identify the different types of chemical compounds Identify the source
Time frame	Length of time of the leak into the water
Source	Spatial extent of the source
Physical properties	Density (lighter or heavier than water; sink or float) Solubility in water Viscosity Other physical properties necessary to the site
Chemical properties	Reaction with oxygen Reaction with rocks/minerals/sediments Other chemical properties necessary to the site
Concentration	At various locations of an extended source? Large industrial sites
Toxicology	Effect of the toxic contaminant on plants Effect on animals Effect on humans Effect on the ecosystem

water-quality testing may be subject to the following problems, especially when the communities or the sampling sites are remote or inaccessible: (i) deterioration of samples during transport to centralized laboratory facilities, (ii) inadequate techniques for sample storage and preservation during prolonged transport, thus limiting the sampling range, and (iii) the need for reporting, which may necessitate further return journeys. If there are delays in sample transport and analysis and, therefore, in reporting remedial action is also likely to be delayed. For these reasons, on-site water testing using portable equipment is appropriate in many remote areas.

2.1 Sampling

Sampling water for subsequent analysis is often considered to be somewhat easier than sampling, say, soils for two main reasons: (i) water tends to be more homogeneous than soils, there is less point-to-point variability between two samples collected within the same vicinity, (ii) it is often physically easier to collect water samples because it can be done with pumps and hose lines, (iii) known volumes of water can be collected from known depths with relative ease. The amounts of water collected depend on the sample being evaluated, but can vary from 100ml for potable drinking water collected from water utilities to 1000l for groundwater samples. Sampling strategy is also less complicated for water samples since, in many cases, the water is mobile and a set number of bulk samples are collected from the same point at various time intervals. For samples of marine water, the samples are often collected sequentially in time within the defined area of interest.

Samples of water from the natural water systems should be routinely taken and analyzed as part of a pre-determined monitoring program by regulatory authorities to ensure that water system remains unpolluted or, if polluted, that the levels of pollution are not increasing or are falling in line with an agreed remediation plan. Typical parameters for ensuring that unpolluted surface waters remain within acceptable chemical standards include (among others that are specific to the water system): pH, major cations and anions, and biochemical oxygen demand. However, the frequency of the sampling protocols will be dependent upon the water system, the type of pollution, and the degree of pollution.

In order to establish remediation protocols for any water system, guidelines and standard test methods must be available to assess the quality of the water. Samples must be taken from locations that are representative of the water and when selecting sampling points, each locality should be considered individually and the following general criteria are suggested guidelines: (i) the sampling points should be selected such that the samples taken are representative of the source from which water is obtained, (ii) the sampling points should include those that yield samples representative of the conditions at the water system, particularly points of possible contamination, (iii) the sampling points should be uniformly distributed throughout a water system, and (iv) the sampling points chosen should generally yield samples that are representative of the system as a whole.

Sampling regimes using variable or random sites have the advantage of being more likely to detect local problems but are less useful for analyzing changes over time. However, he results of analysis are of no value if the samples tested are not properly collected and stored. This has important consequences for sampling regimes, sampling procedures, and methods of sample preservation and storage. Sample bottles must be clean but need not be sterile and preservatives may be required for some analytes. Residual chlorine, pH, and turbidity should be tested immediately after sampling as they will change during storage and transport. Moreover, the time between sampling and analysis should be kept to a minimum. Storage in glass or polyethylene bottles at a low temperature (e.g. 4 °C, 39 °F) in the dark is recommended.

The microbiological examination of water emphasizes assessment of the quality of the water system. This requires the isolation and enumeration of organisms that indicate the presence of any form of microbial contamination. The isolation of specific pathogens in water should be undertaken only by reference laboratories for purposes of investigating and controlling outbreaks of disease.

2.2 Test methods

It is not the intent here to present a complete description of all that test methods that can be applied to water analysis. The standard test methods as documented by ASTM International are instrumental in specifying and evaluating the methods and facilities used in examining the various characteristics of and contaminants in water for health, security, and environmental purposes. These water testing standards allow concerned local government authorities, water distribution facilities, and environmental laboratories to test the quality of water and ensure safe consumption Appendix). More detailed texts are available for that purpose (ASTM, 2019) but it is the intent to present a brief summary of the methods that are available and that should be applied to determine the purity or the degree of contamination of a water sample. Testing procedures and parameters may be grouped into (i) physical property test methods, which are used to determine the properties of the sample that are detectable by the senses such as color, turbidity, odor and taste, (ii) chemical property test methods, which are used to determine the amounts of mineral and organic substances that affect water quality, and (iii) bacteriological test methods, which are used to determine the presence of bacteria, characteristic of fecal pollution.

2.2.1 Physical property test methods

The physical test methods are used to determine color, turbidity, total solids, dissolved solids, suspended solids, odor and taste are recorded.

Color in water may be used to indicate the presence of minerals such as iron and manganese or the presence of substances of vegetable origin such as algae and weeds. Color tests indicate the efficiency of the water treatment system.

Turbidity in water is because of suspended solids and colloidal matter. It may be due to eroded soil caused by dredging or due to the growth of micro-organisms. High turbidity makes filtration expensive. If sewage solids are present, pathogens may be encased in the particles and escape the action of chlorine during disinfection.

Odor and taste are associated with the presence of living microscopic organisms; or decaying organic matter including weeds, algae; or industrial wastes containing ammonia, phenol derivatives, halogen derivatives, and hydrocarbon derivatives. This taste is imparted to fish, rendering them unpalatable. While chlorination dilutes odor and taste caused by some contaminants, it generates a foul odor itself when added to waters polluted with detergents, algae and some other wastes.

2.2.2 Chemical property test methods

The chemical test consists of—but are not necessarily limited to—determination of the following properties: pH, hardness, presence of a selected group of chemical parameters, biocides, highly toxic chemicals, and biochemical oxygen demand (BOD.

The pH is a measure of hydrogen ion concentration and is an indicator of relative acidity or alkalinity of water. A pH of 7.0 indicates that the solution is neutral while values of 7.1 and higher indicate alkalinity while values of 6.9 and lower indicate acidity. Low pH values help in effective chlorination but cause problems with corrosion. Values below 4 generally do not support living organisms in the marine environment. Drinking water should have a pH between 6.5 and 8.5.

The biochemical oxygen demand is an indicator of the amount of oxygen needed by micro-organisms for stabilization of decomposable organic matter under aerobic conditions. A high value for the biochemical oxygen demand is an indication that there is less oxygen available to support life and indicates that there is pollution by organic compounds.

2.2.3 Bacteriological test methods

The bacteriological analysis of a water sample is a method of analyzing water to estimate the numbers (and types) of bacteria present in the sample and represents one aspect of water quality. The method use of a microbiological analytical procedure which uses samples of water and from these samples determines the concentration of bacteria in each sample. From the data, it is possible to draw inferences about the suitability of the water for use from these concentrations. This process is used, for example, to routinely confirm that water is safe for human consumption or that bathing and water for recreational use.

However, analytical test methods procedures for the detection of harmful organisms are impractical for routine water quality surveillance. It must be appreciated that all that bacteriological analysis can prove is that, at the time of examination, contamination or bacteria indicative of fecal pollution, could or could not be demonstrated in a given sample of water using specified culture methods. In addition, the results of routine bacteriological examination must always be interpreted in the light of a thorough knowledge of the water supplies, including their source, treatment, and distribution.

Furthermore, the recognition that microbial infections can be waterborne has led to the development of methods for routine examination to ensure that water intended for human consumption is free from excremental pollution. Although it is now possible to detect the presence of many pathogens in water, the methods of isolation and enumeration are often complex and time-consuming. It is therefore impractical to monitor drinking water for every possible microbial pathogen that might occur with contamination.

2.3 Quality assurance and quality control

Standard methods for water analysis should be tested under local conditions for accuracy and precision, agreed at national level, and applied universally by both water-supply and regulatory agencies. However, the use of standard methods does not in itself ensure that reliable and accurate results will be obtained. In the context of analytical work, the terms quality assurance and quality control are often treated as synonymous. In fact, they are different concepts. Analytical quality control is the generation of data for the purpose of assessing and monitoring how good an analytical method is and how well it is operating. This is normally described in terms of within-day and day-to-day precision.

Quality control (QC) refers to all those analytical processes and procedures designed to ensure that the results of laboratory analysis—as well as analysis at field sites are consistent, comparable, and accurate and within the specified limits of precision. Constituents submitted to the analytical laboratory must be accurately described to avoid faulty interpretations, approximations, or incorrect results. On the other hand, quality assurance (QA) comprises all the steps taken by a laboratory to assure those who receive the data that the laboratory is producing valid results (Stebbing, 1993). Quality assurance thus encompasses analytical quality control but also includes many other aspects such as proving that the individuals who carried out an analysis were competent to do so, and ensuring that the laboratory has, for example: (i) established and documented analytical methods, (ii) equipment calibration procedures, (iii) systems for data retrieval, and (iv) the sample-handling procedures.

The qualitative and quantitative data generated from the laboratory can then be used for decision making. In addition, quality assurance (QA) is a way of preventing mistakes and defects in manufactured products and avoiding problems when delivering products or services to customers. This defect prevention in quality assurance differs subtly from defect detection and rejection in quality control. The terms *quality control* and *quality assurance* are often used interchangeably to refer to ways of ensuring the quality of a service or product.

3 Pollution

Water pollution (also called water contamination) occurs when pollutants are released to the ground and make their way into a water system. This type of water pollution can also occur naturally due to the presence of a minor and unwanted constituent, contaminant or impurity in the water. The pollutant often creates a contaminant plume within a water system, especially in an aquifer. Movement of water and dispersion within the system spreads the pollutant over a wider area. Its advancing boundary, often called a plume edge, can make the water supplies unsafe for humans and wildlife.

Pollution can occur from on-site sanitation systems and landfills, effluent from wastewater treatment plants, leaking sewers, service stations, or from the over-application of fertilizers. Pollution (or contamination) can also occur from naturally occurring contaminants, such as arsenic for fluoride derivatives. Using polluted water causes hazards to public health through poisoning or the spread of disease. Different mechanisms have influence on the transport of pollutants, such as diffusion, adsorption, and precipitation. Thus, in a water ecosystem a spill of a chemical represents the introduction of a non-aqueous phase liquid composed of a large amount of organic matrices in addition to other toxic constituent, such as polynuclear aromatic hydrocarbon derivatives as occur in the higher boiling fractions of crude oil (Parkash, 2003; Gary et al., 2007; Speight, 2014, 2015, 2017; Hsu and Robinson, 2017).

Chemical type, weather, wind and wave conditions, as well as air and sea temperature, all play important roles in ultimate fate of spilled oil in marine environment. After oil is discharged in the environment, a wide variety of physical, chemical and biological processes begin to transform the discharged oil. These chemical and physical processes are collectively referred to as *weathering* and act to change the composition, behavior, route of exposure and toxicity of discharged chemical. The weathering processes are described as follows:

- *Spreading and advection*: Spreading, which dominates the initial stages of the spill and involves the whole oil and is the movement of the entire oil slick horizontally on the surface of water due to effects of gravity, inertia, friction, viscosity and surface tension—on calm water spreading occurs in a circular pattern outward from the center of the point of release. Advection is the movement of oil due to overlying winds or underlying currents, which increase the surface area of the oil, thereby increasing its exposure to air, sunlight and underlying water. The advection effects are not uniform and do not affect the chemical composition of the oil.
- *Evaporation*: Evaporation, which is subject to atmospheric conditions such as temperature, is the preferential transfer of lower-boiling constituents of the petroleum-related spill from the liquid phase to the vapor phase. The chemical composition of the spilled material

is physically altered and although the volume decreases, the remaining components have higher viscosity and specific gravity leading to thickening of the slick and formation of entities such as tar balls.

- *Dissolution*: Dissolution is the transfer of constituents from the spilled material into the aqueous phase. Once dissolved these constituents are bioavailable and if exposed to marine life can cause environmental impacts and injuries. The largest concentration is found near the surface or the release point and hence the effect on marine life may be localized.
- *Natural dispersion*: Natural dispersion is the process of forming small oil droplets that become incorporated in the water in the form of a dilute oil-in-water suspension and occurs when breaking waves mix the oil in the water column. This phenomenon reduces the volume at the spill surface but does not change the physicochemical properties of the spilled material. It is estimated to range from 10% to 60% per day for first three days of the spill, depending on the condition of the ecosystem and may be independent of the type of petroleum or petroleum product.
- *Emulsification*: Emulsification is the mixing of sea water droplets into oil spilled on water surface forming a water-in-oil emulsion.
- *Photo-oxidation*: Photo-oxidation occurs when sunlight in the presence of oxygen transforms hydrocarbon derivatives (by increasing the oxygen content of a hydrocarbon) into new by-products. This results in changing in the interfacial properties of the oil, affecting spreading and emulsion formation, and may result in transfer of toxic by-products into the water column due to the enhanced water solubility of the by-product.
- *Sedimentation and shoreline stranding*: Sedimentation is the incorporation of oil within both bottom and suspended sediments. Shoreline stranding is the visible accumulation of petroleum along the edge of the water system (typically, the shoreline) following a spill. It is affected by proximity of the spill to the affected shoreline, intensity of current and wave action on the affected shoreline and persistence of the spilled product. Sedimentation can begin immediately after the spill but increases and peaks after several weeks, whereas shoreline stranding is a function of the distance of the shoreline from the spill and chemical nature of spilled oil.
- *Biodegradation*: Biodegradation occurs on the water surface, in the water, in the sediments, and at the shoreline. It can begin (depending on the lag time or induction period) after a spill and can continue as long as degradable hydrocarbon derivatives are present.

Weathering processes occur simultaneously—one process does not stop before the other begins. The order in which these processes presented is instantaneous, the relative significance of these processes may change if the spill occurred below the water surface or in tropical or ice conditions. The spill chronology may also vary if the spill is near the shoreline in which case it can contaminate the soil and groundwater even before the weathering or cleanup processes start. Also there are many onshore and offshore operations in a petroleum industry that can cause soil pollution and aquifer contamination.

Aquifer contamination can also take place because of the migration of the oil through the porous media and its subsequent adsorption on the rock surface. Many technologies have been proposed for the treatment of oil contaminated sites; these can be performed by two basic processes in-situ and ex-situ treatment using different cleaning technologies such as thermal treatment, biological treatment, chemical extraction and soil washing, and aeration accumulation techniques (Speight, 1996, 2005; Speight and Lee, 2000).

4 Remediation

Water remediation, especially groundwater remediation, is the process that is used to treat polluted water by removing the pollutants or converting them into benign products. Anthropogenic activities produce innumerable waste materials and by-products that can eventually enter a water system. Historically, the disposal of such waste have not been subject to many regulatory controls and, consequently, waste materials have often been disposed of or stored on land surfaces where they percolate into a water system, usually a groundwater system.

Current practices—such as the over-application of fertilizers and pesticides as well as spills from industrial operations, urban runoff, and leakage from landfills—can still impact a water system. Using contaminated water causes hazards to public health through poisoning and/or the spread of disease. Contaminants that occur in a water system cover a broad range of physical, inorganic chemical, organic chemical, bacteriological, and radioactive parameters. These pollutants and contaminants can be removed from groundwater by applying various techniques, thereby bringing the water to a standard that is in order with the intended use.

The remediation techniques for water span biological, chemical, and physical treatment technologies. Most water treatment techniques utilize a combination of technologies. However, the type and degree of treatment are strongly dependent upon the source and intended use of the water. For example, water for domestic use must be thoroughly disinfected to eliminate disease-causing microorganisms, but may contain appreciable levels of dissolved calcium and magnesium (hardness). On the other hand, water to be used in industrial boilers may contain bacteria that must be removed for the waster to be relatively contaminant free to prevent scale formation on the inside of the boiler and pipes.

4.1 Physical treatment technologies

Previously in this text (Chapter 6), reference has made to the use of physical remediation methods that have been used, and continue to be used, to remediate oil sill in (for the most part) the ocean. There are two main steps in controlling the oil spills are containment and recovery which are mainly used as a barrier to control the spreading oil spill without changing its physical and chemical characteristics. Different equipment are used to control oil spills which are: (i) boomers, (ii) skimmers, (iii) adsorbent materials, and (iv) thermal methods, which do in fact change the character of the oil through the combustion process.

Physical treatment techniques include, but are not limited to: pump and treat, air sparging, and dual-phase extraction (Reddy, 2008). Water purification starts on a physical level, with the removal of the largest particles and obstructions that are inherent in the water to be treated. Air sparging is one physical remediation method that can be used, which involves using pressurized air to strip water clean (Reddy et al., 1995). A more common method is to pump water directly, with filters stripping away and large gravel or rock materials, and then letting the water be further filtered biologically or chemically to ensure that the water is in relatively clean from.

In the physical processes the phase transfer of pollutants is induced and these treatments are typically cost effective and can be completed in short time periods (in comparison with biological treatment). Equipment is readily available and is generally not engineering or energy-intensive (Reddy, 2008).

4.1.1 Booms, skimmers, adsorbents, and thermal methods

Physical remediation methods are mostly to control oil spills in a water environment. There are two main steps in controlling the oil spills are containment and recovery. They are mainly used as a barrier to control the spreading oil spill without changing its physical and chemical characteristics. Different equipment are used to control oil spills which are: (i) booms, (ii) skimmers, and (iii) adsorbent materials.

Booms are used for both containment and recovery. They are floatation device which act like physical barriers which would not allow the oil to spread in water so that oil could be recovered. In the process, the book is typically placed across a narrow entrance to the ocean, such as a stream outlet or small inlet, to close off that entrance so that oil can't pass through into marshland or other sensitive habitat. In places where the boom can deflect oil away from sensitive locations, such as shellfish beds or beaches used by piping plovers as nesting habitat.

There are three main types of boom: (i) a hard boom which is like a floating piece of plastic that has a cylindrical float at the top and is weighted at the bottom so that it has a *skirt* under the water. If the water currents or the winds are not too strong, booms can also be used to send the make the spilled oil in a different direction, which is known as deflection booming, (ii) the sorbent boom looks like a long sausage made out of a material that absorbs oil but a sorbent boom does not have the skirt that are a part of a hard booms and a sorbent book may not be able to contain oil for prolonged period, and (iii) a fire boom which has the appearance metal plates with a floating metal cylinder at the top and thin metal plates that make the skirt in the water; this fire book is constructed to contain oil long enough that the oil can be lit on fire and burned.

During the recovery period, the booms are sailed through the heaviest sections of the spill at slow speed and a shipping vessel traps the spilled oil between the angle of the boom and the vessel hull. They are also characterized into fence boomer, curtain boomer and fire-resistant boomer on the basis of floating tendency, material with which they are made weight and stopping tendency.

Skimmers represent a variety of mechanical equipment used to physically remove floating spills from the water surface. Many designs use a conveyor belt that is placed to carry the spilled oil into a reservoir where it is collected for processing and recovery. Other skimmer technologies use suction to remove spilled material, while weir skimmers use gravity to gather skimmed oil into underwater storage tanks. Skimmers are generally effective only in calm water, and suction skimmers are susceptible to clogging by floating debris.

Oleophilic skimmers try to trap the oil from the surface with help of belts, disks, continuous chain of oleophilic material and then oil is squeezed out in the recovery tank. Weir skimmers use dam for trapping the oil inside and then it can be pumped out through a pipe or hose to storage tank for recycling purpose. The type and the thickness of oil spill determine the success of skimming. Skimmers are effective and work efficiently in calm water but can be adversely affected by floating debris.

Adsorbent materials which are oleophilic and hydrophobic in nature come out as a good controller of oil spills. After skimming operation, adsorbent are used to clean the remaining oil. These adsorbents can be natural organic, inorganic or synthetic materials. Natural organic sorbents includes peat, hay, feathers, ground corncob etc. they can soak up from 3 to 15 times their weight in oil. Natural inorganic sorbents includes perlite vermiculite, glass, clay, wool, sand and volcanic ash which can absorb up to 4 to 20 times their weight in oil. Synthetic absorbents include materials similar to plastic like polyethylene, and nylon fibers. They can absorb up to 70 times weight in oil but cannot be cleaned and reused.

Thermal remediation methods involve burning the oil and by use of this method high rates of oil can be removed by using minimal specialized equipment, such as fire resistant boom or igniters. The thermal technique is more beneficial in calm wind environment and for fresh spills or refined products which can burn quickly without any harm to marine life. Fire resistance booms are used to accumulate oil and concentrate into a slick that is thick enough to burn. The residue that remains is removed by the mechanical means. The thick oil and high supply oxygen condition are important for its success. Factors affecting the thermal methods are water temperature, speed, wave amplitude, wind direction, slick thickness, oil type and amount of weathering and emulsification that has occurred.

The way in which an oil slick breaks up and dissipates depends largely on the chemical and physical persistence of the crude oil or the crude oil product. Low-boiling products such as kerosene tend to evaporate and dissipate quickly and naturally. These are called non-persistent oils. In contrast, persistent oils, such as many crude oils, break up and dissipate more slowly and usually require a clean-up response. Physical properties such as the density, viscosity and pour point of the oil all affect its behavior. However, dissipation does not occur immediately and the time this takes depends on a series of factors, including the amount and type of oil spilled, the weather conditions and whether the oil remains on the water or is washed to the shoreline (Table 8.3).

Table 8.3 Some common factors affecting petroleum hydrocarbon biodegradation

Factor	Comment
Composition	Structure, amount, toxicity
Physical state	spreading, adsorption
Weathered or not	Evaporation, photo-oxidation.
Temperature	Evaporation rate, degradation rate
Mineral nutrients	May be inhibitors
Reaction	low pH may be limiting
Micro-organisms	Inability to degrade polynuclear systems

4.1.2 Pump and treat

Pump and treat is a common method for cleaning up groundwater contaminated with dissolved chemicals, including industrial solvents, metals, and fuel oil. Groundwater is pumped from wells to an above-ground treatment system that removes the contaminants. Pump and treat systems also are used to contain the contaminant plume. Containment of the plume keeps it from spreading by pumping contaminated water toward the wells. This pumping helps keep contaminants from reaching drinking water wells, wetlands, streams, and other natural resources. Pump and treat methods may involve installing one or more wells to extract the contaminated groundwater. Groundwater is pumped from these "extraction wells" to the ground surface, either directly into a treatment system or into a holding tank until treatment can begin. The treatment system may consist of a single cleanup method, such as activated carbon or air stripping, to clean the water.

In the process, free-phase contaminants and/or contaminated groundwater are pumped directly out of the surface. Treatment occurs above ground, and the cleaned groundwater is either discharged into sewer systems or re-injected into the subsurface. Pump and treat systems have been operated at numerous sites for many years. Unfortunately, data collected from these sites reveals that although pump and treat may be successful during the initial stages of implementation, performance drastically decreases at later times. As a result, significant amounts of residual contamination can remain unaffected by continued treatment. Due to these limitations, the pump and treat method is now primarily used for free product recovery and control of contaminant plume migration.

Pump and treat is a safe way to both clean up contaminated groundwater and keep it from moving to other areas where it may affect drinking water supplies, wildlife habitats, or recreational rivers and lakes. Although pumping brings contamination to the ground surface, it does not expose people to that contamination. A pump and treat system is monitored to ensure the extraction wells and treatment units operate as designed. Also, the groundwater is sampled to ensure the plume is decreasing in concentration and is not spreading.

Pump and treat may last from a few years to several decades. The actual cleanup time will depend on several factors, which vary from site to site. For example, it may take longer where: (i) the concentrations of the contaminants are high, or the contamination source has not been completely removed, (ii) the contaminant plume is large, and (iii) the groundwater flow is slow, or the flow path is complex.

Pump and treat is used to remove a wide range of contaminants that are dissolved in groundwater. Pump and treat typically is used once the source of groundwater contamination, such as leaking drums and contaminated soil, has been treated or removed from the site. It also is used to contain plumes so that they do not move offsite or toward lakes, streams, and water supplies. Pump and treat is the most common cleanup method for groundwater.

4.1.3 Air sparging

Air sparging, also known as in situ air stripping and/or in situ volatilization is an in situ remediation technique, used for the treatment of saturated groundwater by volatile organic compounds (VOCs), such as crude oil hydrocarbon derivatives which is a widespread problem for the ground water health. The vapor extraction has manifested itself into becoming very successful and practical when it comes to disposing of volatile organic compounds. It was used as a new development when it came to saturated zone remediation when using air sparging. Being that the act of it was to inject a hydrocarbon-free gaseous medium into the ground where contamination was found.

In situ air sparging is a means by which to enhance the rate of mass removal from contaminated saturated-zone systems. Air sparging involves injecting air into the target contaminated zone, with the expectation that volatile and semi-volatile contaminants will undergo mass transfer (volatilization) from the groundwater to the air bubbles. Because of buoyancy, the air bubbles generally move upward toward the vadose zone, where a soil-venting system is usually employed to capture the contaminated air stream. As the contaminants move into the soil, a soil vapor extraction system is usually used to remove vapors.

In the process, a gas (usually air/oxygen) is injected under pressure into the saturated zone to volatilize groundwater contaminants and to promote biodegradation by increasing subsurface oxygen concentrations. Volatilized vapors migrate into the vadose zone where they are extracted via vacuum, generally by a soil vapor extraction system. The term biosparging is sometimes used interchangeably with air sparging to highlight the bioremediation aspect of the treatment process or can refer to situations where biodegradation is the dominant remedial process, with volatilization playing a secondary role.

The process is based on the principles of air flow dynamics and contaminant transport, transfer and transformation processes (Reddy and Adams, 2001). In the process, injected air moves through aquifer materials in the form of either bubbles or micro-channels. The density of bubbles or micro-channels is found to be depended on the injected air flow rate. Mineral heterogeneity will significantly affect the air flow patterns and the zone of influence. The transport mechanisms include advection, dispersion, and diffusion. The mass transfer mechanisms include volatilization, dissolution, and adsorption/desorption. Besides these, biodegradation is enhanced due to increased dissolved oxygen that can promote aerobic biodegradation.

Air sparging is generally applied for commercial usage. Air sparging contaminant groups are volatile organic compounds and fuels found in groundwater. Air sparging is usually applied to the lighter gasoline constituents such as benzene, ethylbenzene, toluene, and xylene. This method is typically not applied on the higher boiling fuel products such as kerosene and diesel fuel. The usage of air sparging is commonly applied when cleaning up contaminated water under buildings and obstacles to prevent the further contamination of that water source. The usage of air sparging and soil vapor extraction is safe when properly conducted. This makes sure only clean air that meets a certain quality standard is released, therefore it does not pose a threat when the proper sample method is done to make sure that hazardous gases do not exit into the atmosphere. If vapor treatment is required, the extraction wells are connected to an aboveground treatment system (such as a bioreactor or an activated carbon adsorption unit).

Thus, the air sparging system can be adapted to treat the off-gases (referred as contaminated vapors and extracted air). The vapor is treated with granulated activated carbon prior to release to the atmosphere. For example, arsenic-contaminated groundwater were treated by air sparging and what the treatment does is remove arsenic at certain percentage using solution of iron and arsenic only at a molar ratio of 2. Treatment using air sparging is beneficial as groundwater contains high amounts of dissolved iron, which contains the theoretical capacity for the treatment.

4.1.4 Dual phase vacuum extraction

The dual-phase vacuum extraction (DPVE) process (also known as multi-phase extraction process, vacuum-enhanced extraction, or even sometimes as bioslurping) is a technology that uses a high-vacuum system to remove both contaminated groundwater and soil vapor.

In the process, vapor and liquid can be withdrawn together under a vacuum using a down well stinger (called slurping) or separately; extracting the liquid with a submersible pump and the vapor with a soil vapor extraction unit. In recent times the stinger approach is used and has been proven to be generally more efficient. Additional benefits of the process are that the system can be used to lower the water table within the contamination plume, this causes contaminants in the newly exposed capillary fringe to be accessible to soil vapor extraction (SVE). Each system must be designed and operated on a site by site basis with the technique being particularly well suited to medium permeability horizons with generally volatile or semi volatile contaminants or free phase products.

Abstracted water and free phase products are separated, the water is treated and either returned to the site or sent to the sewer for disposal. The free phase product is collected and disposed off-site to a recycling facility abstracted vapors are treated to remove volatile organic compounds. The technique also encourages the ingress of air and thus oxygen into the plume which can enhance microbial activity and thus in situ biological treatment. This is often termed bio-slurping or bioventing and can be useful to reduce dissolved phase contaminants following product removal.

In the actual process, a high-vacuum extraction well is installed with its screened section in the zone of contaminated soils and groundwater. Fluid/vapor extraction systems depress the water table and water flows faster to the extraction well and the dual-phase vacuum extraction process removes contaminants from above and below the water table. As the water table around the well is lowered from pumping, unsaturated soil is exposed. This area (the capillary fringe) is often highly contaminated, as it holds undissolved chemicals, chemicals that are lighter than water, and vapors that have escaped from the dissolved groundwater below. Contaminants in the newly exposed zone can be removed by vapor extraction. Once above ground, the extracted vapors and liquid-phase organics and groundwater are separated and treated. Use of dual-phase vacuum extraction with these technologies can shorten the cleanup time at a site, because the capillary fringe is often the most contaminated area.

4.1.5 Monitoring-well oil skimming

Of particular interest to the crude oil recovery industry is the monitoring-well oil skimming technology. Briefly, an oil skimmer is a device that is designed to remove oil floating on a liquid surface. Depending on the specific design they are used for a variety of applications such as oil spill response, as a part of oily water treatment systems and collecting fats, oils, and greases in wastewater treatment in the food manufacturing industries.

In the crude oil and natural gas industries, monitoring-wells are often drilled for the purpose of collecting ground water samples for analysis. These wells, which are usually six inches or less in diameter, can also be used to remove hydrocarbon derivatives from the contaminant plume within a groundwater aquifer by using a belt-style oil skimmer. Belt oil skimmers, which are simple in design, are commonly used to remove oil and other floating hydrocarbon contaminants from industrial water systems.

A monitoring-well oil skimmer remediates various oils, ranging from light fuel oils such as petrol, light diesel or kerosene to heavy products such as No. 6 oil, creosote and coal tar. It consists of a continuously moving belt that runs on a pulley system driven by an electric motor. The belt material has a strong affinity for hydrocarbon liquids and for shedding water.

In the process, the belt is lowered into the monitoring well past the light non-aqueous phase liquid interphase with water and as the belt moves through this interface, it picks up liquid hydrocarbon contaminant which is removed and collected at ground level as the belt passes through a wiper mechanism. To the extent that the dense light non-aqueous phase liquid hydrocarbon derivatives settle at the bottom of a monitoring well, and the lower pulley of the belt skimmer reaches them, these contaminants can also be removed by a monitoring-well oil skimmer.

Typically, belt skimmers remove very little water with the contaminant, so simple weir-type separators can be used to collect any remaining hydrocarbon liquid, which often makes the water suitable for its return to the aquifer. Because the small electric motor uses little electricity, it can be powered from solar panels or by a wind turbine, making the system self-sufficient. There are many different types of oil skimmer. Each type has different design features and therefore results in different applications and use. It is important to understand the design features before employing a particular skimmer type (Table 8.4).

Weir skimmers function by allowing the oil floating on the surface of the water to flow over a weir. There are two main types of weir skimmer, those that require the weir height to be manually adjusted and those where the weir height is automatic or self-adjusting. Whilst manually adjusted weir skimmer types can have a lower initial cost, the requirement for regular manual adjustment makes self-adjusting weir types more popular in most applications. Weir skimmers will collect water if operating when oil is no longer present. To overcome this limitation most weir type skimmers contain an automatic water drain on the oil collection tank.

Belt oil skimmers is one of the most reliable and economical equipment removing liquid surface floating oil, low electric consumption, no need any consumable, can effective remove all kinds of floating oil (contain machine oil, kerosene, diesel oil, lubricating oil, plant oil and other liquid specific weight less than water, no matter the water surface oil layer thick or thin, all can absorb), can make oil-water separation and waste oil reuse, its principle is adopt different gravity and surface tension etc. characteristic, when the oil skimming belt pass water surface can absorb and take away the surface all kinds of floating oil.

Table 8.4 Factors to consider for the oil skimming process

Factor	Comment
Oil removal flow rate	Alternative skimmer designs have different oil removal flow rates.
	Volume removal rates for oleophilic skimmer types are comparatively low.
	Weir type skimmers are capable of very high oil and water removal rates.
Oil removal concentration	Oil skimmers remove a mixture of oil and water.
	In most situations the 'oil' mixture removed is an emulsion of oil and water more like a 'mousse'.
	Oleophilic and non-oleophilic skimmers can provide a more concentrated oil in the removal stream.
Effectiveness	Oleophilic and non-oleophilic skimmers are not equally effective with all oil.
	Oleophilic skimmers may not work as effectively if there are surfactants in the water that interfere with the oleophilic attraction.
Weir skimmers	Not affected by chemicals.
Effects of trash and debris	May block or interfere with the operation of oil skimmers.
Skimming direction	some skimmers only remove oil from one direction.
Weir skimmer	Allows the oil floating on the surface of the water to flow over a weir.

Oleophilic skimmers function by using an element such as a drum, disc, rope or mop to which the oil adheres. The oil is wiped from the oleophilic surface and collected in a tank. As the oil is adhering to a collection surface the amount of water collected when oil is not present will be limited. On the other hand, non-oleophilic skimmers are distinguished by the component used to collect the oil. A metal disc, belt or drum is used in applications where an oleophilic material is inappropriate, such as in a hot alkaline aqueous parts washer. The skimmer is generally turned off whenever there is no oil to skim thus minimizing the amount of water collected. Metal skimming elements are nearly as efficient as oleophilic skimmers when oil is present.

4.2 Chemical treatment technologies

Chemical remediation can be achieved through a variety of methods, including carbon absorption, ion exchange, oxidation, and chemical precipitation. Chemical remediation is often used alongside physical water treatment to achieve the best results, and can help achieve the cleanest groundwater after the fact.

The chemical treatment of water is the process that removes the majority of the contaminants from water and produces a liquid effluent suitable for disposal to the natural environment (Reddy, 2008). Hover, not all water systems are equal and some waters require different and sometimes specialized treatment methods. At the simplest level, water treatment is carried out through separation of solids and liquids, usually by sedimentation. By progressively converting dissolved material into solids, usually a biological floc, which is then settled out, an effluent stream of increasing purity is produced.

4.2.1 Adsorption/absorption

Adsorption is the adhesion of atoms, ions, or molecules from a gas, liquid or dissolved solid though chemical forces to a surface thereby creating a film of the *adsorbate* on the surface of the *adsorbent*. This process differs from absorption in which a fluid (the *absorbate*) is dissolved or permeates a liquid or solid (the *absorbent*), respectively. Adsorption is a surface phenomenon while absorption involves the whole volume of the material. The term sorption encompasses both adsorption and absorption while desorption is the reverse process.

An adsorbent is any material with a capability to adsorb substance and the amount of surface provided by the adsorbent to appeal the particles of contaminants determine its effectiveness (Table 8.5) Nano-adsorbents are nanomaterials from different organic and inorganic materials of high capacity and capability to absorb certain substances. Nano-adsorbents include nanostructured mixed oxides, metallic nanoparticles, metallic oxide nanoparticles, and magnetic nanoparticles. Some nano-sized metal oxides, such as ferric, aluminum, manganese, biochar, magnesium, cerium, and titanium oxides can provide specific affinity and high surface area to adsorb intensive metals from liquid materials. Nano-adsorbents can be used in different forms, such as thin film and nanocomposite membrane, adsorptive and bioactive nanomaterial, and biomimetic and catalytic membrane). Membrane filters or composite nano-fibrous media can discard bacteria or viruses first and then heavy metals or ions and organic matters in a contaminated mix wastewater at the common form (Thines et al., 2017; Bishoge et al., 2018).

Similar to surface tension, adsorption is a consequence of surface energy. In a bulk material, all the bonding requirements (be the requirement ionic, covalent, or metallic) of the constituent atoms of the material are filled by other atoms in the material. However, atoms on the surface of the adsorbent are not wholly surrounded by other adsorbent atoms and therefore can attract adsorbates. The exact nature of the bonding

Table 8.5 Factors to consider for the adsorption process

1.	Possibility of product recovery.
2.	Excellent control and response to process changes.
3.	No chemical-disposal problem when pollutant (product) recovered and returned to process.
4.	Capability to remove gaseous or vapor contaminants from process streams to extremely low levels.
5.	Product recovery may require distillation or extraction.
6.	Adsorbent may deteriorate as the number of cycles increase.
7.	Adsorbent regeneration.
8.	Prefiltering of gas stream possibly required to remove any particulate materials capable of plugging the adsorbent bed.

depends on the details of the species involved, but the adsorption process is generally classified as physisorption (which is characteristic of weak van der Waals forces) or chemisorption (which is characteristic of covalent bonding). Chemisorption may also occur due to electrostatic attraction.

Adsorption is present in many natural, physical, biological and chemical systems and is widely used in industrial applications such as heterogeneous catalysts, activated charcoal Adsorption chromatography is a sorption processes in which certain adsorbates are selectively transferred from the fluid phase to the surface of insoluble, rigid particles suspended in a vessel or packed in a column.

The most common adsorbent for remediation is activated that is derived from bituminous coal (Speight, 2013). The activate carbon adsorbs/absorbs organic compounds from water by physically or chemically binding them to the carbon atoms.

Granular activated carbon adsorption: a form of activated carbon with a high surface area, adsorbs many compounds including many toxic compounds. Water passing through activated carbon is commonly used in municipal regions with organic contamination, taste or odors. If water is held in the carbon block for longer periods, microorganisms can grow inside which results in fouling and contamination. Silver nanoparticles are excellent anti-bacterial material and they can decompose toxic halo-organic compounds such as pesticides into non-toxic organic products.

Iron oxide adsorption filters the water through a granular medium containing ferric oxide. Ferric oxide has a high affinity for adsorbing dissolved metals such as arsenic. The iron oxide medium eventually becomes saturated, and must be replaced. The sludge disposal is also an issue. In addtion, activated alumina is an adsorbent that effectively removes arsenic. Long-term column performance has been possible through the efforts of community-elected water committees that collect a local water tax for funding operations and maintenance. It has also been used to remove undesirably high concentrations of fluoride.

4.2.2 Ion exchange

Ion exchange is, as the name indicates, an exchange of ions between two electrolytes or between an electrolyte solution and a complex. In most cases the term is used to denote the processes of purification, separation, and decontamination of aqueous and other ion-containing solutions with solid polymer-type ion exchangers or mineral-type ion exchangers.

The process involves passing the water successively over a solid cation exchanger and a solid anion exchanger, which replace cations and anions by hydrogen ion and hydroxide ion, respectively, so that each equivalent of salt is replaced by a mole of water. Demineralization by ion exchange generally produces water of a very high quality. Unfortunately, some organic compounds in wastewater have an adverse effect on ion exchanger units, and microbial growth on the exchangers can diminish their efficiency. In addition, regeneration of the resins is expensive, and the concentrated wastes from regeneration require disposal in a manner that will not damage the environment.

Typical ion exchangers are ion-exchange resins (functionalized porous polymers or gel polymers), zeolites, montmorillonite clay minerals, and soil humus. Ion exchangers are either (i) cation exchangers, which exchange positively charged ions—cations, or (ii) anion exchangers which exchange negatively charged ions—anions. There are also amphoteric exchangers that are able to exchange both cations and anions simultaneously. However, the simultaneous exchange of cations and anions can be more efficiently performed in *mixed beds*, which contain a mixture of anion- exchange resins and cation-exchange resins, or passing the treated solution through several different ion-exchange materials.

Ion exchanges can be unselective or have binding preferences for certain ions or classes of ions, depending on the chemical structure which can be dependent on the size of the ions, their charge, or their structure. Typical examples of ions that can bind to ion exchangers are (without any order of preference): (i) H^+ (proton) and OH^- (hydroxide), (ii) singly charged monatomic ions such as Na+, K+, and Cl^-, (iii) doubly charged monatomic ions such as Ca^{2+} and Mg^{2+}, (iv) polyatomic inorganic ions, such as the sulfate ion, $SO_4{}^{2-}$, and the phosphate ion, $PO_4{}^{3-}$, (v) organic bases that are usually, molecules containing the amine functional group, $-NR_2H^+$, (vi) organic acids, often molecules containing the $-CO_2^-$ functional group, (vii) and biomolecules that that can be ionized, such as amino acids, peptides, and proteins.

Along with adsorption and absorption, ion exchange is a form of sorption and is a reversible insofar as thee ion exchanger can be *regenerated* or *loaded* with desirable ions by washing with an excess of these ions. Ion exchange for water remediation is virtually always carried out by passing the water downward under pressure through a fixed bed of granular medium (either cation exchange media or anion exchange media) or spherical beads. In the process, cations are displaced by certain cations from the solutions and ions are displaced by certain anions from the solution. Ion exchange media most often used for remediation are zeolite derivatives (both natural and synthetic) and synthetic resins.

Ion exchange systems use ion exchange resin-packed columns or zeolite-packed columns to replace unwanted ions. The most common case is water softening which consists of the removal of calcium (Ca^{2+}) ions and magnesium (Mg^{2+}) ions and replacing them with the more benign sodium (Na^+) ions or potassium (K^+) ions. Ion exchange resins are also used to remove toxic ions such as (among others) lead, mercury, and arsenic. Radium can be removed by ion exchange, or by water conditioning. The back flush or sludge that is produced is, however, a low-level radioactive waste.

4.2.3 Oxidation

Oxidation is any chemical reaction that involves the moving of electrons—more specifically, it means the substance that loses electrons is oxidized. Oxidation is the opposite of reduction. A reduction-reaction always comes together with an oxidation-reaction and, when this happens, the reaction is a redox reaction (reduction and oxidation). Oxygen does not have to be present in a reaction for it to be a redox-reaction.

In the oxidation remediation process, oxidizing agents are delivered in the subsurface to destroy (converted to water and carbon dioxide or to nontoxic substances) the organics molecules. The oxidants are introduced as either liquids or gasses. Oxidants include air or oxygen, ozone, and certain liquid chemicals such as hydrogen peroxide, potassium permanganate, and potassium (or sodium) persulfate. In fact, hydrogen peroxide (H_2O_2) and chlorine dioxide (ClO_2) are relatively easy to incorporate into various environmental media under treatment, including water, waste water, leachate, air, and soil. Hydrogen peroxide and chlorine dioxide are frequently used as disinfectants, bleaching, and oxidizing agents. They can oxidize hazardous materials that are either organic or inorganic compounds. Sometimes when these oxidizing agents cannot completely degrade the contaminants, they can transform the contaminants into constituents that are amenable to other forms of degradation, such as biological processes (Gates and Siegrist, 1994).

Chlorine dioxide is an effective water disinfectant that is of particular interest—in the neutral pH range, chlorine dioxide in water remains largely in the molecular form until it contacts a reducing agent with which to react. Chlorine dioxide is a gas that is highly reactive with organic matter and explosive when exposed to light. As a water disinfectant, chlorine dioxide does not chlorinate or oxidize ammonia or other nitrogen-containing compounds.

Ozone (O_3, sometime referred to as tri-oxygen) is an inorganic gas that is formed from dioxygen by the action of ultraviolet light and electrical discharges within the atmosphere of the Earth. Ozone is a powerful oxidant and has many industrial and consumer applications related to oxidation. This same high oxidizing potential, however, causes ozone to damage mucous and respiratory tissues in animals, and also tissues in plants, above concentrations on the order of about 0.1 ppm.

Ozone and oxygen gas can be generated on site from air and electricity and directly injected into soil and groundwater contamination. The process has the potential to oxidize and/or enhance naturally occurring aerobic degradation. The techniques has proven to be effective technique for the removal of dense non-aqueous phase liquids (DNAPL) which are more dense than water and is immiscible in or does not dissolve in water and which tend be below the surface of the water or at the bottom of the water system.

In situ chemical oxidation is a form of advanced oxidation processes and advanced oxidation technology, that is an environmental remediation technique used for groundwater remediation to reduce the concentrations of targeted environmental contaminants to acceptable levels. The process is accomplished by injecting or otherwise introducing strong chemical oxidizers directly into the contaminated medium (soil or groundwater) to destroy chemical contaminants in place. It can be used to remediate a variety of organic compounds, including some that are resistant to natural degradation.

In subterranean arsenic removal (SAR), aerated groundwater is recharged back into the aquifer to create an oxidation zone which can trap iron and arsenic on the soil particles through adsorption process. The oxidation zone created by aerated water boosts the activity of the arsenic-oxidizing microorganisms which can oxidize arsenic from +3 to +5 state. No chemicals are used and almost no sludge is produced during operational stage since iron and arsenic compounds are rendered inactive in the aquifer itself. Thus toxic waste disposal and the risk of its future mobilization is prevented. Also, it has very long operational life, similar to the long lasting tube wells drawing water from the shallow aquifers.

The greatest advantage of this process is there is no need for sludge handling. The arsenic which is trapped into the sand along with the iron flocs constitute an infinitesimal volume of the total volume being handled and hence pose very little environmental threat in its adsorbed form. The whole mass remains down below unlike other processes where there is extra cost of sludge handling and messy disposal problem. The process is chemical free, simple and easy to handle. There is no restriction to the volume it can handle as long as proper time is allowed for the oxygen rich impregnated water to create the adequate oxidizing zone in the deep aquifer.

4.2.4 Permeable reactive barriers

A permeable reactive barrier (PRB, also referred to as a permeable reactive treatment zone, PRTZ), is a technology for in situ groundwater remediation. A permeable reactive barrier is an emplacement of reactive media in the sub-surface designed to intercept a contaminated plume, provide a flow path through the reactive media and transform the contaminant(s) into environmentally acceptable forms to attain remediation concentration goals downgradient of the barrier. The concept behind the permeable reactive barrier technology is that a permanent, semi-permanent or replaceable reactive media is placed in the sub-surface across the flow path of a plume of contaminated groundwater which must move through it under the natural gradient, thereby creating a passive treatment system. Typical reactant media contained in the barriers includes media designed for degrading volatile organics, chelators for immobilizing metals, or nutrients and with a porous material—such as sand—to enhance groundwater flow through the barrier.

In the barrier, treatment walls remove contaminants from groundwater by degrading, transforming, precipitating, adsorbing or adsorbing the target solutes as the water flows through permeable reactive trenches. Permeable reactive barriers are designed to be more permeable than the surrounding aquifer materials so that water can readily flow through it maintaining groundwater hydrogeology while contaminants are treated.

Permeable reactive barriers allow some, but not all, materials to pass through. One definition for a permeable reactive barrier is an in situ treatment zone that passively captures a plume of contaminants and removes or breaks down the contaminants, releasing uncontaminated water. The primary removal methods include: (i) sorption and precipitation, (ii) chemical reaction, and (iii) reactions involving biological mechanisms (Scherer et al., 2000; Hashim et al., 2011; ITRC, 2011). Although often considered, with some justification to be a chemical method of treatment, certain types of permeable reactive barriers utilize biological organisms in order to remediate groundwater.

Thus, a permeable reactive barrier is a subsurface emplacement of reactive materials through which a dissolved contaminant plume must move as it flows, typically under natural gradient. Treated water exits the other side of the permeable reactive barrier. This in situ method for remediating dissolved-phase contaminants in groundwater combines a passive chemical or biological treatment zone with subsurface fluid flow management. The permeable reactive barrier can be installed as permanent unit or as a semi-permanent unit. The most commonly used permeable reactive barrier configuration is that of a continuous trench (perpendicular to and intersects the groundwater plume) in which the treatment material is backfilled. These permeable reactive barriers use collection trenches, funnels, or complete containment to capture the plume and pass the groundwater, by gravity or hydraulic head, through a vessel containing either a single treatment medium or sequential media.

A frequently used configuration is the *funnel and gate* method in which low-permeability walls (the funnel) direct the groundwater plume toward a permeable treatment zone (the gate). A funnel-and-gate system has an impermeable section, called the *funnel*, which directs the captured groundwater flow toward the permeable section (the *gate*). The funnel walls may be aligned in a straight line with the gate, or other geometric arrangements of funnel-and-gate systems can be used depending on the site conditions. This funnel-and-gate configuration allows better control over reactive cell placement and plume capture. At sites where the groundwater flow is very heterogeneous, a funnel-and-gate system can allow the reactive cell to be placed in the more permeable portions of the aquifer. At sites where the contaminant distribution is very non-uniform, a funnel-and-gate system can better homogenize the concentrations of contaminants entering the reactive cell. A system with multiple gates can also be used to ensure sufficient residence times at sites with a relatively wide plume and high groundwater velocity.

Injection or fracturing techniques also have been used to construct subsurface permeable reactive barriers. These techniques use permanent or temporary injection wells to place reactive materials such as zero-valent iron (Fe^0) or edible oils into the subsurface. The injection points are spaced to provide overlapping radii of influence between them to form a free standing treatment zone. The majority of installed trench style permeable reactive barriers s use zero-valent iron for converting contaminants to non-toxic or immobile species. The zero-valent iron is a mild reductant, has the ability to reductively dehalogenate many halogenated hydrocarbon derivatives as well as remove hexavalent chromium, arsenic, and uranium.

There are materials other than zero-valent iron (Fe^0) that can be used to treat contaminants. For example, petroleum hydrocarbon plumes such as those emanating from a variety of sources and which are not amenable to reaction with zero-valent iron, can be treated using a biosparging or slow release oxygen compound containment wall. The aerobic condition created within the wall allows for biodegradation of the dissolved contaminants as they pass through the barrier. Mulch and other vegetative materials can be employed in traditional trench style permeable reactive barriers to treat a number of contaminants—an example is a permeable mulch barrier for remediation of chlorinated solvents and perchlorate.

Reactive chemical agents available for permeable reactive barriers can also be used to precipitate contaminants. These chemicals can modify the pH and redox conditions resulting in metal precipitation as hydroxides (Blowes et al., 2000). The materials that can be used include ferrous salts, lime, limestone, fly ash, phosphate, chemicals such as magnesium hydroxide [$Mg(OH)_2$], magnesium carbonate ($MgCO_3$), calcium chloride ($CaCl_2$), calcium sulfate (CaSO4), and zero-valent metals. The immobilized contaminants and toxic degradation intermediates might be re-mobilized upon environmental condition changes.

Zeolites also have the potential to be used as treatment mineral in the permeable reactive barriers due to high ion exchange, adsorbing, catalytic and molecular sieving capacities (Hashim et al., 2011). Also, fly ash zeolites are byproducts from hydrothermal treatment processes of hard coal fly ash by sol exhibit different dimensions of their channels and cages forming the crystal lattice.

A benefit of a biological permeable reactive barrier system over purely abiotic systems is that the treatment processes may extend downgradient of the constructed treatment zone due to migration of soluble organic carbon, enabling the effects of anaerobic degradation beyond the biowall. Another benefit is the ability of a single system to treat multiple contaminants with different chemical characteristics, including both organic compounds and inorganic compounds (such as e.g., nitrate, sulfate, perchlorate, and metals).

4.2.5 Precipitation

The process of precipitation/coagulation/flocculation transforms dissolved contaminants into insoluble solids, assisting in subsequent removal of the contaminants from the liquid phase through sedimentation or filtration. The process usually uses pH adjustment, addition of a chemical precipitant and flocculation. Description: The precipitation of metals has been the main method of treating metal-impacted industrial wastewaters. In the treatment of ground water, the metal precipitation method is utilized as a pretreatment for other treatment technologies such as chemical oxidation or air stripping, where the occurrence of metals has the potential to hinder with the other treatment processes.

Metal precipitation from impacted water involves the alteration of soluble heavy metal salts to insoluble salts that will precipitate. The precipitate can be removed from the treated water by means of physical methods such as clarification (settling) or filtration. The process typically uses pH adjustment, the supplement of a chemical precipitant and flocculation. Metals then precipitate out from the solution as hydroxide derivatives, sulfide derivatives, or carbonate derivatives. The solubility of each of the metal contaminants and the cleanup standards needed to remove these contaminants dictate which process is used. In some situations, process design brings about the production of sludge that can be sent to recycling units for recovery of any valuable (or unwanted) metals.

Precipitation is commonly used in water treatment to remove metal ions that arise from the so-called heavy metals (Duffus, 2002). Generally, a heavy metal is considered to be a metal with a relatively high density, atomic weight, or atomic number. A density of more than $5\,g/cm^3$ is sometimes quoted as a commonly used criterion. Heavy metals are often assumed to be highly toxic or damaging to the environment—some metals are highly toxic or damaging to the environment while certain others are toxic only if taken in excess or encountered in certain forms.

Heavy metals (and their respective derivatives) occur in the crust of the Earth and may be solubilized in groundwater through natural processes or by change in soil pH. Moreover, groundwater can be contaminated with heavy metals from landfill leachate, sewage, leachate from mine tailings, deep-well disposal of liquid wastes, and seepage from industrial waste lagoons or from industrial spills and leaks. A variety of reactions in soil environment e.g. acid/base, precipitation/dissolution, oxidation/reduction, sorption or ion exchange processes can influence the speciation and mobility of metal contaminants.

In general, the process for removal of heavy metals involves addition of agent to an aqueous waste stream in a stirred reaction vessel, either by a batch process or a continous process with steady flow. Most metals can be converted to insoluble compounds by chemical reactions between the agent and the dissolved metal ions. The insoluble compounds (precipitates) are removed by settling and/or filtering.

Precipitation is becoming the most widely selected means for removing heavy metals from groundwater in pump-and-treat operations. It is used as a pretreatment process with other technologies (such as chemical oxidation or air stripping, where the presence of metals would interfere with treatment. After water is pumped to the surface, precipitation converts soluble heavy metals to insoluble metals that settle and/or are filtered out of the water.

The precipitation process usually adjusts pH, adds chemicals that stimulate precipitation, adds coagulants, and mixes the fluid in a device that promotes coalescence of the particles (flocculation.) The chemical precipitants, coagulants, and flocculants are all used to increase particle size through aggregation. Commonly used precipitants include carbonates, sulfates, sulfides, lime, and other hydroxides. The precipitants generate very fine particles that are held in suspension. Coagulants are often added to aggregate the suspended particles. Mixing in a flocculating unit following the addition of coagulants promotes contact among the particles, which in turn promotes particle growth and settling. In anaerobic environments, bacteria react with soluble metals to form insoluble metal precipitants.

Thus, in the process, chemical precipitants, coagulants, and flocculation agents are used to increase particle size through aggregation. The precipitation process can generate very fine particles that are held in suspension by electrostatic surface charges. These charges cause clouds of counter-ions to form around the particles, giving rise to repulsive forces that prevent aggregation and reduce the effectiveness of subsequent solid-liquid separation processes. Therefore, chemical coagulants are often added to overcome the repulsive forces of the particles. The three main types of coagulants are inorganic electrolytes (such as alum, lime, ferric chloride, and ferrous sulfate), organic polymers, and synthetic polyelectrolytes with anionic or cationic functional groups. The addition of coagulants is followed by low-sheer mixing in a flocculating unit to promote contact between the particles, allowing particle growth through the sedimentation phenomenon (flocculants settling).

Flocculent settling refers to a dilute suspension of particles that flocculate for the duration of the sedimentation process. As flocculation takes place, the particles intensify in mass and settle at a quicker rate. As coalescence or flocculation occurs, the particles increase in mass and settle at a faster rate. The amount of flocculation that occurs depends on the opportunity for contact, which varies with the overflow rate, the depth of the basin, the velocity gradients in the system, the concentration of particles, and the range of particles sizes. The effects of these variables can only be accomplished by sedimentation tests prior to application of the process.

4.2.6 Surfactant-enhanced remediation

Surfactants are unique chemical agents that greatly enhance the solubility of organic contaminants in aqueous media. They are also able to reduce the interfacial tension (IFT, the force existing where two fluids meet that keeps them as separate fluids) between the aqueous and organic phases to mobilize the organic phase.

The primary objective in surfactant-enhanced process design is to remove the maximum amount of contaminant with a minimum amount of chemicals and in minimal time while maintaining hydraulic control over the injected chemicals and contaminant. Each step in the design process must keep this in mind. Design challenges include precisely locating the dense non-aqueous phase liquid, finding the optimum surfactant solution for a given dense non-aqueous phase liquid composition and type, and fully characterizing the hydraulic properties of the aquifer, particularly the heterogeneities typically present in the subsurface environment. Because it is impossible to know with certainty the variations in aquifer properties over the treatment zone, numerical modeling tools are used to simulate how the system may respond in the presence of these unknown factors. Numerical modeling is also necessary to understand the dynamics of the flooding process under the hydrogeologic conditions at the site.

Surfactant-enhanced remediation—like surfactant enhanced crude oil recovery (Speight, 2014)—increases the mobility and solubility of the contaminants present as dense non-aqueous phase liquids (DNAPLs). Thus, the process is used most often when the groundwater is contaminated by dense non-aqueous phase liquids. These dense compounds, such as trichloroethylene (TCE), sink in groundwater because they have a higher density than water. They then act as a continuous source for contaminant plumes that can stretch for miles within an aquifer. These compounds may biodegrade very slowly. They are commonly found in the vicinity of the original spill or leak where capillary forces have trapped them.

A surfactant flood can be designed to remove contaminants either primarily by solubilization or primarily by mobilization. Surfactant mobilization can remove denser non-aqueous phase liquids in less time; however, there is greater risk of uncontrolled downward movement of dense non-aqueous phase liquids, as the dense non-aqueous phase liquids is being physically displaced by the surfactant solution. Thus, to conduct a mobilization flood, it is necessary to have an aquitard as a barrier to prevent vertical migration of the dense non-aqueous phase liquids. It is important to identify from the outset whether solubilization or mobilization of the dense non-aqueous phase liquids is desired, because not all surfactants can be used to conduct a mobilization flood.

Surfactant enhanced aquifer remediation (SEAR), in its most basic form, could thus be considered a chemical enhancement to pump and treat. A chemical solution is pumped across a contaminated zone by introduction at an injection point and removal from an extraction point. To cover the entire contaminated zone, a number of injection and extraction wells are used; the well configuration is determined by the subsurface distribution of non-aqueous phase liquids and the hydrogeologic properties of the aquifer.

In the process, surfactants are injected into the contaminated water. A typical system uses an extraction pump to remove groundwater downstream from the injection point. The extracted water is treated aboveground to separate the injected surfactants from the contaminants and groundwater. Once the surfactants have separated from the water they are re-used. The surfactants used are non-toxic, food-grade, and biodegradable.

4.3 Biological treatment technologies

If a physical or chemical treatability method show little or no degradation of the contaminant(s) in contamination contained in the water, inoculation with strains known to be capable of degrading the contaminant(s) may be helpful. A bioremediation process generally requires a mechanism for stimulating and maintaining the activity of these microorganisms. This mechanism is usually a delivery system for providing one or more of the following: An electron acceptor (oxygen, nitrate); nutrients (nitrogen, phosphorus); and an energy source (carbon). Generally, electron acceptors and nutrients are the two most critical components of any delivery system. In biological water remediation, biologic materials help to break down unwanted chemicals that are not easily separated from the water, particularly in industrial waste that forms in groundwater. An advantage to this method is that physically pumping groundwater out is not required to treat it.

Biological treatment processes (or biotreatment processes) are those processes which remove dissolved and suspended organic chemical constituents through biodegradation, as well as suspended matter through physical separation. Biotreatment demands that the appropriate reactor conditions prevail in order to maintain sufficient levels of viable (i.e. living) micro-organisms (or, collectively, biomass) to achieve removal of organics. The latter are normally measured as biochemical oxygen demand (BOD) or chemical oxygen demand (COD) which are indirect measurements of organic matter levels since both refer to the amount of oxygen utilized for oxidation of the organics. The microorganisms that grow on the organic substrate on which they feed derive energy and generate cellular material from oxidation of the organic matter, and can be aerobic (oxygen-dependent) or anaerobic (oxygen-independent). They are subsequently separated from the water to leave a relatively clean, clarified effluent.

In a typical in-situ bioremediation system, groundwater is extracted using one or more wells and, if necessary, treated to remove residual dissolved constituents. The treated groundwater is then mixed with an electron acceptor and nutrients, and other constituents if required, and re-injected up-gradient of or within the contaminant source. Infiltration galleries or injection wells may be used to re-inject treated water. In an ideal configuration, a "closed-loop" system would be established. All water extracted would be reinjected without treatment and all remediation would occur in situ. This ideal system would continually recirculate the water until cleanup levels had been achieved. If a local regulation does not allow re-injection of extracted groundwater, it may be feasible to mix the electron acceptor and nutrients with fresh water instead.

The most attractive feature of biological processes is the very high chemical conversion efficiency achievable. Unlike chemical oxidation processes, aerobic processes are capable of quantitatively mineralizing large organic molecules that is, converting them to the end mineral constituents of carbon dioxide, water, and inorganic nitrogen products, at ambient temperatures without significant onerous byproduct formation. In doing so a variety of materials are released from the biomass in the reactor which are collectively referred to as extracellular polymeric substances and which contain a number of components which contribute to membrane fouling in a membrane reactor. The relative and overall concentrations of the various components are determined both by the feed characteristics and operational facets of the system, such as microbial speciation. Anaerobic processes generate methane as an end product, a possible thermal energy source, and similarly generate extracellular polymeric substances. Biotreatment processes are generally robust to variable organic loads, create little odor (if aerobic) and generate a waste product (sludge) which is readily processed. On the other hand, they are slower than chemical processes, susceptible to toxic shock and consume energy associated with aeration in aerobic systems and mixing in all biotreatment systems.

Biological treatment processes for water systems depend on supplying colonies of micro-organisms with optimum quantities of air and nutrients to achieve the same reactions that occur in natural self-purification processes. The three changes that occur during self-purification include coagulation of colloidal solids passing through the primary sedimentation stage; oxidation of carbon, hydrogen, nitrogen, and phosphorus; and nitrification.

By way of clarification, biodegradation is a microbial process that occurs when all of the nutrients and physical conditions involved are suitable for growth. Bacteria and fungi—the inhabitants of various water systems—are the microorganisms responsible for biodegradation. Most biodegradation processes occur at temperatures between 10 and 35 °C (50 and 95 °F). Biodegradation can occur under aerobic conditions where oxygen is the electron acceptor and under anaerobic conditions where nitrate, sulfate, or another compound is the electron acceptor. The process of biodegradation as might occur in a water system can be divided into three stages: (i) biodeterioration, (ii) biofragmentation, and (iii) assimilation.

Biodeterioration is a surface-level degradation that modifies the mechanical, physical, and chemical properties of the contaminant. This stage occurs when the material is exposed to abiotic factors in the outdoor environment and allows for further degradation by weakening the structure of the contaminant. Briefly, abiotic factors are factors such as climate or habitat that influence or affect an ecosystem and the organisms in it. Some other abiotic factors that influence these initial changes are compression (mechanical), light, temperature, and chemicals in the environment. While biodeterioration typically occurs as the first stage of biodegradation, it can in some cases be parallel to biofragmentation.

Biofragmentation of a contaminant is the process in which bonds within the contaminant are cleaved thereby generating lower molecular weight products. The steps taken to fragment the contaminant differ based on the presence of oxygen in the system. The breakdown of materials by microorganisms when oxygen is present is aerobic digestion, and the breakdown of materials when is oxygen is not present is anaerobic digestion. The main difference between these processes is that anaerobic reactions produce methane, while aerobic reactions do not (however, both reactions produce carbon dioxide, water, some type of residue, and a new biomass). In addition, aerobic digestion typically occurs more rapidly than anaerobic digestion, while anaerobic digestion does a better job reducing the volume and mass of the material. Due to the ability of the anaerobic digestion process to reduce the volume and mass of waste materials and produce a natural gas, anaerobic digestion technology is widely used for waste management systems and as a source of local, renewable energy.

The resulting products from biofragmentation are then integrated into microbial cells (the *assimilation* stage). Some of the products from fragmentation are easily transported within the cell by membrane carriers. However, others still have to undergo biotransformation reactions to yield products that can then be transported inside the cell. Once inside the cell, the products enter catabolic pathways that either lead to the production of adenosine triphosphate (ATP) or elements of the cells structure.

The biodegradation of organic matter in the aquatic and terrestrial environments is a crucial environmental process. Some organic pollutants are biocidal. For example, effective fungicides must be antimicrobial in action and, therefore, in addition to killing harmful fungi, fungicides frequently harm beneficial saprophytic fungi (fungi that decompose dead organic matter) and bacteria. Herbicides are designed for plant control and insecticides are used to control insects and can also be harmful when introduced into a water system.

The biodegradation of crude oil is essential to the elimination of oil spills. This oil is degraded by both marine bacteria and filamentous fungi. The physical form of crude oil makes a large difference in its potential for degradation. Degradation in water occurs at the water-oil interface. Therefore, thick layers of crude oil prevent contact with bacterial enzymes and oxygen. Apparently, bacteria synthesize an emulsifier that keeps the oil dispersed in the water as a fine colloid and therefore accessible to the bacterial cells.

Some of the most important microorganism-mediated chemical reactions in aquatic and soil environments are those involving nitrogen compounds and the cycle of such compounds throughout the biosphere. Among the biochemical transformations in the nitrogen cycle (Fig. 8.1) are: (i) nitrogen fixation, in which molecular nitrogen is fixed as organic nitrogen, (ii) nitrification, which is the process of oxidizing ammonia to nitrate, (iii) nitrate reduction, in which nitrogen in nitrate ions is reduced to nitrogen in a lower oxidation state; and (iv) denitrification, which is the reduction of nitrate and nitrite to ammonia.

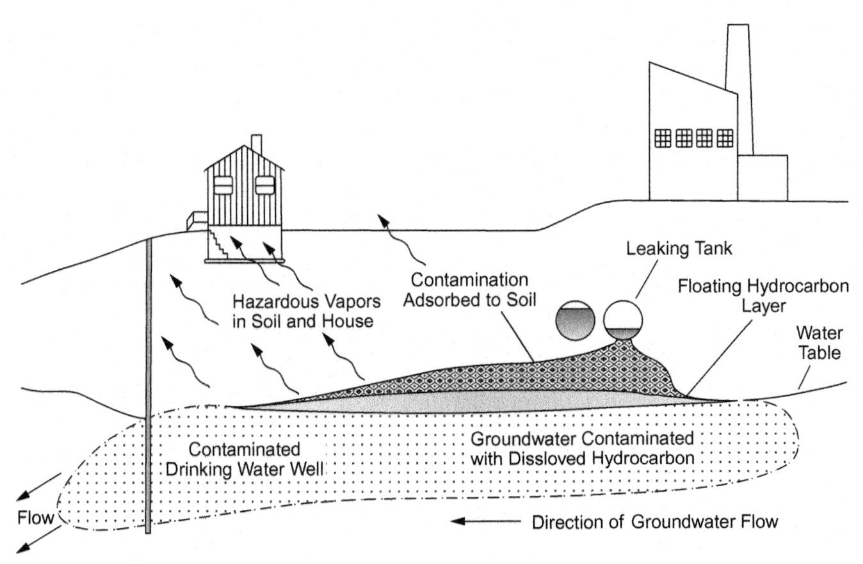

Fig. 8.1 General schematic of a contaminant plume.

Biodegradation of phosphorus compounds is important in the environment for two reasons: (i) provision of a source of algal nutrient orthophosphate from the hydrolysis of polyphosphates and (ii) the biodegradation and, hence, the deactivation of highly toxic organophosphate compounds, such as the organophosphate insecticides. The organophosphorus compounds of greatest environmental concern tend to be sulfur-containing phosphorothionate and phosphorodithioate ester insecticides. These are used because they exhibit higher ratios of insect: mammal toxicity than do their non-sulfur analogs. The biodegradation of these compounds is an important environmental chemical process. Fortunately, unlike the organic halogen insecticides that they largely displaced, the organophosphates readily undergo biodegradation and do not accumulate.

Sulfur compounds are very common in water—sulfate ions (SO_4^{2-}) occur in varying concentration in practically all natural waters. Organic sulfur compounds, both those of natural origin and pollutant species, are very common in natural aquatic systems, and the degradation of these compounds is an important microbial process. Sometimes the degradation products, such as the odorous and toxic hydrogen sulfide, cause serious problems with water quality.

One consequence of bacterial action on metal compounds is the occurrence of drainage of acidic aqueous solutions from mines. Acid mine drainage (the outflow of acidic water from metal mines or coal mines) is one of the most-common and damaging problems in the waters flowing from coal mines and draining from the spoil piles (mine tippage, gob piles) left over from coal processing and washing are highly acidic and have the ability to sterilize the surrounding land and water systems with the ensuing serious (often fatal) effects on the flora and fauna. Acidic mine water results from the presence of sulfuric acid produced by the oxidation of pyrite (FeS_2). Microorganisms are closely involved in the overall process. The prevention and cure of acid mine water is one of the major challenges facing the environmental chemist.

Selenium is also subject to bacterial oxidation and reduction. These transitions are important because selenium is a crucial element in nutrition, particularly of livestock. Microorganisms are closely involved with the selenium cycle, and microbial reduction of oxidized forms of selenium has been known for some time. A soil dwelling strain of *Bacillus megaterium* has been found to be capable of oxidizing elemental selenium to selenite (SeO).

In terms of bioremediation, the basic biological treatment processes used in the system include waste stabilization ponds and constructed wetland systems, trickling (or percolating) filter systems, and activated sludge systems. An aerobic stabilization pond is a large and shallow excavation in the ground, where the treatment of the waste occurs by natural processes involving the use of both bacteria and algae. In aerobic ponds, oxygen is supplied by natural surface re-aeration and by algal photosynthesis. Higher animals such as rotifers and protozoa are also present in the pond and the main function of these higher species is to act as predators on the bacteria, and to a lesser extent on algae, which helps in controlling the suspended solids (SS) concentration in the effluent. Water systems, such as ponds, in which the stabilization of wastes is brought about by a combination of aerobic, anaerobic, and facultative bacteria are known as facultative stabilization ponds.

Three zones can be defined in facultative ponds: (i) a surface zone where aerobic bacteria and algae exist in a symbiotic relationship, (ii) an intermediate zone that is partly aerobic and partly anaerobic—usually on a time basis—in which the decomposition of organic matter is carried out by facultative bacteria, and (iii) an anaerobic bottom zone in which accumulated solids are decomposed by anaerobic bacteria.

Thus, the key parameters that determine the effectiveness of in-situ groundwater bioremediation are: (i) hydraulic conductivity of the aquifer, which controls the distribution of electron acceptors and nutrients in the subsurface, (ii) biodegradability of the petroleum constituents, which determines both the rate and degree to which constituents will be degraded by microorganisms,; and (iii) the location of the contamination in the subsurface. Contaminants must be dissolved in groundwater or adsorbed onto more permeable sediments within the aquifer.

In general, the aquifer medium will determine hydraulic conductivity. Fine-grained media (e.g., clay minerals, silt minerals) have lower intrinsic permeability than coarse-grained media (e.g., sands, gravels). Bioremediation is generally effective in permeable (e.g., sandy, gravelly) aquifer media. However, depending on the extent of contamination, bioremediation also can be effective in less permeable silt or clay media. In general, an aquifer medium of lower permeability will require longer to clean up than a more permeable medium. Soil structure and stratification are important to in-situ groundwater bioremediation because they affect groundwater flowrates and patterns when water is extracted or injected. Structural characteristics such as microfracturing can result in higher permeability than expected for certain soils (e.g., clay minerals). In this case, however, flow will increase in the fractured media but not in the unfractured media. The stratification of soils with different permeability can dramatically increase the lateral flow of groundwater in the more permeable strata while reducing the flow through less permeable strata. This preferential flow behavior can lead to reduced effectiveness and extended remedial times for less-permeable strata.

4.3.1 Bioaugmentation

Bioaugmentation (biological augmentation) is the addition of archaea (single-celled microorganisms) or bacterial cultures required to speed up the rate of degradation of a contaminant. Organisms that originate from contaminated areas may already be able to break down waste, but perhaps inefficiently and slowly. Before applying the process, it is often necessary to study the indigenous varieties of microorganisms present to determine if biostimulation is viable. If the indigenous variety of microorganisms do not have the metabolic capability to perform the remediation process, exogenous varieties with such sophisticated pathways are introduced. Bioaugmentation is commonly used in the treatment of municipal waste water before discharge.

If a treatability study shows no degradation (or an extended lab period before significant degradation is achieved) in contamination contained in the groundwater, then inoculation with strains known to be capable of degrading the contaminants may be helpful. This process increases the reactive enzyme concentration within the bioremediation system and subsequently may increase contaminant degradation rates over the non-augmented rates, at least initially after inoculation.

At sites where groundwater is contaminated with chlorinated ethylene derivatives, such as tetrachloroethylene ($Cl_2C{=}CCl_2$, also called perchloroethylene) and trichloroethylene ($Cl_2C{=}CHCl$), bioaugmentation can be used to ensure that the in situ microorganisms can completely degrade these contaminants to ethylene ($CH_2{=}CH_2$) and chloride derivatives, which are non-toxic. Bioaugmentation is typically only applicable to bioremediation of chlorinated ethylene derivatives, although there are emerging cultures with the potential to biodegrade other compounds including benzene, toluene, ethyl benzene and the xylene isomers (known collectively as BTEX because of the frequent occurrence of these chemicals in mixtures), chloroethane derivatives, chloromethane derivatives, and methyl t-butyl ether (MTBE). Bioaugmentation is typically performed in conjunction with the addition of electron donor (biostimulation) to achieve geochemical conditions in groundwater that favor the growth of the dechlorinating microorganisms in the bioaugmentation culture.

4.3.2 Bioslurping

Bioslurping is an in situ remediation technology that combines the elements of bioventing and vacuum-enhanced pumping to recover free product from the groundwater while promoting the aerobic bioremediation of hydrocarbon contaminants. Vacuum extraction/recovery removes free product along with some groundwater; vapor extraction removes high volatility vapors from the vadose zone; and bioventing enhances biodegradation in both the vadose zone and the capillary fringe.

Thus, bioslurping combines elements of bioventing and vacuum-enhanced pumping of free-product that is lighter than water (light non-aqueous phase liquid, LNAPL) to recover free-product from the water. The biological processes in the term bioslurping refer to aerobic biological degradation of the hydrocarbon derivatives when air is introduced into the unsaturated zone contaminated soil. This process minimizes changes in the water table elevation which minimizes the creation of a smear zone. Bioventing of vadose zone soils is achieved by drawing air into the soil due to withdrawing soil gas via the recovery well.

The bioslurping system is made up of a well into which an adjustable length slurp tube is installed. The slurp tube, connected to a vacuum pump, is lowered into the light non-aqueous phase liquid layer, and pumping begins to remove free product along with some groundwater (vacuum enhanced extraction/recovery). The vacuum-induced negative pressure zone in the well promotes flow of the light non-aqueous phase liquid toward the well and also draws the light non-aqueous phase liquid trapped in small pore spaces above the water table. When the level of the light non-aqueous phase liquid declines slightly in response to pumping, the slurp tube begins to draw in and extract vapors (vapors extraction). This removal of vapors promotes air movement through the unsaturated zone, increasing oxygen content and enhancing aerobic bioremediation (bioventing).

When mounding due to the introduced vacuum causes a slight rise in the water table, the slurp cycles back to removing the light non-aqueous phase liquid and groundwater. This cycling minimizes water table fluctuations, reducing smearing associated with other recovery techniques. Liquid (product and groundwater) removed through the slurp tube is sent to an oil/water separator, and vapors are sent to a liquid vapor separator. Aboveground water and vapor treatment systems may also be included, if required. However, in some cases, system design modifications have allowed discharge of groundwater and vapor extracted via bioslurping without treatment. Results of field tests of bioslurping systems have shown that LNAPL and vapor recovery are directly correlated with the degree of vacuum. A comparison of bioslurping to conventional methods of LNAPL recovery reported that bioslurping achieved the greater recovery rates than either skimming or dual-pump methods.

4.3.3 Biosparging

Bio-sparging is a technique in which air is injected beneath the groundwater table and, by way of this injection, the water is enriched with oxygen and the biological break down is stimulated. The volatilization of pollutants also takes place. It is typical for the injected air quantities to be lower than those in sparging due to the volatilization of pollutants.

The processes uses indigenous microorganisms to biodegrade organic constituents in the saturated zone. Such as the area in an aquifer, below the water table, in which relatively all pores and fractures are saturated with water. In biosparging, air (or oxygen) and nutrients (if needed) are injected into the saturated zone to increase the biological activity of the indigenous microorganisms. Biosparging can be used to reduce concentrations of chemical constituents (such as crude oil constituents and crude products) that are dissolved in the water. Thus, in biosparging, air (or oxygen) and nutrients (if needed) are injected into the saturated zone to increase the biological activity of the indigenous microorganisms. Biosparging can be used to reduce concentrations of petroleum constituents that are dissolved in groundwater, adsorbed to soil below the water table, and within the capillary fringe.

The biosparging process is similar to air sparging but, while air sparging removes constituents primarily through volatilization, biosparging promotes biodegradation of constituents rather than volatilization (generally by using lower flow rates than are used in air sparging). In practice, some degree of volatilization and biodegradation occurs when either air sparging or biosparging is used. When volatile constituents are present, biosparging is often combined with bioventing and can also be used with other remedial technologies. When biosparging is combined with vapor extraction, the vapor extraction system creates a negative pressure in the vadose zone through a series of extraction wells that control the vapor plume migration.

When an area is contaminated with crude oil, crude oil products, or aromatic compounds, such as benzene, toluene, ethylbenzene, xylene and naphthalene (BTEXN), these pollutants can be removed by means of biosparging. This technique is suitable for on-site (in situ) remediation of both soil and groundwater and is the best option for mineral oil removal there is.

The time a remediation project takes varies from one to several years and greatly depends on concentrations, the type of contaminants present and the extent to which the soil can be aerated. After remediation of, for example, crude oil there is usually some residual contamination, this consist of the immobile heavier oil fractions with no risk of diffusion. In comparison to physical in situ techniques the amount of residual contaminants is much smaller, since these physical methods predominantly target the volatile and/or soluble fraction of the oil.

Air sparging for remediation of dense non-aqueous phase liquids involves injecting air or other gases directly into the groundwater to vaporize and recover the contaminants. Volatile components of the dense non-aqueous phase liquids will vaporize and move upward to the atmosphere or to a vapor extraction system installed in the vadose zone.

The vaporization of volatile organic chemicals, and even mixtures of such chemicals, is well understood. Injected air moves laterally, driven by the injection pressure, and upward, due to the buoyancy of air. As the injected air moves through a formation and comes in contact with non-aqueous phase liquids, contaminated soil, or water containing dissolved-phase contamination, the volatile contaminants partition into the air. Partitioning from the dissolved phase is described by a compound's Henry's law constant; partitioning from dense non-aqueous phase liquids is described by its vapor pressure.

Henry's law is one of the gas laws states that: *at a constant temperature, the amount of a given gas that dissolves in a given type and volume of liquid is directly proportional to the partial pressure of that gas in equilibrium with that liquid.* An equivalent way of stating the law is that the solubility of a gas in a liquid is directly proportional to the partial pressure of the gas above the liquid:

$$C = kP_{gas}$$

In this equation, C is the solubility of a gas at a fixed temperature in a particular solvent (in units of M or mL gas/L), k is Henry's law constant (often in units of M/atm), and P_{gas} is the partial pressure of the gas (often in units of atm). In addition, oxygen present in the injected air will dissolve in the water, promoting in situ biodegradation of nonvolatile contaminants or those located downgradient of the sparging zone.

4.3.4 Bioventing

Bioventing is an in situ remediation technology that uses microorganisms to biodegrade organic constituents in a water system. Bioventing enhances the activity of indigenous bacteria and archaea and stimulates the natural in situ biodegradation of hydrocarbon derivatives (as, for example, form a spill of crude oil or crude oil products) by inducing air or oxygen flow into the unsaturated zone and, if necessary, by adding nutrients. During bioventing, oxygen may be supplied through direct air injection into residual contamination in soil. Bioventing primarily assists in the degradation of adsorbed fuel residuals, but also assists in the degradation of volatile organic compounds (VOCs) as vapors move slowly through biologically active soil.

In the process for subsurface water, recovery wells are used to remove contaminated groundwater, which is treated aboveground, in this case using a bioreactor containing microorganisms that are acclimated to the contaminant. This would be considered ex situ treatment. Following bioreactor treatment, the clean water is supplied with oxygen and nutrients, and then it is reinjected into the site. The reinjected water provides oxygen and nutrients to stimulate in situ biodegradation. In addition, the reinjected water flushes the vadose zone to aid in removal of the contaminant for aboveground bioreactor treatment. This remediation scheme is a very good example of a combination of physical, chemical, and biological treatments being used to maximize the effectiveness of the remediation treatment.

4.3.5 Phytoremediation

Phytoremediation refers to the technologies that use living plants to clean up water that is contaminated with hazardous contaminants. The description can be refined to the use of green plants and the associated microorganisms, along with proper soil amendments and agronomic techniques to either contain, remove or render toxic environmental contaminants harmless.

Phytoremediation may be applied wherever the static water environment has become polluted or is suffering ongoing chronic pollution. Contaminants such as metals, pesticides, solvents, explosives, and crude oil and its derivatives, have been mitigated in phytoremediation projects worldwide. Many plants such as mustard plants, alpine pennycress, hemp, and pigweed have proven to be successful at hyper-accumulating contaminants at toxic waste sites.

Thus, the phytoremediation process makes use of specific types of plants and trees that have roots capable of absorbing contaminants from water over time. This process can be carried out in areas where the roots can tap the ground water. Growing and, in some cases, harvesting plants on a contaminated site as a remediation method is a passive technique that can be used to clean up sites with shallow, low to moderate levels of contamination. Phytoremediation can be used to clean up metals, pesticides, solvents, explosives, crude oil, polyaromatic hydrocarbon derivatives, and landfill leachates. The process can also be used for river basin management through the hydraulic control of contaminants.

5 Metals in water systems

Metals and their derivatives are introduced in aquatic systems as a result of the weathering of soils and rocks, from volcanic eruptions, and from a variety of human activities involving the mining, processing, or use of metals and/or substances that contain metal pollutants (Fergusson, 1990; Bradl, 2002; Tchounwou et al., 2012; Chowdhury et al., 2016). There are different types of sources of pollutants: (i) point sources, which is localized pollution where the pollutants come from single, identifiable sources (ii) non-point sources, where pollutants come from dispersed sources that are often difficult to identify. There are only a few examples of localized metal pollution, like the natural weathering of ore bodies and the little metal particles coming from coal-burning power plants via smokestacks in air, water and soils around the factory. The most common metal pollution in freshwater comes from mining companies that usually use an acid mine drainage mine drainage system to release metals from their respective ores—metals are soluble in an acid solution—and after the drainage process the acid can contaminant a water system.

In addition, metals can be deposited in bottom sediments, where they remain for many years. Streams coming from draining mining areas are often very acidic and contain high concentrations of dissolved metals with little aquatic life. Both localized and dispersed metal pollution cause environmental damage because metals are non-biodegradable. Unlike some organic pesticides, metals cannot be broken down into less harmful components in the environment. Generally the ionic form of a metal is more toxic, because it can form toxic compounds with other ions. Electron transfer reactions that are connected with oxygen can lead to the production of toxic oxy-radicals, a toxicity mechanism now known to be of considerable importance in both animals and plants. Some oxy-radicals, such as superoxide anion (O_2^-) and the hydroxyl radical (OH^-), can cause serious cellular damage.

The metals that are most important in this respect are the so-called heavy metals. However, the term *heavy metal* is somewhat imprecise, but includes most metals with an atomic number greater than 20 (Table 8.6), and excludes alkali metals, alkaline earths, lanthanides and actinides. More generally, the term heavy metal generally refers to any metallic chemical element that has a relatively high density and is toxic or poisonous at low concentrations. Examples of heavy metals include mercury (Hg), cadmium (Cd), arsenic (As), chromium, (Cr), thallium (Tl), and lead (Pb). Heavy metals are natural components of the Earth's crust and cannot be degraded or destroyed. As trace elements, some heavy metals (for example: copper, Cu, selenium, Se, and zinc, Zn) are essential to maintain the metabolism of the human body. However, at

Table 8.6 The periodic table of the elements showing the groups and periods including the lanthanide elements and the actinide elements

Group→ ↓Period	1	2		3	4	5	6	7	8	9	10	11	12	13	14	15	16	17	18
1	1 H																		2 He
2	3 Li	4 Be												5 B	6 C	7 N	8 O	9 F	10 Ne
3	11 Na	12 Mg												13 Al	14 Si	15 P	16 S	17 Cl	18 Ar
4	19 K	20 Ca		21 Sc	22 Ti	23 V	24 Cr	25 Mn	26 Fe	27 Co	28 Ni	29 Cu	30 Zn	31 Ga	32 Ge	33 As	34 Se	35 Br	36 Kr
5	37 Rb	38 Sr		39 Y	40 Zr	41 Nb	42 Mo	43 Tc	44 Ru	45 Rh	46 Pd	47 Ag	48 Cd	49 In	50 Sn	51 Sb	52 Te	53 I	54 Xe
6	55 Cs	56 Ba	*	71 Lu	72 Hf	73 Ta	74 W	75 Re	76 Os	77 Ir	78 Pt	79 Au	80 Hg	81 Tl	82 Pb	83 Bi	84 Po	85 At	86 Rn
7	87 Fr	88 Ra	**	103 Lr	104 Rf	105 Db	106 Sg	107 Bh	108 Hs	109 Mt	110 Ds	111 Rg	112 Cn	113 Nh	114 Fl	115 Mc	116 Lv	117 Ts	118 Og
			*	57 La	58 Ce	59 Pr	60 Nd	61 Pm	62 Sm	63 Eu	64 Gd	65 Tb	66 Dy	67 Ho	68 Er	69 Tm	70 Yb		
			**	89 Ac	90 Th	91 Pa	92 U	93 Np	94 Pu	95 Am	96 Cm	97 Bk	98 Cf	99 Es	100 Fm	101 Md	102 No		

* Lanthanide elements.
** Actinide elements.

higher concentrations they can lead to poisoning. Heavy metal poisoning could result, for instance, from drinking-water contamination (e.g. lead pipes), high ambient air concentrations near emission sources, or intake via the food chain.

Heavy metals are extremely toxic for living beings and they are highly persistent pollutants. Once they get into the groundwater, it becomes extremely difficult to handle these metals because of the complex speciation chemistry coming into play. However, many techniques have been devised over the past few decades to remediate heavy metal contaminated groundwater. In an ecosystem, the heavy metals can become mobile in soils depending on soil pH and their speciation. Thus, a fraction of the total mass can leach to aquifer or can become bioavailable to living organisms (Santona et al., 2006). The diffusion phenomenon of contaminants through soil layers and the change in mobility of heavy metals in aquifers with intrusion of organic pollutants are being studied in more details in recent years (Hashim et al., 2011; Satyawali et al., 2010).

When the pH in water decreases (i.e. the water becomes more acidic), metal solubility increases and the metal particles become more mobile. That is why metals are more toxic in soft waters. Metals can become 'locked up' in bottom sediments, where they remain for many years. Streams coming from draining mining areas are often very acidic and contain high concentrations of dissolved metals with little aquatic life. Both localized and dispersed metal pollution cause environmental damage because metals are non-biodegradable. Unlike some organic pesticides, metals cannot be broken down into less harmful components in the environment.

Some metals, such as manganese, iron, copper, and zinc are essential micronutrients. They are essential to life in the right concentrations, but in excess, these chemicals can be poisonous. At the same time, chronic low exposures to heavy metals can have serious health effects in the long term. Many rivers are polluted with heavy metals from old mine workings and some species of algae become very tolerant to polluted conditions. Heavy metals are dangerous because they tend to bioaccumulate—that is, they increase in the concentration of a chemical in a biological organism over time, compared to the concentration in the environment. Heavy metals can enter a water system as part of industrial waste and/or domestic consumer waste, or even from acid rain breaking down soils and releasing heavy metals into streams, lakes, rivers, and groundwater.

Although all contaminated sites are unique and a site-specific approach to remediation is often required, heavy metal contamination creates even more complex challenges. Depending on the site characteristics (such as geographical location, types of co-contaminants, climate, depth, soil type, pH levels, water content, particle size, and the presence of clay minerals), site-specific alternatives must be evaluated carefully. Remediation methods in general use include isolation, immobilization, toxicity reduction, physical separation and extraction. The ideal and sustainable alternative is permanent removal or recovery to natural levels so there will never be a risk of return to the ecosystem. Recoverable remediation alternatives include isolation, physical separation, and extraction from both solid (soil) and liquid (water) matrices. Existing or emerging technologies include: precipitation, ion exchange, and reverse osmosis.

Groundwater contaminants are often dispersed in plumes over large areas, deep below the surface, making conventional types of remediation technologies difficult to apply. In those cases, chemical treatment technologies may be the best choice. Chemicals are used to decrease the toxicity or mobility of metal contaminants by converting them to inactive states. Oxidation, reduction and neutralization reactions can be used for this purpose.

Chemical precipitation is a widely used technology for the removal of heavy metals from water. It has long been the primary method of treating metal-laden industrial wastewater. The process involves the transformation of dissolved contaminants into insoluble solids, thereby facilitating the contaminant's subsequent removal from the liquid phase by physical methods, such as clarification and filtration. In a precipitation process, chemical precipitants (also known as coagulants and flocculants) are used to increase particle size through aggregation. The amount of chemical that is required during treatment is dependent on pH and alkalinity of the water. Usually, heavy metals in water are precipitated by adding sodium hydroxide or lime during neutralization.

In-situ treatment by using *reductants* is another technology for the remediation of heavy metal pollutants. Thus, when groundwater is passed through a reductive zone or a purpose-built barrier, reduction of the metal may occur. Manipulation of sub-surface redox conditions can be implemented by injection of liquid reductants, gaseous reductants or reduced colloids. Several soluble reductants such as sulfite, thiosulfate, hydroxylamine, dithionite, hydrogen sulfide and also the colloidal reductants, such as zero-valent iron (Fe^0) and ferrous iron (Fe^{2+}) in clay minerals for soil remediation purpose (Hashim et al., 2011)

Ion exchange is a reversible chemical reaction wherein an ion from water or wastewater solution is exchanged for a similarly charged ion attached to an immobile solid particle. These solid ion exchange particles are either naturally occurring inorganic zeolites or synthetically produced organic resins. It is a process that is very similar to biosorption whereby the latter is known to actually function predominantly on the basis of ion exchange. Ion exchange uses mainly hydrocarbon-derived polymeric resins.

An important class of ion-exchange resins includes solvent-impregnated resins which combine the advantages of liquid-liquid extraction and ion exchange involving a separate solid phase. Solvent-impregnated resins remove very low concentration of contaminants in the presence of high concentration of microelements (e.g. calcium, magnesium, sodium, potassium and chloride) present in water at nearly neutral pH and the presence of other anions, all of which compete for available sites on the solvent-impregnated resins. Commercial ion-exchange resins and synthetically prepared type II solvent-impregnated resins can be used for groundwater remediation as a material for use in permeable reactive barriers. In the type II solvent-impregnated resins, extractant molecules were bound to a functional matrix due to acid-base interactions. A major advantage of ion-exchange resins over other adsorbents is that they can be effectively regenerated up to 100% efficiency (Hashim et al., 2011).

Reverse osmosis is a membrane process that acts as a molecular filter to remove over 99% of all dissolved minerals (Fig. 8.2). In this process, water passes through the membrane while the dissolved and particulate matter is left behind. In the process, mechanical pressure is applied to an impure solution to force pure water through

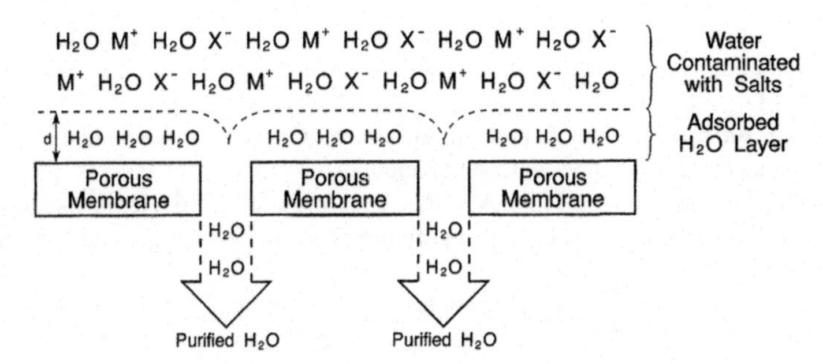

Fig. 8.2 Schematic representation of the reverse osmosis process.

a semi-permeable membrane. Reverse osmosis is theoretically the most thorough method of large scale water purification available, although perfect semi-permeable membranes are difficult to create. Unless membranes are well-maintained, algae and other microbial life forms can colonize the membranes.

The process is very effective for removal of ionic species from solution. The resulting concentrated by-product solutions make eventual recovery of metals more feasible. Despite the effectiveness, the membranes are relatively expensive both to procure and operate. The use of elevated pressures makes this technique costly and sensitive to operating conditions. A significant advantage of reverse osmosis over other traditional water treatment technologies is the ability to reduce the concentration of other ionic contaminants, as well as dissolved organic compounds. Reverse osmosis had been applied in heavy metal removal both in small and large scales.

Phytodegradation is the degradation or breakdown of organic contaminants by internal and external metabolic processes driven by the plant. The process involves the use of plants to uptake, store and degrade contaminants within the plant tissue. During the process, plants actually metabolize and destroy contaminants within their tissues. Some contaminants can be absorbed by the plant and are then broken down by plant enzymes (Newman and Reynolds, 2004; Akpor and Muchie, 2010; Akpor et al., 2014). During phytodegradation, the plants are able to take-up metal contaminants directly from the soil water or release exudates that help to degrade pollutants via cometabolism in the rhizophere (the narrow region of soil that is directly influenced by root secretions and associated soil microorganisms). For environmental application, it is vital that the metabolites which accumulate in vegetation are non-toxic or at least significantly less toxic than the parent compound.

References

Akpor, O., Muchie, M., 2010. Remediation of heavy metals in drinking water and wastewater treatment systems: Processes and applications. Int. J. Phys. Sci. 5 (12), 1807–1817.

Akpor, O., Otohinoyi, B.I., Olaolu, T.D., Aderiye, B.I., 2014. Pollutants in wastewater effluents: impacts and remediation processes. Int. J. Environ. Res. Earth Sci. 3 (3), 050–059.

ASTM, 2019. Annual book of standards. ASTM International, West Conshocken, PA.

Bishoge, O.K., Zhang, L., Suntu, S.L., Jin, H., Zewde, A.A., Qi, Z., 2018. Remediation of water and wastewater by using engineered nanomaterials: a review. J. Environ. Sci. Health A 53 (6), 537–554.

Blowes, D.W., Ptacek, C.J., Benner, S.G., McRae, C.W.T., Bennett, T.A., Puls, R.W., 2000. Treatment of inorganic contaminants using permeable reactive barriers. J. Contam. Hydrol. 45, 123–137.

Bradl, H. (Ed.), 2002. Heavy Metals in the Environment: Origin, Interaction and Remediation. Vol. 6. Academic Press, London, United Kingdom.

Chowdhury, S., Jafar Mazumder, M.A., Al-Attas, O., Husain, t., 2016. Heavy metals in drinking water: Occurrences, implications and future needs in developing countries. Sci. Total Environ. 569–570, 476–488.

Duffus, J.H., 2002. Heavy metals—a meaningless term? Pure Appl. Chem. 74 (5), 793–807.

Fergusson, J.E. (Ed.), 1990. The Heavy Elements: Chemistry, Environmental Impact and Health Effects. Pergamon Press, Oxford, United Kingdom.

Gary, J.G., Handwerk, G.E., Kaiser, M.J., 2007. Petroleum Refining: Technology and Economics, fifth ed. CRC Press, Taylor & Francis Group, Boca Raton, FL.

Gates, D.D., Siegrist, R.L., 1994. In-situ chemical oxidation of trichloroethylene using hydrogen peroxide. J. Environ. Eng. 121 (9), 639–644.

Hashim, M.A., Mukhopadhyay, S., Narayan Sahu, J., Sengupta, N., 2011. Remediation Technologies for Heavy Metal Contaminated Groundwater. J. Environ. Manage. 92, 2355–2388.

Hsu, C.S., Robinson, P.R. (Eds.), 2017. Handbook of Petroleum Technology. Springer International Publishing AG, Cham, Switzerland. https://clu-in.org/download/techfocus/prb/PRB-5-ITRC.pdf.

ITRC, 2011. Permeable reactive barrier: Technology Update. Interstate Technology & Regulatory Council, Washington, DC.

Khan, F.I., Husain, T., Hejazi, R., 2004. An overview and analysis of site remediation technologies. J. Environ. Manage. 71, 95–122.

Newman, L.A., Reynolds, C.M., 2004. Phytodegradation of organic compounds. Curr. Opin. Biotechnol. 15 (3), 225–230.

Parkash, S., 2003. Refining Processes Handbook. Gulf professional publishing. In: Elsevier. Amsterdam, Netherlands.

Reddy, K.R., 2008. Physical and chemical groundwater remediation technologies. In: Darnault, C.J.G. (Ed.), Overexploitation and Contamination of Shared Groundwater Resources. Springer Science, Dordrecht, Berlin, Netherlands. Chapter 12.

Reddy, K.R., Adams, J.A., 2001. Cleanup of chemical spills using air Sparging. In: Handbook of Chemical Spill Technologies. M. Fingas (Editor). McGraw-Hill, New York. (Chapter 14. Page 41.1-14.29).

Reddy, K.R., Kosgi, S., Zhou, J., 1995. A review of in situ air Sparging for the remediation of VOC-contaminated saturated soils and groundwater. Hazard. Waste Hazard. Mater. 12 (2), 97–118.

Richardson, S.D., 2007. Water analysis: emerging contaminants and current issues. Anal. Chem. 79, 4295–4324.

Santona, L., Castaldi, P., Melis, P., 2006. Evaluation of the interaction mechanisms between red muds and heavy metals. J. Hazard. Mater. 136, 324–329.

Satyawali, Y., Schols, E., Van Roy, S., Dejonghe, W., Diels, L., Vanbroekhoven, K., 2010. Stability investigations of zinc and cobalt precipitates immobilized by in situ bioprecipitation (ISBP) process. J. Hazard. Mater. 181, 217–225.

Scherer, M.M., Richter, S., Valentine, R.L., Alvarez, P.J.J., 2000. Chemistry and microbiology of permeable reactive barriers for in situ groundwater cleanup. Crit. Rev. Environ. Sci. Technol. 30, 363–411.

Sharma, H.D., Reddy, K.R., 2004. Geoenvironmental Engineering: Site Remediation, Waste Containment, and Emerging Waste Management Technologies. John Wiley & Sons, Hoboken, NJ.

Speight, J.G., 1996. Environmental Technology Handbook. Taylor & Francis, Washington, DC.

Speight, J.G., 2005. Environmental Analysis and Technology for the Refining Industry. John Wiley & Sons Inc, Hoboken, NJ.

Speight, J.G., 2013. The Chemistry and Technology of Coal, third ed. CRC Press, Taylor & Francis Group, Boca Raton, FL.

Speight, J.G. 2014. The Chemistry and Technology of Petroleum 5th Edition. CRC Press, Taylor & Francis Group, Boca Raton, Florida.

Speight, J.G., 2015. Handbook of Petroleum Product Analysis, second ed. John Wiley & Sons Inc, Hoboken, NJ.

Speight, J.G., 2017. Handbook of Petroleum Refining. CRC Press, Taylor & Francis Group, Boca Raton, FL.

Speight, J.G., Lee, S., 2000. Environmental Technology Handbook, second ed Taylor & Francis, New York.

Stebbing, L., 1993. Quality Assurance: The Route to Efficiency and Competitiveness, third ed. Prentice Hall, Upper Saddle River, NJ.

Tchounwou, P.B., Yedjou, C.G., Patlolla, A.K., Sutton, D.J., 2012. Heavy metal toxicity and the environment. Experientia Suppl. 101, 133–164.

Thines, R.K., Mubarak, N.M., Nizamuddin, S., Sahu, J.N., Abdullah, E.C., Ganesan, P., 2017. Application potential of carbon nanomaterials in water and wastewater treatment: a review. J. Taiwan Inst. Chem. Eng. 72, 116–133.

Pollution prevention

Chapter outline

1 Introduction

In the United States, the Pollution Prevention Act (which became law in 1990—similar acts are in law in many other countries) required the Environmental Protection Agency must establish a source reduction program and to focus industry, government, and public attention on reducing the amount of pollution through cost-effective changes in production, operation, and raw materials use. Opportunities for source reduction are often not realized because of existing regulations and the industrial resources required for compliance, focus on treatment and disposal. Source reduction is fundamentally different and more desirable than waste management or pollution control.

Pollution prevention is any practice that reduces, eliminates, or prevents pollution at its source, also known as source reduction. Source reduction is fundamentally different

Natural Water Remediation. https://doi.org/10.1016/B978-0-12-803810-9.00009-7

and more desirable than recycling, treatment and disposal and focus on (i) reducing the amount of any hazardous substance, pollutant, or contaminant entering any waste stream or otherwise released into the environment (including fugitive emissions); prior to recycling, treatment or disposal; and (ii) reducing the hazards to public health and the environment associated with the release of such substances, pollutants or contaminants. There are significant opportunities for industry to reduce or prevent pollution at the source through cost-effective changes in production, operation, and raw materials use. The opportunities for source reduction are often not realized because existing regulations focus upon treatment and disposal.

Source reduction refers to practices that reduce hazardous substances from being released into the environment prior to recycling, treatment or disposal. The term includes equipment or technology modifications, process or procedure modifications, reformulation or redesign of products, substitution of raw materials, and improvements in housekeeping, maintenance, training, or inventory control. Pollution prevention includes practices that increase efficiency in the use of energy, water, or other natural resources, and protect the resource base through conservation. Pollution prevention reduces both financial costs (waste management and cleanup) and environmental costs (health problems and environmental damage). Pollution prevention protects the environment by conserving and protecting natural resources while strengthening economic growth through more efficient production in industry and less need for households, businesses and communities to handle waste.

Finally, the Pollution Prevention Act establishes a national policy that requires that (i) pollution should be prevented or reduced at the source whenever feasible, (ii) pollution that cannot be prevented should be recycled in an environmentally safe manner whenever feasible, (iii) pollution that cannot be prevented or recycled should be treated in an environmentally safe manner whenever feasible; and (iv) disposal or other release into the environment should be employed only as a last resort and should be conducted in an environmentally safe manner.

Pollution prevention as it pertains to the prevention of water pollution represents one of the most important problems with which any country is to confront as regarding the environment. In addition, waste management can be divided into two important sub-categories: (i) public waste and (ii) production waste, which encompasses waste from commerce, industries and institution, waste from constructions, demolitions and waste water treatment sludge.

In addition to the Pollution Prevention Act there are several other regulatory laws that complement the Act and are presented in the following section.

2 Environmental regulations

Water is one of the renewable resources essential for sustaining all forms of life, food production, economic development, and for general well-being. It is impossible to substitute for most of its uses, difficult to de pollute, expensive to transport, and it is truly a unique gift to mankind from nature. Water is also one of the most manageable natural resources as it is capable of diversion, transport, storage, and recycling. All these properties impart to water its great utility for human beings. The surface water and groundwater resources of the country play a major role in agriculture, hydropower generation, livestock production, industrial activities, forestry, fisheries, navigation, recreational activities etc. The freshwater ecosystems of the world comprise only about 0.5% of the surface of the Earth. Rivers constitute an insignificant amount (0.1%) of the land surface. Only 0.01% v/v of the waters of the Earth occur in river channels. In spite of these low quantities, running waters are of enormous significance (Wetzel, 2001; Kumar et al., 2005).

In the last few decades, there has been a tremendous increase in the demand for freshwater due to rapid growth of population and the accelerated pace of industrialization. Human health is threatened by most of the agricultural development activities particularly in relation to excessive application of fertilizers and unsanitary conditions (Okeke and Igboanua, 2003). Anthropogenic activities related to extensive urbanization, agricultural practices, industrialization, and population expansion have led to water quality deterioration in many parts of the world (Baig et al., 2009; Mian et al., 2010; Wang et al., 2010). In addition, deficient water resources have increasingly restrained water pollution control and water quality improvement (Bu et al., 2010). Water pollution has been a research focus for government and scientists. Therefore, protecting river water quality is extremely urgent because of serious water pollution and global scarcity of water resources.

Environmental issues permeate everyday life. These issues range from the effects on the lives of workers in various occupations where hazards can result from exposure to chemical agents to the influence of these agents on the lives of the population at large (Lipton and Lynch, 1994; Speight, 1996; Boyce, 1997).

Environmental regulations pertaining to water quality laws govern the release of pollutants into water systems, including surface water Lakes, rivers, and oceans), groundwater, and stored drinking water. Some water quality laws, such as regulations pertaining to drinking water regulations, may be designed solely with reference to human health. Other regulations, including restrictions on the alteration of the chemical, physical and biological; characteristics of water systems may also reflect efforts to protect aquatic ecosystems more broadly. Regulatory efforts may include identifying and categorizing water pollutants, dictating acceptable pollutant concentrations in water systems, and limiting pollutant discharges from effluent sources. Regulatory areas include sewage treatment and disposal, industrial industries (including the natural gas and crude oil industries), agricultural industries, agricultural waste water management, and control of surface runoff from construction sites and urban environments, such as landfills. In fact, when the regulations enacted in the United States are considered, there is not one regulation that cannot be applied to the protection of the various water systems.

There are a variety of regulations (Table 9.1) that apply to maintain a non-polluted environment and most have been amended several times since they were first enacted. In each case, the most recent amendments provide stricter regulations for the establishment and enforcement of national ambient air quality standards for, as an example, sulfur dioxide. These standards do not stand alone and there are many national standards for sulfur emissions.

In this section, reference is made to the various environmental laws, which have been enacted in the United States, and are listed below, in alphabetical order rather than in order or preference.

Table 9.1 Environmental regulations that have an influence on water systems

Regulation	First enacted
Clean Air Act	1970
Clean Water Act (Water Pollution Control Act)	1948
Comprehensive Environmental Response, Compensation and Liability Act	1980
Hazardous Material Transportation Act	1974
Occupational Safety and Health Act	1970
Oil Pollution Act	1924
Resource Conservation and Recovery Act	1976
Safe Drinking Water Act	1974[8]
Superfund Amendments and *Re*-authorization Act (SARA)	1986
Toxic Substances Control Act	1976

2.1 Clean air act

The Clean Air Act (CAA) is the comprehensive federal law that regulates air emissions from stationary and mobile sources. Among other things, this law authorizes EPA to establish National Ambient Air Quality Standards (NAAQS) to protect public health and public welfare and to regulate emissions of hazardous air pollutants.

One of the goals of the Act was to set and achieve NAAQS in every state by 1975 in order to address the public health and welfare risks posed by certain widespread air pollutants. The setting of these pollutant standards was coupled with directing the states to develop state implementation plans (SIPs), applicable to appropriate industrial sources in the state, in order to achieve these standards. The Act was amended in 1977 and 1990 primarily to set new goals (dates) for achieving attainment of NAAQS since many areas of the country had failed to meet the deadlines.

Section 112 of the Clean Air Act addresses emissions of hazardous air pollutants. Prior to 1990, CAA established a risk-based program under which only a few standards were developed. The 1990 Clean Air Act Amendments revised Section 112 to first require issuance of technology-based standards for major sources and certain area sources. "Major sources" are defined as a stationary source or group of stationary sources that emit or have the potential to emit 10 tons per year or more of a hazardous air pollutant or 25 tons per year or more of a combination of hazardous air pollutants. An "area source" is any stationary source that is not a major source.

For major sources, Section 112 requires that EPA establish emission standards that require the maximum degree of reduction in emissions of hazardous air pollutants. These emission standards are commonly referred to as "maximum achievable control technology" or "MACT" standards. Eight years after the technology-based MACT standards are issued for a source category, EPA is required to review those standards to determine whether any residual risk exists for that source category and, if necessary, revise the standards to address such risk.

The first *Clean Air Act* of 1970 and the 1977 Amendments consisted of three titles. *Title I* dealt with stationary air emission sources, *Title II* with mobile air emission sources, and *Title III* with definitions of appropriate terms as well as applicable standards for judicial review. The *Clean Air Act Amendments of 1990* contain extensive provisions for control of the accidental release of toxic substances from storage or transportation as well as the formation of acid rain (acid deposition). In addition, the requirement that the standards be technology based removes much of the emotional perception that all chemicals are hazardous as well as the guesswork from legal enforcement of the legislation. The requirement also dictates environmental and health protection with an ample margin of safety.

The relevance of the Clean Air Act to water systems is the reduction of the emissions of the oxides of sulfur and the oxides of nitrogen that will react with moisture in the atmosphere (or the water in a water system) leading to the production of acid rain that will pollute water systems and cause the death (and extinction) of aquatic life. Thus:

$$SO_2 + H_2O \rightarrow H_2SO_3 \text{ (sulfurous acid)}$$

$$SO_3 + H_2O \rightarrow H_2SO_4 \text{ (sulfuric acid)}$$

$$NO + H_2O \rightarrow HNO_2 \text{ (nitrous acid)}$$

$$3NO_2 + 2H_2O \rightarrow HNO_3 \text{ (nitric acid)}$$

Thus, acid rain is rain—or any other form of precipitation—that is unusually acidic insofar as the rain (or other form of precipitation), meaning that it has elevated levels of hydrogen ions (low pH). This type of precipitation can have harmful effects on the flora and fauna (including humans) wherever it falls to the surface of the Earth. Of particular importance in the current context is the acidification of water systems, such as lakes, by the rain as well as acid-related damage to the infrastructure.

2.2 Comprehensive environmental response, compensation, and liability act

The Comprehensive Environmental Response, Compensation, and Liability Act (CERCLA) that is generally known as *Superfund*, was first signed into law in 1980. The central purpose of this Act is to provide a response mechanism for cleanup of any hazardous substance released, such as an accidental spill, or of a threatened release of a chemical. While the Resource Conservation Recovery Act (RCRA) regulates the storage, transportation, treatment, and disposal of solid and hazardous wastes to prevent contaminants from leaching into groundwater and, basically, deals basically with the management of wastes that are generated, treated, stored, or disposed of, CERCLA provides a response to the environmental release of various pollutants or contaminants into the air, water, or land.

Under this Act, a hazardous substance is any substance requiring (i) special consideration due to its toxic nature under the Clean Air Act, the Clean Water Act, or the Toxic

Substances Control Act and (ii) defined as hazardous waste under RCRA. Additionally, a pollutant or contaminant can be any other substance not necessarily designated by or listed in the Act but that *will or may reasonably* be anticipated to cause any adverse effect in organisms and/or their offspring.

The Superfund Amendments and Reauthorization Act (SARA) addresses closed waste disposal sites that may release hazardous substances into any environmental medium. The most revolutionary part of SARA is the Emergency Planning and Community Right-to-Know Act (EPCRA), which for the first time mandated public disclosure. It is covered under Title III of SARA.

In the context of water pollution, this Act authorizes the government to clean up contamination caused by chemical spills or hazardous waste that do or could pose threats to the environment (groundwater). The Act guards against the potential for runoff (also known as overland flow) which occurs when excess storm water, melt water, or other sources flows over the surface of the Earth and into a water system. Runoff that occurs on the ground surface that can drain to a common point is called a (a drainage basin). When runoff flows along the ground, it can pick up soil contaminants including pesticides or fertilizers that discharge into a water system thereby causing water erosion and pollution.

2.3 Federal insecticide, fungicide, and rodenticide act

The Federal Insecticide, Fungicide, and Rodenticide Act (FIFRA) is a regulatory statute governing the licensing, distribution, sale, and use of pesticides, including insecticides, fungicides, rodenticides, and other designated classes of chemicals (Copeland, 2012). Its objective is to protect human health and the environment from unreasonable adverse effects of pesticides. To that end, it establishes a nationally uniform pesticide labeling system requiring the registration of all pesticides and herbicides sold in the United States, and requiring users to comply with conditions of use included on the national label. A FIFRA label encompasses the terms on which a chemical is registered, and its requirements become part of the regulatory scheme under this Act. In registering the chemical, the Environmental Protection Agency makes a finding that when a chemical is used in accordance with widespread and commonly recognized practice, the chemical will not generally cause unreasonable adverse effects on the environment.

Pesticides used to control weeds, insects, and other pests receive public attention because of potential impacts on humans and the environment. Depending on the chemical, possible health effects from overexposure to pesticides include cancer, reproductive or nervous-system disorders, and acute toxicity. Similar effects are possible in the aquatic environment. However, many pesticides and their breakdown products do not have standards or guidelines, and current standards and guidelines do not yet account for exposure to mixtures and seasonal pulses of high concentrations. The effects of pesticides on aquatic life are a concern, because more than one-half of streams sampled had concentrations of at least one pesticide that exceeded an Environmental Protection Agency guideline for the protection of aquatic life. Whereas most toxicity and exposure assessments of pesticides are based on controlled experiments with a single contaminant, most contamination of waterbodies occurs as pesticide mixtures.

Thus, the Act authorizes the US Environmental Protection Agency to control the availability of pesticides that have the ability to leach into groundwater by governing the labeling, distribution, sale, and use of pesticides, including insecticides and herbicides. Its objective is to protect human health and the environment from unreasonable adverse effects of pesticides. It also establishes a nationally uniform labeling system requiring the registration of all pesticides sold in the United States, and requiring users to comply with the national label.

2.4 Hazardous materials transportation act

The Hazardous Materials Transportation Act authorizes the establishment and enforcement of hazardous material regulations for all modes of transportation by highway, water, and rail. The purpose of the Act is to ensure safe transportation of hazardous materials. The Act prevents any person from offering or accepting a hazardous material for transportation anywhere within this nation if that material is not properly classified, described, packaged, marked, labeled, and authorized for shipment pursuant to the regulatory requirements.

Under Department of Transportation regulations, a hazardous material is defined as any substance or material, including a hazardous substance and hazardous waste, which is capable of posing an unreasonable risk to health, safety, and property during transportation. The Act also imposes restrictions on the packaging, handling, and shipping of hazardous materials. For shipping and receiving of hazardous chemicals, hazardous wastes, and radioactive materials, the appropriate documentation, markings, labels, and safety precautions are required.

A hazardous material is any item or chemical which is a health hazard or a physical hazard, including the following: (i) chemicals that are carcinogens, toxic or highly toxic agents, reproductive toxins, irritants, corrosives, hepatotoxins, nephrotoxins, neurotoxins, agents that act on the hematopoietic system, and agents that damage the lungs, skin, eyes, or mucous membranes, (ii) chemicals that are combustible liquids, compressed gases, explosives, flammable liquids, flammable solids, organic peroxides, oxidizers, pyrophorics, unstable (reactive) or water-reactive; and (iii) chemicals that, in the course of normal handling, use or storage, may produce or release dusts, gases, fumes, vapors, mists or smoke having any of the above characteristics.

The Act is the principal federal law in the United States that regulates the transportation of hazardous materials and the purpose of the law is to protect the environment the risks meant that are inherent in the transportation of hazardous material in intrastate, interstate, and foreign commerce. The Act was passed as a means to improve the uniformity of existing regulations for transporting hazardous materials and to prevent spills and illegal dumping endangering the public and the environment, a problem exacerbated by uncoordinated and fragmented regulations. In any of these instances, a spill onto the surface can lead to runoff of the hazardous material into a water system. Road and railway accidents that occur in the transportation of hazardous materials can result in injury, death, and the destruction of property as well as the environment including, related to the current context, the water systems. In concert with this Act, the United States Coast Guard enforces regulations governing the bulk transportation of hazardous materials by vessels and regulations issued under other laws.

2.5 Occupational safety and health act

Occupational health hazards are those factors arising in or from the occupational environment that adversely impact health. Thus, the *Occupational Safety and Health Administration* (*OSHA*) came into being in 1970 and is responsible for administering the Occupational Safety and Health Act.

The goal of the Act is to ensure that employees do not suffer material impairment of health or functional capacity due to a lifetime occupational exposure to chemicals. The statute imposes a duty on employers to provide employees with a safe workplace environment, free of known hazards that may cause death or serious bodily injury.

The Act is also responsible for the means by which chemicals are contained. Workplaces are inspected to ensure compliance and enforcement of applicable standards under the Act. In keeping with the nature of the Act, there is also a series of standard tests relating to occupational health and safety as well as the general recognition of health hazards in the workplace. The Act is also the means by which guidelines have evolved for the management and disposition of chemicals used in chemical laboratories.

This Act has some (but limited) jurisdiction of over-the-road vehicle operation. In the instance of spills occurring while the material is on the vehicle or otherwise in transportation, the Hazardous Waste Operations and Emergency Response (HAZWOPER) standard does not cover the operator per se but does, however, cover emergency response personnel who respond to the incident.

2.6 Oil pollution act

The Oil Pollution Act of 1990 deals with pollution of waterways by crude oil and is intended to avoid oil spills from vessels and facilities by enforcing removal of spilled oil and assigning liability for the cost of cleanup and damage, requires specific operating procedure. The Act also defines responsible parties and financial liability; implements processes for measuring damages; specifies damages for which violators are liable; and establishes a fund for damages, cleanup, and removal costs. The Act specifically deals with crude oil vessels and onshore and offshore facilities and imposes strict liability for oil spills on their owners and operators.

Initially, the Act only limited liability for *deliberate* discharge of oil into marine waters. In 1970, the Congress of the United States placed oil pollution under the authority of the Federal Water Pollution Act (FWPA) of 1965, which later became the Clean Water Act of 1972 and had previously only covered sewage and Industrial discharge. The Federal Water Pollution Act set specific liability limitations.

In the current form, this Act streamlines and strengthens the ability of the Environmental Protection Agency to prevent and respond to catastrophic spillage of oil. A trust fund financed by a tax on oil is available to clean up spills when the responsible party is incapable or unwilling to do so. One of top priorities of the Environmental Protection Agency is to prevent, prepare for, and respond to oil spills that occur in and around inland waters of the United States. EPA is the lead federal response agency for oil spills occurring in inland waters. The Act requires oil storage facilities and vessels to submit to the Federal government plans detailing how they will respond to large discharges that endanger the environment, including water systems. The environmental Protection Agency has published regulations for aboveground storage facilities; the Coast Guard has done so for oil tankers.

2.7 Resource conservation and recovery act

Since its initial enactment in 1976, the Resource Conservation and Recovery Act (RCRA) continues to promote safer solid and hazardous waste management programs. Besides the regulatory requirements for waste management, the Act specifies the mandatory obligations of generators, transporters, and disposers of waste as well as those of owners and/or operators of waste treatment, storage, or disposal facilities. The Act also defines solid waste as: garbage, refuse, sludge from a treatment plant, from a water supply treatment plant, or air pollution control facility and other discarded material, including solid, liquid, semisolid, or contained gaseous material resulting from industrial, commercial, mining, and agricultural operations and from community activities.

The Act also states that solid waste does not include solid, or dissolved, materials in domestic sewage, or solid or dissolved materials in irrigation return flows or industrial discharges. A solid waste becomes a hazardous waste if it exhibits any one of four specific characteristics: (i) ignitability, (ii) reactivity, (iii) corrosivity, or (iv) toxicity. Certain types of solid wastes (e.g., household waste) are not considered to be hazardous, irrespective of their characteristics. Hazardous waste generated in a product or raw-material storage tank, transport vehicles, or manufacturing processes and samples collected for monitoring and testing purposes are exempt from the regulations.

Hazardous waste management is based on a beginning-to-end concept so that all hazardous wastes can be traced and fully accounted for. All generators and transporters of hazardous wastes as well as owners and operators of related facilities in the United States must file a notification with the Environmental Protection Agency. The notification must state the location of the facility and a general description of the activities as well as the identified and listed hazardous wastes being handled. Thus all regulated hazardous waste facilities must exist and/or operate under valid, activity-specific permits.

Regulations pertaining to companies that generate and/or transport wastes require that detailed records be maintained to ensure proper tracking of hazardous wastes through transportation systems. Approved containers and labels must be used, and wastes can only be delivered to facilities approved for treatment, storage, and disposal. The Act is the public law that creates the framework for the proper management of hazardous and non-hazardous solid waste (Table 9.2).

Under this Act, industrial wastewater must be managed (and not discharged into a water system) as solid or hazardous waste while they are being collected, stored, or treated before discharge). In addition, any sludge generated by treating the industrial wastewater must be managed as solid or hazardous waste.

Table 9.2 Categories of solid waste that can pollute a water system

Category	Comment
Garbage	Also known as municipal solid waste.
Refuse	For example: metal scrap, wall board, and empty containers
Sludge	From water and waste treatment plants or pollution control facilities
Industrial waste	Manufacturing process (wastewater, sludge, and solids)
Other materials	Solid, semisolid, liquid, or contained gaseous materials resulting from industrial, commercial, mining, agricultural, and community activities (e.g., boiler slag).

2.8 *Safe drinking water act*

The Safe Drinking Water Act, first enacted in 1974, was amended several times in the 1970s and 1980s to set national drinking water standards. The Act calls for regulations that (i) apply to public water systems, (ii) specify contaminants that may have any adverse effect on the health of persons, and (iii) specify contaminant levels. In addition, the difference between primary and secondary drinking water regulations is defined, and a variety of analytical procedures are specified. Statutory provisions are included to cover underground injection control systems. The Act also requires maximum levels at which a contaminant must have no known or anticipated adverse effects on human health, thereby providing an *adequate margin of safety*.

The Superfund Amendments and Reauthorization Act (SARA) set standards the same for groundwater as for drinking water in terms of necessary cleanup and remediation of an inactive site that might be a former crude oil refinery. Under the Act, all underground injection activities must comply with the drinking water standards as well as meet specific permit conditions that are in unison with the provisions of the Clean Water Act. However, under the Resource Conservation and Recovery Act, class IV injection wells are no longer permitted and there are several restrictions on underground injection wells that may be used for storage and disposal of hazardous wastes.

The Act includes mandatory levels for contaminants and the standards are organized into six groups: (i) microorganisms, (ii) disinfectants, (iii) disinfection byproducts, (iv) inorganic chemicals, (v) organic chemicals, and (vi) radionuclides. The Act requires that a public water system must notify its customers when it violates drinking water regulations or is providing drinking water that may pose a health risk. Such notifications are provided either immediately, as soon as possible (but within 30 days of the violation) or annually, depending on the health risk associated with the violation.

2.9 *Toxic substances control act*

The Toxic Substances Control Act (TSCA) was first enacted in 1976 and was designed to provide controls for those chemicals that may threaten human health or the environment. The Act authorizes the US Environmental Protection Agency to control the manufacture, use, storage, distribution, or disposal of toxic chemicals that have the potential to leach into groundwater.

Particularly hazardous chemicals are the cyclic nitrogen species that may be produced when crude oil is processed and that often occur in residua and cracked residua. The objective of the Act is to provide the necessary control before a chemical is allowed to be mass-produced and enter the environment. The term *chemical substance*, as defined by the Act, is any organic or inorganic substance of a particular molecular identity, including any combination of these substances occurring in whole or in part as a result of a chemical reaction or occurring in nature, and any element or uncombined radical.

The Act specifies a *premanufacture notification* requirement by which any manufacturer must notify the Environmental Protection Agency at least 90 days prior to the production of a new chemical substance. Notification is also required even if there is a new use for the chemical that can increase the risk to the environment. No notification is required for chemicals that are manufactured in small quantities solely for scientific research and experimentation. A *new chemical substance* is defined as a chemical that is not listed in the Environmental Protection Agency Inventory of Chemical Substances or is an unlisted reaction product of two or more chemicals. In addition, the term *chemical substance* means any organic or inorganic substance of a particular molecular identity, including any combination of such substances occurring in whole or in part as a result of a chemical reaction or occurring in nature, and any element or uncombined radical. The term *mixture* means any combination of two or more chemical substances if the combination does not occur in nature and is not, in whole or in part, the result of a chemical reaction.

The Act provides authority to the Environmental Protection Agency to require reporting, record-keeping and testing requirements, and restrictions relating to chemical substances and/or mixtures. The Act also requires that any person who manufactures (including imports), processes, or distributes in commerce a chemical substance or mixture and who obtains information which reasonably supports the conclusion that such substance or mixture presents a substantial risk of injury to health or the environment to immediately inform the Environmental Protection Agency, except where the Agency EPA has been adequately informed of such information. Also, the Agency screens all submissions related to the Act as well as voluntary submissions. The latter are not required by law, but are submitted by industry and public interest groups for a variety of reasons.

Contrary to what the name implies, the Act does not separate chemicals into categories of toxic and non-toxic but it does prohibit the manufacture or importation of chemicals that are not on the Act Inventory or subject to one of many exemptions. Chemicals listed on the inventory are referred to as *existing chemicals* while chemicals not listed are referred to as *new chemicals*.

However, spillage of many any of these chemicals and/or mixtures that fall under the Act could cause irreparable harm to a water system and render the water unsuitable for consumption.

2.10 Water pollution control act

Several acts are related to the protection of the waterways in the United States. Of particular interest in the present context is the *Water Pollution Control Act* (*Clean Water Act*) which is one of the first and most influential modern environmental laws in the United States. The objective of the Act is to restore and maintain the chemical, physical, and biological integrity of water systems and authorizes development of surface water protection strategies and authorizes a number of programs to prevent water pollution from a variety of potential sources.

The Water Pollution Control Act of 1948 and The Water Quality Act of 1965 were generally limited to control of pollution of interstate waters and the adoption of water-quality standards by the states for interstate water within their borders. The first comprehensive water-quality legislation in the United States came into being in 1972 as the Water Pollution Control Act. This Act was amended in 1977 and became the Clean Water Act. Further amendments in 1978 were enacted to deal more effectively with spills of crude oil. Other amendments followed in 1987 under the new name Water Quality Act and were aimed at improving water quality in those areas where there were insufficiencies in compliance with the discharge standards.

Section 311 of the Clean Water Act includes elaborate provisions for regulating intentional or accidental discharges of crude oil and of hazardous substances. Included are response actions required for oil spills and the release or discharge of toxic and hazardous substances. As an example, the person in charge of a vessel or an onshore or offshore facility from which any designated hazardous substance is discharged, in quantities equal to or exceeding its reportable quantity, must notify the appropriate federal agency as soon as such knowledge is obtained.

The Act is the primary federal law in the United States that governs water pollution and it is the objective of the Act to restore and maintain the chemical, physical, and biological integrity of the waters systems by preventing point and nonpoint pollution sources (though it does not enable the control of nonpoint sources); recognizing the responsibilities of the states in addressing pollution and providing assistance to states to do so, including funding for publicly owned treatment works for the improvement of wastewater treatment and maintaining the integrity of wetlands.

3 Pollution prevention

Pollution prevention is the responsibility of everyone. In fact, making the public aware of the problem is the first step to prevent water pollution. Hence, importance of water and pollution prevention measures should be a part of awareness and education program. Polluter pays principle should be adopted so that the polluters will be the first people to suffer by way of paying the cost for the pollution. Ultimately, the polluter pays principle should be designed to prevent people from polluting and making them behave in an environmentally responsible manner.

Preventing pollution may be a new role for production-oriented managers and workers, but their cooperation is crucial. It will be the workers themselves who must make pollution prevention succeed in the workplace. Several options have been identified that refineries can undertake to reduce pollution. These include pollution prevention options, recycling options, and waste treatment options. Furthermore, pollution prevention options are often is presented in four different categories, viz.: (i) pollution prevention options, (ii) waste recycling, and (iii) waste treatment. Either one or the other or any combination of the three options may be in operation in any industrial operation.

Contamination of water can result in (i) poor drinking water quality, (ii) loss of water supply, (iii) degraded surface water systems, (iv) high cleanup costs, (v) high costs for alternative water supplies, and/or (vi) potential health problems. The consequences of contaminated water or degraded surface water are often serious. For example, estuaries that have been impacted by high nitrogen from water sources have lost critical shellfish habitats. In terms of water supply, in some instances, water contamination is so severe that the water supply must be abandoned as a source of drinking water.

In other cases, the water can be cleaned up and used again, if the contamination is not too severe and if the municipality is willing to spend a good deal of money. Follow-up water quality monitoring is often required for many years. Because water generally moves slowly, contamination often remains undetected for long periods of time. This makes cleanup of a contaminated water supply difficult, if not impossible. If a cleanup is undertaken, it can cost anywhere in the range from thousands to millions of dollars. Once the contaminant source has been controlled or removed, the contaminated water can be treated in one of several ways: (i) containing the contaminant to prevent migration, (ii) pumping the water, treating it, and returning it to the aquifer, (iii) leaving the water in place an d treating either the water or the contaminant, and (iv) allowing the contaminant to attenuate (reduce) naturally (with monitoring), following the implementation of an appropriate source control.

Selection of the appropriate remedial technology is based on site-specific factors and often takes into account cleanup goals based on potential risk that are protective of human health and the environment. The technology selected is one that will achieve those cleanup goals. Different technologies are effective for different types of contaminants, and several technologies are often combined to achieve effective treatment but are site specific and spill specific (Chapter 8). The effectiveness of treatment depends in part on local hydrogeological conditions, which must be evaluated prior to selecting an appropriate treatment option.

Given the difficulty and high costs of cleaning up a contaminated aquifer, some communities choose to abandon existing wells and use other water sources, if available. Using alternative supplies is probably more expensive than obtaining drinking water from the original source. A temporary and expensive solution is to purchase bottled water, but it is not a realistic long-term solution for the drinking water supply problem of a community. A community might decide to install new wells in a different area of the aquifer. In this case, appropriate siting and monitoring of the new wells are critical to ensure that contaminants do not move into the new water supplies.

The key challenges to better management of the water quality comprise of temporal and spatial variation of rainfall, uneven geographic distribution of surface water resources, persistent droughts, overuse of ground water and contamination, drainage and salinization and water quality problems due to treated, partially treated and untreated wastewater from urban settlements, industrial establishments and runoff from irrigation sector besides poor management of municipal solid waste and animal dung in rural areas. Improper use of fertilizers, herbicides and pesticides in farming should be stopped and organic methods of farming should be adopted. Cropping practices in riparian zone should be banned to protect the riparian vegetation growing there.

The conditions of the contaminated area plays a major role on whether bioremediation is the appropriate method of cleanup for the given oil spill. The success of bioremediation is dependent upon physical conditions and chemical conditions. Physical parameters include temperature, surface area of the oil, and the energy of the water. Chemical parameters include oxygen and nutrient content, pH, and the composition of the oil. Temperature affects bioremediation by changing the properties of the oil and also by influencing the crude oil-degrading microbes. When the temperature is lowered, the viscosity of the crude oil is increased which changes the toxicity and solubility of the oil, depending upon its composition. Temperature also has an effect on the growth rate of the microorganisms, as well as the degradation rate of the hydrocarbon derivatives, depending upon their characteristics.

Biostimulation, the addition of nutrients, is practiced for cleanup of crude oil spills in seawater when there is an existing population of oil degrading microbes present. When an oil spill occurs, the result is a large increase in carbon and this also stimulates the growth of the indigenous crude oil-degrading microorganisms. However, these microorganisms are limited in the amount of growth and remediation that can occur by the amount of available nitrogen and phosphorus.

By adding these supplemental nutrients in the proper concentrations, the hydrocarbon degrading microbes are capable of achieving their maximum growth rate and hence the maximum rate of pollutant uptake. It has been found that when using nitrogen for the supplemental nutrient, a maximum growth rate is achieved by the oil degrading microorganisms. Biostimulation has been proven to be an effective way of achieving increased hydrocarbon degradation by the indigenous microbial population.

Supplemental nutrients tended to move downward during rising tides and seaward during falling tides. This is very useful information in determining the proper timing to add nutrients in order to allow for the maximum residence time of the nutrients in the contaminated areas. The results of this experiment concluded that the nutrients should be applied during low tide at the high tide line, which results in maximum contact time of the nutrients with the oil and hydrocarbon degrading microorganisms.

Waves also have an effect on the distribution and movement of the water and dissolved nutrients, which determine the residence time of the nutrients in the oil affected area. The role of waves on solute movement varies whether there is a tide or not. When Research performed by, showed that when a wave is present there is a sharp seaward hydraulic gradient in the backwash zone, and a gentle gradient landward of this area. Furthermore, the contact time of the nutrients is increased when the waves break seaward of their location and the waves increased the dispersion and washout of the nutrients in the tidal zone, and residence time was approximately 75% when a wave was present with a tide, as compared to a tide with no waves.

The marine environment encompassing the vast majority of surface of the Earth is a large repertoire of a variety of microorganisms. The environmental roles of the biosurfactants produced by many such marine microorganisms have been reported earlier.

3.1 Options

Having defined the process products and emission pollution prevention is the operational guideline for industrial operators, process engineers, process chemists, and, for that matter, anyone who handles industrial products, especially crude oil and/or crude oil products. It is in this area that environmental analysis plays a major role (EPA, 2004).

Pollution prevention is, simply, reduction or elimination of discharges or emissions to the environment. The limits of pollutants emitted to the atmosphere, the land, and water are defined by various pieces of legislation that have been put into place over the past four decades (Chapter 5) (Speight, 1996; Woodside, 1999). This includes all pollutants such as hazardous and non-hazardous wastes, regulated and unregulated chemicals from all sources.

Pollution associated with crude oil refining typically includes volatile organic compounds (volatile organic compounds), carbon monoxide (CO), sulfur oxides (SO_x), nitrogen oxides (NO_x), particulates, ammonia (NH_3), hydrogen sulfide (H_2S), metals, spent acids, and numerous toxic organic compounds. Sulfur and metals result from the impurities in crude oil. The other wastes represent losses of feedstock and crude oil products. These pollutants may be discharged as air emissions, wastewater, or solid waste. All of these wastes are treated. However, air emissions are more difficult to capture than wastewater or solid waste. Thus, air emissions are the largest source of untreated wastes released to the environment.

Pollution prevention can be accomplished by reducing the generation of wastes at their source (source reduction) or by using, reusing or reclaiming wastes once they are generated (environmentally sound recycling). However, environmental analysis plays a major role in determining if emissions-effluents (air, liquid or solid) fall within the parameters of the relevant legislation. For example, issues to be addressed are the constituents of gaseous emissions, the sulfur content of liquid fuels, and the potential for leaching contaminants (through normal rainfall or through the agency of acid rain) from solid products such as coke.

Pollution prevention options are usually subdivided into four areas: (i) good operating practices, (ii) processes modification, (iii) feedstock modification, and (iv) product reformulation (Lo, 1991). The options described here include only the first three of these categories since product reformulation is not an option that is usually available to the environmental analyst, scientist or engineer.

When pollution prevention and recycling options are not economically viable, pollution can still be reduced by treating wastes so that they are transformed in to less environmentally harmful wastes or can be disposed of in a less environmentally harmful media.

The toxicity and volume of some de-oiled and dewatered sludge can be further reduced through thermal treatment. Thermal sludge treatment units use heat to vaporize the water and volatile components in the feed and leave behind a dry solid residue. The vapors are condensed for separation into the hydrocarbon and water components. Non-condensable vapors are either flared or sent to the gas-cleaning amine unit (Mokhatab et al., 2006; Speight, 2019) for treatment and use as a fuel gas.

Furthermore, because oily sludge makes up a large portion of many industrial process solid wastes, any improvement in the recovery of oil from the sludge can significantly reduce the volume of waste. There are a number of technologies currently in use to mechanically separate oil, water and solids, including: belt filter presses, recessed chamber pressure filters, rotary vacuum filters, scroll centrifuges, disc centrifuges, shakers, thermal driers and centrifuge-drier combinations.

Waste material such as tank bottoms from crude oil storage tanks constitute a large percentage of industrial solid waste and pose a particularly difficult disposal problem due to the presence of heavy metals. Tank bottoms are comprised of high molecular weight hydrocarbon derivatives, solids, water, rust and scale. Minimization of tank bottoms is carried out most cost effectively through careful separation of the oil and water remaining in the tank bottom. Filters and centrifuges can also be used to recover the oil for recycling.

Spent clay from industrial filters often contains significant amounts of entrained hydrocarbon derivatives and, therefore, must be designated as hazardous waste. Back washing spent clay with water or steam can reduce the hydrocarbon content to levels so that it can be reused or handled as a non-hazardous waste. Another method used to regenerate clay is to wash the clay with naphtha, dry it by steam heating and then feed it to a burning kiln for regeneration. In some cases clay filtration can be replaced entirely with hydrotreating process options.

Decant oil sludge from the fluidized bed catalytic cracking unit can (and often does) contain significant concentrations of catalyst fines. These fines often prevent the use of decant oil as a feedstock or require treatment which generates an oily catalyst sludge. Catalyst fines in the decant oil can be minimized by using a decant oil catalyst removal system. One system incorporates high voltage electric fields to polarize and capture catalyst particles in the oil. The amount of catalyst fines reaching the decant oil can be minimized by installing high efficiency cyclones in the reactor to shift catalyst fines losses from the decant oil to the regenerator where they can be collected in the electrostatic precipitator.

3.2 Operating practices

Good operating practices (Table 9.3) prevent waste by better handling of feedstocks and products without making significant modifications to current production technology. If feedstocks are handled appropriately, they are less likely to become wastes inadvertently through spills or outdating. If products are handled appropriately, they can be managed in the most cost-effective manner. For example, a significant portion

Table 9.3 A selection of good operating practices

- Specify sludge and water content for feedstock Minimize carryover to API separator
- Use recycled water for desalter
- Replace desalting with chemical treatment system
- Collect catalyst fines during delivery
- Recover coke fines

of industrial waste arises from oily sludge found in combined process/storm sewers. Segregation of the relatively clean rainwater runoff from the process streams can reduce the quantity of oily sludge generated. Furthermore, there is a much higher potential for recovery of oil from smaller, more concentrated process streams.

Solids released to the industrial wastewater sewer system can account for a large portion of the sludge from an industrial process. Solids entering the sewer system (primarily soil particles) become coated with oil and are deposited as oily sludge in the API oil/water separator. Because a typical sludge has a solids content of five to 30% by weight, preventing one pound of solids from entering the sewer system can eliminate several pounds 3–20 pounds of oily sludge.

Methods used to control solids include using a street sweeper on paved areas, paving unpaved areas, planting ground cover on unpaved areas, re-lining sewers, cleaning solids from ditches and catch basins, and reducing heat exchanger bundle cleaning solids by using anti-foulants in cooling water. Benzene and other solvents in wastewater can often be treated more easily and effectively at the point at which they are generated rather than at the wastewater treatment plant after it is mixed with other wastewater.

3.3 Process modifications

The crude oil industry requires very large, capital-intensive process equipment. Expected lifetimes of process equipment are measured in decades. This limits economic incentives to make capital-intensive process modifications to reduce wastes generation. However, some process modifications (Table 9.4) or process improvement (Table 9.5) reduce waste generation. The crude oil industry has made many improvements in the design and modification of processes and technologies to recover product and unconverted raw materials. In the past, they pursued this strategy to the point that the cost of further recovery could not be justified. Now the costs of end-of-pipe treatment and disposal have made source reduction a good investment. Greater reductions are possible when process engineers trained in pollution prevention plan to reduce waste at the design stage. For example, although barge loading is not a factor for all

Table 9.4 Options for process modifications

- Add coking operations.
- Certain industrial hazardous wastes can then be used as coker feedstock, reducing the quantity of sludge for disposal.
- Install secondary seals on floating roof tanks.
- Where appropriate, replace with fixed roofs to eliminate the collection of rainwater, contamination of crude oil or finished products, and oxidation of crude oil.
- Where feasible,
  - Replace clay filtration with hydrotreating.
  - Substitute air coolers or electric heaters for water heat exchangers to reduce sludge production.
  - Install tank agitators. This can prevent solids from settling out.
  - Concentrate similar wastewater streams through a common dewatering system.

Table 9.5 Process improvement

• Segregate wastes to reduce the quantity of oily sludge generated and increase the potential for oil recovery. • Reuse rinse waters where possible. • Use optimum pressures, temperatures and mixing ratios. • Sweep or vacuum streets and paved process areas to reduce solids going to sewers. • Use water softeners in cooling water systems to extend the useful life of the water.

refineries, it is an important emissions source for many facilities. One of the largest sources of volatile organic carbon emissions is the fugitive emissions from loading of tanker barges. These emissions could be reduced by >90% by installing a vapor loss control system that consists of vapor recovery or the destruction of the volatile organic carbon emissions in a flare.

Fugitive emissions are one of the largest sources of industrial hydrocarbon emissions. A leak detection and repair (LDAR) program consists of using a portable detecting instrument to detect leaks during regularly scheduled inspections of valves, flanges, and pump seals. Older industrial boilers may also be a significant source of emissions of sulfur oxides (SO_x), nitrogen oxides (NO_x), and particulate matter. It is possible to replace a large number of old boilers with a single new cogeneration plant with emissions controls.

Since storage tanks are one of the largest sources of VOC emissions, a reduction in the number of these tanks can have a significant impact. The need for certain tanks can often be eliminated through improved production planning and more continuous operations. By minimizing the number of storage tanks, tank bottom solids and decanted wastewater may also be reduced. Installing secondary seals on the tanks can significantly reduce the losses from storage tanks containing gasoline and other volatile products.

Solids entering the crude distillation unit are likely to eventually attract more oil and produce additional emulsions and sludge. The amount of solids removed from the desalting unit should, therefore, be maximized. A number of techniques can be used such as: using low shear mixing devices to mix desalter wash water and crude oil; using lower pressure water in the desalter to avoid turbulence; and replacing the water jets used in some refineries with mud rakes which add less turbulence when removing settled solids.

Purging or blowing down a portion of the cooling water stream to the wastewater treatment system controls the dissolved solids concentration in the recirculating cooling water. Solids in the blowdown eventually create additional sludge in the wastewater treatment plant. However, minimizing the dissolved solids content of the cooling water can lower the amount of cooling tower blowdown. A significant portion of the total dissolved solids in the cooling water can originate in the cooling water makeup stream in the form of naturally occurring calcium carbonates. Such solids can be controlled either by selecting a source of cooling tower makeup water with less dissolved solids or by removing the dissolved solids from the makeup water stream. Common treatment methods include: cold lime softening, reverse osmosis, or electrodialysis.

In many refineries, using high-pressure water to clean heat exchanger bundles generates and releases water and entrained solids to the industrial wastewater treatment system. Exchanger solids may then attract oil as they move through the sewer system and may also produce finer solids and stabilized emulsions that are more difficult to remove. Solids can be removed at the heat exchanger cleaning pad by installing concrete overflow weirs around the surface drains or by covering drains with a screen. Other ways to reduce solids generation are by using anti-foulants on the heat exchanger bundles to prevent scaling and by cleaning with reusable cleaning chemicals that also allow for the easy removal of oil.

Surfactants entering the industrial wastewater streams will increase the amount of emulsions and sludge generated. Surfactants can enter the system from a number of sources including: washing unit pads with detergents; treating gasoline with an end point over 200 °C (>392 °F) thereby producing spent caustics; cleaning tank truck tank interiors; and using soaps and cleaners for miscellaneous tasks. In addition, the overuse and mixing of the organic polymers used to separate oil, water, and solids in the wastewater treatment plant can actually stabilize emulsions. The use of surfactants should be minimized by educating operators, routing surfactant sources to a point downstream of the DAF unit and by using dry cleaning, high pressure water or steam to clean oil surfaces of oil and dirt.

Replacing 55-gal drums with bulk storage facilities can minimize the chances of leaks and spills. And, just as 55-gal drums can lead to leaks, underground piping can be a source of undetected releases to the soil and groundwater. Inspecting, repairing or replacing underground piping with surface piping can reduce or eliminate these potential sources.

Finally, open ponds used to cool, settle out solids and store process water can be a significant source of volatile organic carbon emissions. Wastewater from coke cooling and coke volatile organic carbon removal is occasionally cooled in open ponds where volatile organic carbon easily escape to the atmosphere. In many cases, open ponds can be replaced with closed storage tanks.

3.4 Material substitution options

The concept of raw materials substitution implies effective and efficient use of raw materials (to minimize losses along the process system) as well as using different raw materials that will not generate waste during processing. This concept also further implies re-using materials or using recycled materials and is, essentially: (i) changing the source of raw gas feed and substituting the feed with a feed that will produce less waste in the process system; (ii) changing chemicals for other chemical reactions in the process by substituting them with different chemicals that will not generate waste and that are more environmentally friendly and safe to process or use. This ultimately translates to reformulating and redesigning products that will be environment friendly.

An example is that spent conventional degreaser solvents can be reduced or eliminated through substitution with less toxic and/or biodegradable products. In addition, chromate containing wastes can be reduced or eliminated in cooling tower and heat exchanger sludge by replacing chromates with less toxic alternatives such as phosphates. Also, using catalysts of a higher quality will lead in increased process efficiency while the required frequency of catalyst replacement can be reduced. Similarly, the replacement of **ceramic catalyst support with activated alumina supports** presents the opportunity for recycling the activated alumina supports with the spent alumina catalyst.

However, waste problems that cannot be solved by simple procedural adjustments or improvements in process practices require more substantial long-term changes. It is necessary to develop possible prevention options for the waste problems. The process or production changes that may increase production efficiency and reduce waste generation include: (i) changes in the production process, continuous versus batch, (ii) equipment and installation changes, (iii) changes in process control, automation, (iv) changes in process conditions, such as retention times, temperatures, agitation, pressure, catalysts, (v) use of dispersants in place of organic solvents where appropriate, (v) reduction in the quantity or type of raw materials used in production, (vi) raw material substitution through the use of wastes as raw materials or the use of different raw materials that produce less waste or less hazardous waste, and (vii) process substitution with cleaner technology.

Waste reuse often can be implemented if materials of sufficient purity can be concentrated or purified. Technologies such as reverse osmosis, ultrafiltration, electrodialysis, distillation, electrolysis, and ion exchange may enable materials to be reused and reduce or eliminate the need for waste treatment. Where waste treatment is necessary, a variety of technologies should be considered. These include physical, chemical, and biological treatment processes. In some cases the treatment method can also recover valuable materials for reuse. Another industry or factory may be able to use or treat a waste that you cannot treat on-site. It may be worth investigating the possibility of setting up a waste exchange bureau as a structure for sharing treatment and also reuse facilities. There may also be the need to reconsider the possibilities for product improvements or changes yielding cleaner, more environmentally-friendly products, both for existing products and in the development of new products.

Finally, the substitution of currently-used chemicals with less hazardous chemicals is one of the most effective ways of eliminating or reducing exposure to materials that are toxic or pose other hazards as well as reducing the amount of hazardous waste produced in the process.

A hazardous chemical (or material) is the source of danger or injury and can include any chemical or material that has the ability or a property that can cause an adverse health effect or harm to a person under certain conditions or harm to the environment. On the other hand, risk from such chemicals or materials is the probability or chance that exposure to a chemical hazard will actually cause harm to a person or cause an adverse effect.

Other methods for controlling the production and emissions of chemicals include elimination, isolation, enclosure, local exhaust ventilation, process or equipment modification, good housekeeping, administrative controls and personal protective equipment. All these methods reduce or eliminate the risk of injury or harm by interrupting the path of exposure between the hazardous material and the worker. Substitution removes the hazard at the source.

3.5 Recycling

Recycling is the use, reuse or reclamation of a waste after it is generated. At present the crude oil industry is focusing on recycling and reuse as the best opportunities for pollution prevention (Table 9.6). Although pollution is reduced more if wastes are prevented in the first place, a next best option for reducing pollution is to treat wastes so that they can be transformed into useful products.

Caustic substances used to absorb and remove hydrogen sulfide and phenol contaminants from intermediate and final product streams can often be recycled. Spent caustics may be saleable to chemical recovery companies if concentrations of phenol or hydrogen sulfide are high enough. Process changes in the industrial may be needed to raise the concentration of phenols in the caustic to make recovery of the contaminants economical. Caustics containing phenols can also be recycled on-site by reducing the pH of the caustic until the phenols become insoluble thereby allowing physical separation. The caustic can then be treated in the industrial wastewater system.

Oily sludge can be sent to a coking unit or the crude distillation unit where it becomes part of the industrial products. Sludge sent to the coker can be injected into the coke drum with the quench water, injected directly into the delayed coker, or injected into the coker blowdown contactor used in separating the quenching products. Use of sludge as a feedstock has increased significantly in recent years and is currently carried out by most refineries. The quantity of sludge that can be sent to the coker is restricted by coke quality specifications that may limit the amount of sludge solids in the coke. Coking operations can be upgraded, however, to increase the amount of sludge that they can handle.

Significant quantities of catalyst fines are often present around the catalyst hoppers of fluid catalytic cracking reactors and regenerators. Coke fines are often present around the coker unit and coke storage areas. The fines can be collected and recycled before being washed to the sewers or migrating off-site via the wind. Collection techniques include dry sweeping the catalyst and coke fines and sending the solids to be recycled or disposed of as non-hazardous waste. Coke fines can also be recycled for fuel use. Another collection technique involves the use of vacuum ducts in dusty areas (and vacuum hoses for manual collection) that run to a small baghouse for collection.

An issue that always arises relates to the disposal of laboratory sample from any process control or even environmental laboratory that is associated with an industrial process. Samples from such a laboratory can be recycled to the chemical waste recovery and/or disposal system.

Table 9.6 Options for recycling

- Use phenols and caustics as chemical feedstocks.
- Use oily waste sludge as feedstock in coking operations.
- Regenerate catalysts. Extend useful life. Recover valuable metals from spent catalyst. Possibly use catalyst as a concrete admixture or as a fertilizer.
- Regenerate filtration clay.
- Wash clay with naphtha
- Dry by steam heating
- Use a burning kiln for regeneration.
- Recover valuable product from sludge with solvent extraction.

4 Wastes and treatment

Waste elimination is common sense and provides several obvious benefits (Table 9.7). Yet waste elimination continues to elude many companies in every sector, including the industrial section, and activity from industrial waste (that is a function of their production system design). It may not matter how a refiner categorizes the waste or how the refiner chooses to pursue waste elimination, one thing remains constant and that is once identified, waste could be eliminated. There are models and structures that allow a refiner to identify and eliminate waste to increase productivity, and hence cost structures, that have a direct impact on industrial operations and, more than all else, profitability. Waste elimination though identification (by judicious analysis) and treatment subscribe to the smooth operation of an industrial process.

Generally process wastes (emissions) are categorized as gaseous, liquid, and solid. This does not usually include waste from or from accidental spillage of a crude oil feedstock a product. Also, the efficiency of a pollutant removal technology varies widely depending on (i) the treatment technology, (ii) site-specific considerations such as local weather patterns, (iii) maintenance of the system, and (iv) design constraints. Each of these items should be given serious consideration before the technology is applied.

In addition, any waste that is generated from a domestic and/or an industrial source, if not sent for disposal by a recognized (legal) method has the potential to eventually enter a water system. As defined elsewhere in this text, water is uniquely vulnerable to pollution (Chapter 5). Known as a universal solvent, water is able to dissolve more substances than any other liquid on earth (Chapter 3). This is the reason why water is so easily polluted. Toxic substances from farms, towns, and factories readily dissolve into and mix with water, causing water pollution.

An example, in the case of groundwater, is that when rain falls and seeps deep into the earth, filling the cracks, crevices, and porous spaces of an aquifer (basically an underground storehouse of water), it becomes groundwater which is one of the least visible but most important natural resources. Groundwater becomes polluted when contaminants (such as from pesticides and fertilizers, waste leached from landfills and septic systems) enter an aquifer, rendering it unsafe for human use. Once polluted, an aquifer may be unusable for decades, or even thousands of years. Furthermore, groundwater can also spread contamination far from the original polluting source as it seeps into streams, lakes, and oceans.

Table 9.7 Benefits of waste elimination

- Solve the waste disposal problems created by land bans
- Reduce waste disposal costs
- Reduce costs for energy, water and raw materials
- Reduce operating costs
- Protect workers, the public and the environment
- Reduce risk of spills, accidents and emergencies
- Reduce vulnerability to lawsuits and improve its public image
- Generate income from wastes that can be sold.

Surface water (such as is in lakes, rivers, and oceans) is in peril due to the influx of contaminants such as nitrates and phosphates. While plants and animals need these nutrients to grow, they have become major pollutants due to farm waste and fertilizer runoff. Municipal and industrial waste discharges contribute their fair share of toxins as well. There is also all the random junk that industry and individuals dump directly into waterways. Contaminants such as chemicals, nutrients, and heavy metals are carried from farms, factories, and cities by streams and rivers into our bays and estuaries; from there they travel out to sea. Meanwhile, marine debris—particularly—is blown in by the wind or washed in via storm drains and sewers. The oceans are also sometimes spoiled by spills of crude oil and/or crude oil products consistently soaking up carbon pollution from the atmosphere.

Hence, in all senses of the word *pollution*, water systems must be protected from contamination from pollutants by (i) air emissions, (ii) *water and treatment, as well as* (iii) *other forms of waste.*

4.1 Air emissions

Air emissions include point and non-point sources (Chapter 4). Point sources are emissions that exit stacks and flares and, thus, can be monitored and treated. Non-point sources are *fugitive emissions* that are difficult to locate and capture. Fugitive emissions occur throughout refineries and arise from the thousands of valves, pumps, tanks, pressure relief valves, flanges, etc. While individual leaks are typically small, the sum of all fugitive leaks from an industrial process can be one of largest emission sources from the process.

The numerous process heaters used in refineries to heat process streams or to generate steam (boilers) for heating or steam stripping, can be potential sources of emissions for sulfur oxides (SO_x,), nitrogen oxides NO_x,), carbon monoxide (CO), particulate matter, and hydrocarbon derivatives. When operating properly and when burning cleaner fuels such as industrial fuel gas, fuel oil or natural gas, these emissions are relatively low. If, however, combustion is not complete, or heaters are fired with industrial fuel pitch or residuals, emissions can be significant.

The majority of gas streams exiting an industrial process contain varying amounts of industrial fuel gas, hydrogen sulfide and ammonia. These streams are collected and sent to the gas treatment and sulfur recovery units to recover the industrial fuel gas and sulfur though a variety of add-on technologies (Speight, 1993, 1996). Emissions from the sulfur recovery unit typically contain some hydrogen sulfide, sulfur oxides, and nitrogen oxides. Other emissions sources from industrial processes arise from periodic regeneration of catalysts. These processes generate streams that may contain relatively high levels of carbon monoxide, particulates and volatile organic compounds. Before being discharged to the atmosphere, such off-gas streams may be treated first through a carbon monoxide boiler to burn carbon monoxide and any volatile organic compounds, and then through an electrostatic precipitator or cyclone separator to remove particulates.

Sulfur is removed from a number of industrial process off-gas streams (sour gas) in order to meet the sulfur oxide emissions limits of the Clean Air Act and to recover saleable elemental sulfur. Process off-gas streams, or sour gas, from the coker, catalytic cracking unit, hydrotreating units and hydroprocessing units can contain high concentrations of hydrogen sulfide mixed with light industrial fuel gases.

Before elemental sulfur can be recovered, the fuel gases (primarily methane and ethane) need to be separated from the hydrogen sulfide. This is typically accomplished by dissolving the hydrogen sulfide in a chemical solvent. Solvents most commonly used are amines, such as diethanolamine (DEA, $HOCH_2CH_2NHCH_2CH_2OH$). Dry adsorbents such as molecular sieves, activated carbon, iron sponge (Fe_2O_3) and zinc oxide (ZnO) are also used (Speight, 1993). In the amine solvent processes, diethanolamine solution or similar ethanolamine solution is pumped to an absorption tower where the gases are contacted and hydrogen sulfide is dissolved in the solution. The fuel gases are removed for use as fuel in process furnaces in other industrial operations. The amine-hydrogen sulfide solution is then heated and steam stripped to remove the hydrogen sulfide gas.

Current methods for removing sulfur from the hydrogen sulfide gas streams are typically a combination of two processes in which the primary process is the Claus Process followed by either the Beavon Process or the SCOT Process or the Wellman-Lord Process.

In the Claus process (Fig. 9.1) (Speight, 1993 and references cited therein), the hydrogen sulfide, after separation from the gas stream using *amine extraction,* is fed to the Claus unit, where it is converted in two stages. The first stage is a **thermal step:** in which the hydrogen sulfide is partially oxidized with air in a reaction furnace at high temperatures (1000–1400 °C, 1830–2550 °F). Sulfur is formed, but some hydrogen sulfide remains unreacted, and some sulfur dioxide is produced. The second stage is a catalytic stage in which the remaining hydrogen sulfide is reacted with the sulfur dioxide at lower temperatures (200–350 °C, 390–660 °F) over a catalyst to produce more sulfur. The overall reaction is the conversion of hydrogen sulfide and sulfur dioxide to sulfur and water:

$$2H_2S + SO_2 \rightarrow 3S + 2H_2O$$

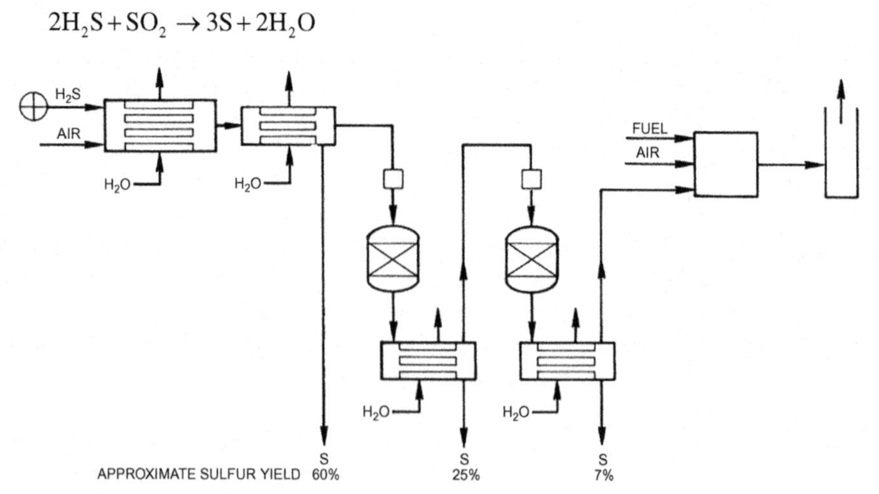

Fig. 9.1 The Claus process.

The catalyst is necessary to ensure that the components react with reasonable speed but, unfortunately, the reaction does not always proceed to completion. For this reason two or three stages are used, with sulfur being removed between the stages. For the analysts, it is valuable to know that carbon disulfide (CS_2) is a by-product from the reaction in the high-temperature furnace. The carbon disulfide can be destroyed catalytically before it enters the catalytic section proper.

Generally, the Claus process may only remove about 90% of the hydrogen sulfide in the gas stream and, as already noted, other processes such as the Beavon process, the SCOT process, or Wellman-Lord processes are often used to further recover sulfur.

In the Beavon process, the hydrogen sulfide in the relatively low concentration gas stream from the Claus process can be almost completely removed by absorption in a quinone solution. The dissolved hydrogen sulfide is oxidized to form a mixture of elemental sulfur and hydroquinone. The solution is injected with air or oxygen to oxidize the hydroquinone back to quinone. The solution is then filtered or centrifuged to remove the sulfur and the quinone is then reused. The Beavon process is also effective in removing small amounts of sulfur dioxide, carbonyl sulfide, and carbon disulfide that are not affected by the Claus process. These compounds are first converted to hydrogen sulfide at elevated temperatures in a cobalt molybdate catalyst prior to being fed to the Beavon unit. Air emissions from sulfur recovery units will consist of hydrogen sulfide, sulfur oxides, and nitrogen oxides in the process tail gas as well as fugitive emissions and releases from vents.

The SCOT process is also widely used for removing sulfur from the Claus tail gas. The sulfur compounds in the Claus tail gas are converted to hydrogen sulfide by heating and passing it through a cobalt-molybdenum catalyst with the addition of a reducing gas. The gas is then cooled and contacted with a solution of di-isopropanolamine (DIPA) that removes all but trace amounts of hydrogen sulfide. The sulfide-rich di-isopropanolamine is sent to a stripper where hydrogen sulfide gas is removed and sent to the Claus plant. The di-isopropanolamine is returned to the absorption column.

The Wellman-Lord process is divided into two main stages: (i) absorption and (ii) regeneration. In the absorption section, hot flue gases are passed through a pre-scrubber where ash, hydrogen chloride, hydrogen fluoride and sulfur trioxide are removed. The gases are then cooled and fed into the absorption tower. A saturated solution of sodium sulfite is then sprayed into the top of the absorber onto the flue gases; the sodium sulfite reacts with the sulfur dioxide forming sodium bisulfite. The concentrated bisulfate solution is collected and passed to an evaporation system for regeneration. In the regeneration section, sodium bisulfite is converted, using steam, to sodium sulfite that is recycled back to the flue gas. The remaining product, the released sulfur dioxide, is converted to elemental sulfur, sulfuric acid or liquid sulfur dioxide.

Most industrial process units and equipment are sent into a collection unit, called the blowdown system. Blowdown systems provide for the safe handling and disposal of liquid and gases that are either automatically vented from the process units through pressure relief valves, or that are manually drawn from units. Recirculated process streams and cooling water streams are often manually purged to prevent the continued buildup of contaminants in the stream. Part or all of the contents of equipment can also be purged to the blowdown system prior to shut down before normal or emergency shutdowns. Blowdown systems utilize a series of flash drums and condensers to separate the blowdown into its vapor and liquid components. The liquid is typically composed of mixtures of water and hydrocarbon derivatives containing sulfides, ammonia, and other contaminants, which are sent to the wastewater treatment plant. The gaseous component typically contains hydrocarbon derivatives, hydrogen sulfide, ammonia, mercaptans, solvents, and other constituents, and is either discharged directly to the atmosphere or is combusted in a flare. The major air emissions from blowdown systems are hydrocarbon derivatives in the case of direct discharge to the atmosphere and sulfur oxides when flared.

4.2 Water treatment

In addition, many aquifers and isolated surface waters are high in water quality and may be pumped from the supply and transmission network directly to any number of end uses, including human consumption, irrigation, industrial processes, or fire control. However, clean water sources are the exception in m a y parts of the world, particularly regions where the population is dense or where there is heavy agricultural use. In these places, the water supply must receive varying degrees of treatment before distribution. Impurities enter water as it moves through the atmosphere, across the surface of the Earth, and between soil particles in the ground. These background levels of impurities are often supplemented by human activities.

Generally, and without many exceptions, the characteristics of raw water determine the treatment method to be applied. Most public water systems **are** relied on for drinking water as well **as** for industrial consumption and firefighting, so that human consumption, the highest use of the water, defines the degree of treatment. Thus, the water from industrial sources contain a varied mix contaminants (Chapter 4) and each water must requires analysis (Chapter 8) before a treatment technology is applied. Chemicals from industrial discharges and pathogenic organisms of human origin, if allowed to enter the water distribution system, may cause health problems. Excessive silt and other solids may make water aesthetically unpleasant and unsightly. Heavy metal pollution, including lead, zinc, and copper, may be caused by corrosion of the very pipes that carry water from the source to the consumer.

Water used in processing operations accounts for a significant portion of the total wastewater. Process wastewater arises from desalting crude oil, steam-stripping operations, pump gland cooling, product fractionator reflux drum drains and boiler blowdown. Because process water often comes into direct contact with oil, it is usually highly contaminated. Most cooling water is recycled over and over. Cooling water typically does not come into direct contact with process oil streams and therefore contains less contaminants than process wastewater. However, it may contain some oil contamination due to leaks in the process equipment. Storm water (i.e., surface water runoff) is intermittent and will contain constituents from spills to the surface, leaks in equipment and any materials that may have collected in drains. Runoff surface water also includes water coming from crude and product storage tank roof drains. Sewage water needs no further explanation of its origins but must be treated as opposed to discharge on to the land or into ponds.

Primary wastewater treatment consists of the separation of oil, water and solids in two stages. During the first stage, an API separator, a corrugated plate interceptor, or other separator design is used. Wastewater moves very slowly through the separator allowing free oil to float to the surface and be skimmed off, and solids to settle to the bottom and be scraped off to a sludge collection hopper. The second stage utilizes physical or chemical methods to separate emulsified oils from the wastewater. Physical methods may include the use of a series of settling ponds with a long retention time, or the use of dissolved air flotation (DAF). In DAF, air is bubbled through the wastewater, and both oil and suspended solids are skimmed off the top. Chemicals, such as ferric hydroxide or aluminum hydroxide, can be used to coagulate impurities into a froth or sludge that can be more easily skimmed off the top. Some wastes associated with the primary treatment of wastewater at crude oil refineries may be considered hazardous and include API separator sludge, primary treatment sludge, sludge from other gravitational separation techniques, float from DAF units, and wastes from settling ponds.

After primary treatment, the wastewater can be discharged to a publicly owned treatment works (POTW) or undergo *secondary treatment* before being discharged directly to surface waters under a National Pollution Discharge Elimination System (NPDES) permit. In secondary treatment, microorganisms may consume dissolved oil and other organic pollutants biologically. Biological treatment may require the addition of oxygen through a number of different techniques, including activated sludge units, trickling filters, and rotating biological contactors. Secondary treatment generates biomass waste that is typically treated anaerobically and then dewatered.

Some refineries employ an additional stage of wastewater treatment called *polishing* to meet discharge limits. The polishing step can involve the use of activated carbon, anthracite coal, or sand to filter out any remaining impurities, such as biomass, silt, trace metals and other inorganic chemicals, as well as any remaining organic chemicals.

Certain industrial wastewater streams are treated separately, prior to the wastewater treatment plant, to remove contaminants that would not easily be treated after mixing with other wastewater. One such waste stream is the sour water drained from distillation reflux drums. Sour water contains dissolved hydrogen sulfide and other organic sulfur compounds and ammonia which are stripped in a tower with gas or steam before being discharged to the wastewater treatment plant.

Wastewater treatment plants are a significant source of industrial air emissions and solid wastes. Air releases arise from fugitive emissions from the numerous tanks, ponds and sewer system drains. Solid wastes are generated in the form of sludge from a number of the treatment units.

Many refineries unintentionally release, or have unintentionally released in the past, liquid hydrocarbon derivatives to ground water and surface waters. At some refineries, contaminated ground water has migrated off-site and resulted in continuous seeps to surface waters. While the actual volume of hydrocarbon derivatives released in such a manner are relatively small, there is the potential to contaminate large volumes of ground water and surface water possibly posing a substantial risk to human health and the environment.

4.3 Other waste and treatment

Solid wastes are generated from many of the refining processes, crude oil handling operations, as well as wastewater treatment (Chapter 4). Both hazardous and non-hazardous wastes are generated, treated and disposed. Solid wastes from an industrial process are typically in the form of sludge (including sludge from wastewater treatment), spent process catalysts, filter clay, and incinerator ash. Treatment of these wastes includes incineration, land treating off-site, land filling onsite, land filling off-site, chemical fixation, neutralization, and other treatment methods (Speight, 1996; Woodside, 1999).

A significant portion of the non-crude oil product outputs of refineries is transported off-site and sold as by-products. These outputs include sulfur, acetic acid, phosphoric acid, and recovered metals. Metals from catalysts and from the crude oil that have deposited on the catalyst during the production often are recovered by third party recovery facilities.

Storage tanks are used throughout the refining process to store crude oil and intermediate process feeds for cooling and further processing. Finished crude oil products are also kept in storage tanks before transport off site. Storage tank bottoms are mixtures of iron rust from corrosion, sand, water, and emulsified oil and wax, which accumulate at the bottom of tanks. Liquid tank bottoms (primarily water and oil emulsions) are periodically drawn off to prevent their continued build up. Tank bottom liquids and sludge are also removed during periodic cleaning of tanks for inspection. Tank bottoms may contain amounts of tetraethyl or tetramethyl lead (although this is increasingly rare due to the phase out of leaded products), other metals, and phenols. Solids generated from leaded gasoline storage tank bottoms are listed as a hazardous waste.

5 Management, mismanagement, and the future

Although numerous cases have been documented where many industries have simultaneously reduced pollution and operating costs, there are often barriers to doing so. The primary barrier to most pollution reduction projects is cost. Many pollution reduction options simply do not pay for themselves. Corporate investments typically must earn an adequate return on invested capital for the shareholders and some pollution prevention options at some facilities may not meet the requirements set by the companies. In addition, the equipment used in the crude oil refining industry are very capital intensive and have very long lifetimes. This reduces the incentive to make process modifications to (expensive) installed equipment that is still useful. It should be noted that pollution prevention techniques are, nevertheless, often more cost-effective than pollution reduction through end-of-pipe treatment.

Of course, facility training programs that emphasize the importance of keeping solids out of the sewer systems will help reduce that portion of wastewater treatment plant sludge arising from the everyday activities of industrial personnel. For example, educating personnel on how to avoid leaks and spills can reduce contaminated soil.

A systematic approach will produce better results than piecemeal efforts. An essential first step is a comprehensive waste audit (Table 9.8). The waste audit should systematically evaluate opportunities for improved operating procedures, process modifications, process redesign and recycling.

Crude oil industrial wastes result from processes designed to remove naturally occurring contaminants in the crude oil, including water, sulfur, nitrogen and heavy metals (Table 9.9). Setting up a pollution prevention program does not require exotic or expensive technologies. Some of the most effective techniques are simple and inexpensive. Others require significant capital expenditures, however many provide a return on that investment.

The conditions of the contaminated area plays a major role on whether bioremediation is the appropriate method of cleanup for the given oil spill. The success of bioremediation is dependent upon physical conditions and chemical conditions. Physical parameters include temperature, surface area of the oil, and the energy of the water. Chemical parameters include oxygen and nutrient content, pH, and the composition of the oil. Temperature affects bioremediation by changing the properties of the oil and also by influencing the crude oil-degrading microbes. When the temperature is lowered, the viscosity of the crude oil is increased which changes the toxicity and solubility of the oil, depending upon its composition. Temperature also has an effect on the

Table 9.8 Elements of a waste audit

• List all generated waste Identify the composition of the waste and the source of each substance • Identify options to reduce the generation of these substances in the production or manufacturing process • Focus on wastes that are most hazardous and techniques that are most easily implemented • Compare the technical and economic feasibility of the options identified • Evaluate the results and schedule periodic reviews of the program so it can be adapted to reflect changes in regulations, technology, and economic feasibility

Table 9.9 Options for waste reduction

• Segregate process (oily) waste streams from relatively clean rainwater runoff in order to reduce the quantity of oily sludge • Generated and increased the potential for oil recovery. Significant portion of the industrial waste comes from oily sludge found in combined process/storm sewers. • Conduct inspection of crude oil industrial systems for leaks. For example, check hoses, pipes, valves, pumps and seals. Make necessary repairs where appropriate. • Conserve water. Reuse rinse waters if possible. Reduce equipment-cleaning frequency where beneficial in reducing net waste generation. • Use correct pressures, temperatures and mixing ratios for optimum recovery of product and reduction in waste produced. • Employ street sweeping or vacuuming of paved process areas to reduce solids to the sewers. • Pave runoff areas to reduce transfer of solids to waste systems. Use water softeners in cooling water systems to extend useful cycling time of the water.

growth rate of the microorganisms, as well as the degradation rate of the hydrocarbon derivatives, depending upon their characteristics.

Biostimulation, the addition of nutrients, is practiced for cleanup of crude oil spills in seawater when there is an existing population of oil degrading microbes present. When an oil spill occurs, the result is a large increase in carbon and this also stimulates the growth of the indigenous crude oil-degrading microorganisms. However, these microorganisms are limited in the amount of growth and remediation that can occur by the amount of available nitrogen and phosphorus.

By adding these supplemental nutrients in the proper concentrations, the hydrocarbon degrading microbes are capable of achieving their maximum growth rate and hence the maximum rate of pollutant uptake. It has been found that when using nitrogen for the supplemental nutrient, a maximum growth rate is achieved by the oil degrading microorganisms. Biostimulation has been proven to be an effective way of achieving increased hydrocarbon degradation by the indigenous microbial population.

Supplemental nutrients tended to move downward during rising tides and seaward during falling tides. This is very useful information in determining the proper timing to add nutrients in order to allow for the maximum residence time of the nutrients in the contaminated areas. The results of this experiment concluded that the nutrients should be applied during low tide at the high tide line, which results in maximum contact time of the nutrients with the oil and hydrocarbon degrading microorganisms.

Waves also have an effect on the distribution and movement of the water and dissolved nutrients, which determine the residence time of the nutrients in the oil affected area. The role of waves on solute movement varies whether there is a tide or not. When Research performed by, showed that when a wave is present there is a sharp seaward hydraulic gradient in the backwash zone, and a gentle gradient landward of this area. Furthermore, the contact time of the nutrients is increased when the waves break seaward of their location and the waves increased the dispersion and washout of the nutrients in the tidal zone, and residence time was approximately 75% when a wave was present with a tide, as compared to a tide with no waves.

The marine environment encompassing the vast majority of surface of the Earth is a repertoire of a large number of microorganisms. The environmental roles of the biosurfactants produced by many such marine microorganisms have been reported earlier.

Controlling the source of pollution in a water system is currently one of the biggest challenges facing scientists, engineers, and regulators. For example, many nonpoint pollutants originate from common anthropogenic activities. Also, because of its widespread nature, nonpoint source pollution is difficult to contain, even harder to eliminate, and costly to mitigate (Table 9.10). Source control through public education, community planning, and regulatory guidelines can be very effective, but often requires substantial changes in human behavior. Furthermore, with the rapid changes occurring in nonpoint source pollution treatment technology, it is important to continue gathering information on what works, what does not work, and what factors contribute to successful nonpoint source pollution reduction.

Table 9.10 Characteristics of point and nonpoint sources of chemicals

Point sources
Wastewater effluent (municipal and industrial)
Runoff and leachate from waste disposal sites
Runoff and infiltration from animal feedlots
Runoff from mines, oil fields, industrial sites lacking a sewer system
Storm sewer output from cities with a population > 100,000
Overflows of combined storm and sanitary sewers
Runoff from construction sites
Nonpoint sources
Runoff from agriculture (including return flow from irrigated agriculture)
Runoff from pasture and range
Septic tank leachate and runoff from failed septic systems
Runoff from construction sites
Runoff from abandoned mines
Atmospheric deposition over a water surface
Activities on land that generate contaminants
Logging
Wetland conversion
Construction
Development of land
Development of waterways

References

Baig, J.A., Kazi, T.G., Arain, M.B., Afridi, H.I., Kandhro, G.A., Sarfraz, R.A., Jamali, M.K., Shah, A.Q., 2009. Evaluation of arsenic and other physico-chemical parameters of surface and ground water of Jamshoro, Pakistan. J. Hazard. Mater. 166, 662–669.

Boyce, A., 1997. Introduction to Environmental Technology. Van Nostrand Reinhold, New York.

Bu, H., Tan, X., Li, S., Zhang, Q., 2010. Water quality assessment of the Jinshui River (China) using multivariate statistical techniques. Environ. Earth Sci. 60, 1631–1639.

Copeland, C., 2012. Pesticide Use and Water Quality: Are the Laws Complementary or in Conflict? Congressional Research Service. Untied States Congress, Washington, DC (July 13). https://fas.org/sgp/crs/misc/RL32884.pdf.

EPA. 2004. Environmental Protection Agency, Washington, DC. Web site: http://www.epa.gov.

Kumar, R., Singh, R.D., Sharma, K.D., 2005. Water resources of India. Curr. Sci. 85 (5), 794–811.

Lipton, S., Lynch, J., 1994. Handbook of Health Hazard Control in the Chemical Process Industry. John Wiley and Sons Inc, New York.

Lo, P., 1991. Wastewater and Solid Waste Management. County Sanitation District of Los Angeles County, Whittier, CA.

Mian, I.A., Begum, S., Riaz, M., Ridealgh, M., McClean, C.J., Cresser, M.S., 2010. Spatial and temporal trends in nitrate concentrations in the river Derwent, North Yorkshire, and its need for NVZ status. Sci. Total Environ. 408, 702–712.

Mokhatab, S., Poe, W.A., and Speight, J.G. 2006. Handbook of Natural Gas Transmission and Processing. Elsevier, Amsterdam, Netherlands.

Okeke, C.O., Igboanua, A.H., 2003. Characteristics and quality assessment of surface water and groundwater resources of Akwa Town, Southeast, Nigeria. J. Niger. Assoc. Hydrol. Geol. 14, 71–77.

Speight, J.G., 1993. Gas Processing: Environmental Aspects and Methods. Butterworth-Heinemann, Oxford, England.

Speight, J.G., 1996. Environmental Technology Handbook. Taylor & Francis, Philadelphia, Pennsylvania.

Speight, J.G., 2019. Natural Gas: A Basic Handbook, second ed. Gulf Publishing Company, Elsevier, Cambridge, MA.

Wang, X., Han, J., Xu, L., Zhang, Q., 2010. Spatial and seasonal variations of the contamination within water body of the Grand Canal, China. Environ. Pollut. 158, 1513–1520.

Wetzel, G.W., 2001. Limnology: Lake and River Ecosystems. Academic Press, New York, pp.15–42.

Woodside, G., 1999. Hazardous Materials and Hazardous Waste Management. John Wiley & Sons Inc, Hoboken, NJ.

Further reading

Water Management Forum, 2003. Inter-basin Transfer of Water in India-Prospects and Problems. The Institution of Engineers (India), New Delhi.

Conversion tables

1 Area

1 square centimeter (1 cm^2) = 0.1550 square inches
1 square meter 1 (m^2) = 1.1960 square yards
1 hectare = 2.4711 acres
1 square kilometer (1 km^2) = 0.3861 square miles
1 square inch (1 $inch^2$) = 6.4516 square centimeters
1 square foot (1 ft^2) = 0.0929 square meters
1 square yard (1 yd^2) = 0.8361 square meters
1 acre = 4046.9 square meters
1 square mile (1 mi^2) = 2.59 square kilometers

2 Concentration conversions

1 part per million (1 ppm) = 1 microgram per liter (1 μg/L)
1 microgram per liter (1 μg/L) = 1 milligram per kilogram (1 mg/kg)
1 microgram per liter (μg/L) × 6.243×10^8 = 1 lb per cubic foot (1 lb/ft^3)
1 microgram per liter (1 μg/L) × 10^{-3} = 1 milligram per liter (1 mg/L)
1 milligram per liter (1 mg/L) × 6.243×10^5 = 1 pound per cubic foot (1 lb/ft^3)
I gram mole per cubic meter (1 g mol/m^3) × 6.243×10^5 = 1 pound per cubic foot (1 lb/ft^3)
10,000 ppm = 1% w/w
1 ppm hydrocarbon in soil × 0.002 = 1 lb of hydrocarbons per ton of contaminated soil

3 Nutrient conversion factor

1 pound, phosphorus × 2.3 (1 lb P × 2.3) = 1 pound, phosphorous pentoxide (1 lb P_2O_5)
1 pound, potassium × 1.2 (1 lb K × 1.2 = 1 pound, potassium oxide (1 lb K_2O)

4 Temperature conversions

$^oF = (^oC \times 1.8) + 32$
$^oC = (^oF - 32)/1.8$
$(^oF - 32) \times 0.555 = {^oC}$
Absolute zero = −273.15°C
Absolute zero = −459.67 °F

5 Sludge conversions

1,700 lbs wet sludge = 1 yd^3 wet sludge
1 yd^3 sludge = wet tons/0.85
Wet tons sludge × 240 = gallons sludge
1 wet ton sludge × % dry solids/100 = 1 dry ton of sludge

6 Various constants

Atomic mass	$mu = 1.6605402 \times 10^{-27}$
Avogadro's number	$N = 6.0221367 \times 10^{23}\ mol^{-1}$
Boltzmann's constant	$k = 1.380658 \times 10^{-23}\ J\ K^{-1}$
Elementary charge	$e = 1.60217733 \times 10^{-19}\ C$
Faraday's constant	$F = 9.6485309 \times 104\ C\ mol^{-1}$
Gas (molar) constant	$R = k \cdot N \,\tilde{}\, 8.314510\ J\ mol^{-1}\ K^{-1}$
	$= 0.08205783\ L\ atm\ mol^{-1}\ K^{-1}$
Gravitational acceleration	$g = 9.80665\ m\ s^{-2}$
Molar volume of an ideal gas at 1 atm and 25 °C	$V_{ideal\ gas} = 24.465\ L\ mol^{-1}$
Planck's constant	$h = 6.6260755 \times 10^{-34}\ J\ s$
Zero, Celsius scale	0 °C = 273.15°K

7 Volume conversion

Barrels (petroleum, U. S.) to Cu feet multiply by 5.6146
Barrels (petroleum, U. S.) to Gallons (U. S.) multiply by 42
Barrels (petroleum, U. S.) to Liters multiply by 158.98
Barrels (US, liq.) to Cu feet multiply by 4.2109
Barrels (US, liq.) to Cu inches multiply by 7.2765 × 103
Barrels (US, liq.) to Cu meters multiply by 0.1192
Barrels (US, liq.) to Gallons multiply by (U. S., liq.) 31.5
Barrels (US, liq.) to Liters multiply by 119.24
Cubic centimeters to Cu feet multiply by 3.5315×10^{-5}
Cubic centimeters to Cu inches multiply by 0.06102
Cubic centimeters to Cu meters multiply by 1.0×10^{-6}
Cubic centimeters to Cu yards multiply by 1.308×10^{-6}
Cubic centimeters to Gallons (US liq.) multiply by 2.642×10^{-4}
Cubic centimeters to Quarts (US liq.) multiply by 1.0567×10^{-3}
Cubic feet to Cu centimeters multiply by 2.8317×10^{4}
Cubic feet to Cu meters multiply by 0.028317
Cubic feet to Gallons (US liq.) multiply by 7.4805
Cubic feet to Liters multiply by 28.317
Cubic inches to Cu cm multiply by 16.387

Cubic inches to Cu feet multiply by 5.787×10^{-4}
Cubic inches to Cu meters multiply by 1.6387×10^{-5}
Cubic inches to Cu yards multiply by 2.1433×10^{-5}
Cubic inches to Gallons (US liq.) multiply by 4.329×10^{-3}
Cubic inches to Liters multiply by 0.01639
Cubic inches to Quarts (US liq.) multiply by 0.01732
Cubic meters to Barrels (US liq.) multiply by 8.3864
Cubic meters to Cu cm multiply by 1.0×10^{6}
Cubic meters to Cu feet multiply by 35.315
Cubic meters to Cu inches multiply by 6.1024×10^{4}
Cubic meters to Cu yards multiply by 1.308
Cubic meters to Gallons (US liq.) multiply by 264.17
Cubic meters to Liters multiply by 1000
Cubic yards to Bushels (Brit.) multiply by 21.022
Cubic yards to Bushels (US) multiply by 21.696
Cubic yards to Cu cm multiply by 7.6455×105
Cubic yards to Cu feet multiply by 27
Cubic yards to Cu inches multiply by 4.6656×10^{4}
Cubic yards to Cu meters multiply by 0.76455
Cubic yards to Gallons multiply by 168.18
Cubic yards to Gallons multiply by 173.57
Cubic yards to Gallons multiply by 201.97
Cubic yards to Liters multiply by 764.55
Cubic yards to Quarts multiply by 672.71
Cubic yards to Quarts multiply by 694.28
Cubic yards to Quarts multiply by 807.90
Gallons (US liq.) to Barrels (US liq.) multiply by 0.03175
Gallons (US liq.) to Barrels (petroleum, US) multiply by 0.02381
Gallons (US liq.) to Bushels (US) multiply by 0.10742
Gallons (US liq.) to Cu centimeters multiply by 3.7854×10^{3}
Gallons (US liq.) to Cu feet multiply by 0.13368
Gallons (US liq.) to Cu inches multiply by 231
Gallons (US liq.) to Cu meters multiply by 3.7854×10^{-3}
Gallons (US liq.) to Cu yards multiply by 4.951×10^{-3}
Gallons (US liq.) to Gallons (wine) multiply by 1.0
Gallons (US liq.) to Liters multiply by 3.7854
Gallons (US liq.) to Ounces (US fluid) multiply by 128.0
Gallons (US liq.) to Pints (US liq.) multiply by 8.0
Gallons (US liq.) to Quarts (US liq.) multiply by 4.0
Liters to Cu centimeters multiply by 1000
Liters to Cu feet multiply by 0.035315
Liters to Cu inches multiply by 61.024
Liters to Cu meters multiply by 0.001
Liters to Gallons (US liq.) multiply by 0.2642
Liters to Ounces (US fluid) multiply by 33.814

8 Weight conversion

1 ounce (1 ounce) = 28.3495 grams (18.2495 g)
1 pound (1 lb) = 0.454 kilogram
1 pound (1 lb) = 454 grams (454 g)
1 kilogram (1 kg) = 2.20462 pounds (2.20462 lb)
1 stone (English) = 14 pounds (14 lb)
1 ton (US; 1 short ton) = 2,000 lbs
1 ton (English; 1 long ton) = 2,240 lbs
1 metric ton = 2204.62262 pounds
1 tonne = 2204.62262 pounds

9 Other approximations

14.7 pounds per square inch (14.7 psi) – 1 atmosphere (1 atm)
1 kilopascal (kPa) $\times 9.8692 \times 10^{-3}$ = 14.7 pounds per square inch (14.7 psi)
1 $yd^3 = 27\ ft^3$
1 US gallon of water = 8.34 lbs
1 imperial gallon of water – 10 lbs
1 ft^3 = 7.5 gallon = 1728 cubic inches = 62.5 lbs.
1 yd^{3}= 0.765 m^3
1 acre-inch of liquid = 27,150 gallons = 3.630 ft^3
1-foot depth in 1 acre (in-situ) = 1613 × (20 to 25 % excavation factor) = ~2000 yd^3
1 yd^3 (clayey soils-excavated) = 1.1 to 1.2 tons (US)
1 yd^3 (sandy soils-excavated) = 1.2 to 1.3 tons (US)
Pressure of a column of water in psi = height of the column in feet by 0.434.

Glossary

Absolute permeability The ability of a rock to conduct a fluid when only one fluid is present in the pores of the rock

Accuracy a measure of how close the test result will be to the true value of the property being measured; a relative term in the sense that systematic errors or biases can exist but be small enough to be inconsequential

Acid catalyst a catalyst having acidic character; alumina is an example of such a catalyst

Acid deposition Acid rain; a form of pollution depletion in which pollutants, such as nitrogen oxides and sulfur oxides, are transferred from the atmosphere to soil or water; often referred to as atmospheric self-cleaning. The pollutants usually arise from the use of fossil fuels

Acid gas a corrosive gas such as hydrogen sulfide or carbon dioxide that forms an acid with water

Acidity the capacity of an acid to neutralize a base such as a hydroxyl ion (OH^-)

Acidic mine water (acid mine water) The outflow of acidic water from metal mines or from coal mines

Acid number a measure of the reactivity of petroleum with a caustic solution and given in terms of milligrams of potassium hydroxide that are neutralized by one gram of petroleum

Acid rain The precipitation phenomenon that incorporates anthropogenic acids and other acidic chemicals from the atmosphere to the land and water (see Acid deposition)

Acid sludge The residue left after treating petroleum oil with sulfuric acid for the removal of impurities; a black, viscous substance containing the spent acid and impurities

Acid treating a process in which unfinished petroleum products, such as gasoline, kerosene, and lubricating-oil stocks, are contacted with sulfuric acid to improve their color, odor, and other properties

Additive A substance added to petroleum products (such as lubricating oils) to impart new or to improve existing characteristics

Adsorption The transfer of a substance from a solution to the surface of a solid resulting in relatively high concentration of the substance at the place of contact; see also Chromatographic adsorption

Aerosol A suspension of fine solid (usually micron-size particles) particles or liquid droplets in air or in another gas

Air pollution the discharge of toxic gases and particulate matter introduced into the atmosphere, principally as a result of human activity

Air quality A measure of the amount of pollutants emitted into the atmosphere and the dispersion potential of an area to dilute those pollutants

Air toxics Hazardous air pollutants

Alicyclic hydrocarbon a hydrocarbon that has a cyclic structure (e.g., cyclohexane); also collectively called naphthenes

Aliphatic hydrocarbon a hydrocarbon in which the carbon-hydrogen groupings are arranged in open chains that may be branched. The term includes *paraffins* and *olefins* and provides a distinction from *aromatics* and *naphthenes*, which have at least some of their carbon atoms arranged in closed chains or rings

Aliquot that quantity of material of proper size for measurement of the property of interest; test portions may be taken from the gross sample directly, but often preliminary operations such as mixing or further reduction in particle size are necessary

Alkalinity the capacity of a base to neutralize the hydrogen ion (H^+)

Alkanes hydrocarbons that contain only single carbon-hydrogen bonds. The chemical name indicates the number of carbon atoms and ends with the suffix "ane"

Alkenes hydrocarbons that contain carbon-carbon double bonds. The chemical name indicates the number of carbon atoms and ends with the suffix "ene"

Alkylate the product of an alkylation process

Alkylation in the petroleum industry, a process by which an olefin (e.g., ethylene) is combined with a branched-chain hydrocarbon (e.g., *iso*-butane); alkylation may be accomplished as a thermal or as a catalytic reaction

Alkyl groups a group of carbon and hydrogen atoms that branch from the main carbon chain or ring in a hydrocarbon molecule. The simplest alkyl group, a methyl group, is a carbon atom attached to three hydrogen atoms

Alluvial aquifer A water-bearing deposit of unconsolidated material (e.g., sand, gravel) left behind by a river or other flowing water; see Artesian aquifer, Aquifer

Alumina (Al_2O_3) A chemical used in separation methods as an adsorbent and in refining as a catalyst

Amphoteric having both basic and acidic properties

Anaerobic bacteria bacteria that thrive in oxygen-poor environments

Analytical equivalence the acceptability of the results obtained from the different laboratories; a range of acceptable results

Analyte The chemical for which a sample is tested, or analyzed. *Antibody* A molecule having chemically reactive sites specific for certain other molecules

Anisotropic having some physical property that varies with direction from a given location

Anthrosphere The part of the biosphere made or modified by humans and used for their activities

Antibody A molecule having chemically reactive sites specific for certain other molecules

Anticline A structural configuration of a package of folding rocks and in which the rocks are tilted in different directions from the crest

API gravity a measure of the *lightness* or *heaviness* of petroleum that is related to density and specific gravity$^{o}API = (141.5/sp\ gr\ @\ 60^{\circ}F) - 131.5$

Aquasphere See Hydrosphere

Aquiclude A rock formation that is too impermeable to yield groundwater

Aquifer a subsurface rock interval that will produce water; often the underlay of a petroleum reservoir; a body of permeable rock or sediment that is saturated with water and yields useful amounts of water; a body of rock that is sufficiently permeable to conduct groundwater and to yield economically significant quantities of water to wells and springs; see Artesian aquifer, Alluvial aquifer, Confined aquifer

Areic region A region which lacks surface streams because of low rainfall or lithologic conditions; an areic region contributes little or no surface drainage, as in a desert region

Artesian aquifer A confined aquifer containing groundwater that will flow upwards out of a well without the need for pumping; see Artesian well, Alluvial aquifer, Aquifer

Artesian well A well drilled into an artesian aquifer; see Artesian aquifer, Alluvial aquifer, Aquifer

Aromatic hydrocarbon a hydrocarbon characterized by the presence of an aromatic ring or condensed aromatic rings; benzene and substituted benzene, naphthalene and substituted naphthalene, phenanthrene and substituted phenanthrene, as well as the higher condensed ring systems; compounds that are distinct from those of aliphatic compounds or alicyclic compounds

Asphalt the nonvolatile product obtained by distillation and further processing of an asphaltic crude oil; a manufactured product

Asphaltene (asphaltenes) the brown to black powdery material produced by treatment of petroleum, petroleum residua, or bituminous materials with a low-boiling liquid hydrocarbon, e.g. pentane or heptane; soluble in benzene (and other aromatic solvents), carbon disulfide, and chloroform (or other chlorinated hydrocarbon solvents)

ASTM International The official organization in the United States for designing standard tests for petroleum and other industrial products; formerly the American Society for Testing and Materials (ASTM)

Atmosphere The layer or a set of layers of gases surrounding the Earth that is held in place by gravity

Atmospheric residuum a residuum obtained by distillation of a crude oil under atmospheric pressure and which boils above 350 °C (660 °F).

Atmospheric equivalent boiling point (AEBP) a mathematical method of estimating the boiling point at atmospheric pressure of non-volatile fractions of petroleum

Attainment area a geographical area that meets NAAQS for criteria air pollutants (See also Non-attainment area)

Attapulgus clay see Fuller's earth

Aulacogen a long, narrow rift in a continent, often filled with thick sediments

BACT best available control technology

Baghouse a filter system for the removal of particulate matter from gas streams; so called because of the similarity of the filters to coal bags

Barrel the unit of measurement of liquids in the petroleum industry; equivalent to 42 US standard gallons or 33.6 imperial gallons

Base fluid a liquid or foam substance into which additives are mixed or added to comprise a fracturing fluid system – the base fluid for many hydraulic fracturing systems is water; in certain other applications, the base fluid may also be a carbon dioxide-based or nitrogen-based foam

Base number the quantity of acid, expressed in milligrams of potassium hydroxide per gram of sample that is required to titrate a sample to a specified end-point

Basic nitrogen nitrogen (in petroleum) that occurs in pyridine form

Basic sediment and water (bs&w, bsw) the material that collects in the bottom of storage tanks usually composed of oil, water, and foreign matter; also called bottoms, bottom settlings

Baumé gravity the specific gravity of liquids expressed as degrees on the Baumé (°B or °Bé) scale. For liquids lighter than waterSp gr 60 F = 140/(130 + °BJ)For liquids heavier than waterSp gr 60 F = 145/(145 – °BJ)

Bbl see Barrel

Benzene a colorless aromatic liquid hydrocarbon (C_6H_6)

Benzin refined light naphtha used for extraction purposes

Benzine an obsolete term for light petroleum distillates covering the gasoline and naphtha range; see Ligroine (Ligroin)

Benzol the general term which refers to commercial or technical (not necessarily pure) benzene; also the term used for aromatic naphtha

Bioaccumulation is the build-up of toxic substances in a food chain. A common example in aquatic systems is the accumulation of heavy metals such as mercury (Hg) in fish. At the start of the chain, mercury is absorbed by algae in the form of methyl mercury (CH_3Hg^+). Fish then eat the algae and absorb the methylmercury and since they are absorbing it at a faster rate than it can be excreted, it accumulates in the body of the fish

Biocide a chemical substance capable of destroying some life forms

Biogenic material derived from bacterial or vegetation sources

Biological lipid any biological fluid that is miscible with a nonpolar solvent. These materials include waxes, essential oils, chlorophyll, etc

Biological oxidation the oxidative consumption of organic matter by bacteria by which the organic matter is converted into gases

Biomagnification The process whereby predatory animals such as fish and birds absorb a toxin (such as mercury) from the food source that they consume, which then accumulates in their bodies leading to a higher concentration of the toxin in their own bodies than in the species they have eaten

Biomass biological organic matter

Biosphere That part of the environment made or modified by floral and faunal life forms and used for their activities

Bitumen a semi-solid to solid hydrocarbonaceous material found filling pores and crevices of sandstone, limestone, or argillaceous sediments

Bituminous containing bitumen or constituting the source of bitumen

Bituminous rock see Bituminous sand

Bituminous sand a formation in which the bituminous material (see Bitumen) is found as a filling in veins and fissures in fractured rocks or impregnating relatively shallow sand, sandstone, and limestone strata; a sandstone reservoir that is impregnated with a heavy, viscous black petroleum-like material that cannot be retrieved through a well by conventional production techniques

Black acid(s) a mixture of the sulfonates found in acid sludge that is insoluble in naphtha, benzene, and carbon tetrachloride; very soluble in water but insoluble in 30 per cent sulfuric acid; in the dry, oil-free state, the sodium soaps are black powders

Black carbon The name given to small particles of carbon left over from the combustion of fossil fuels and biomass; found in atmospheric particulate matter, in soil, and in sediments

Black oil any of the dark-colored oils; a term now often applied to heavy oil

Blender The equipment used to prepare the slurries and gels commonly used in stimulation treatment

Bog A term generally implies a water-logged area that does not have a solid foundation; see Swamp

Boiling point a characteristic physical property of a liquid at which the vapor pressure is equal to that of the atmosphere and the liquid is converted to a gas

Boiling range the range of temperature, usually determined at atmospheric pressure in standard laboratory apparatus, over which the distillation of oil commences, proceeds, and finishes

Boundary layer The term used to indicate the boundary where a fluid (water) is in direct contact with a mineral surface

Brine rejection A process that occurs when salt water freezes

Brine rejection point The temperature at which salt water starts to freeze

Bromine number the number of grams of bromine absorbed by 100 g of oil which indicates the percentage of double bonds in the material

Brown acid oil-soluble petroleum sulfonates found in acid sludge that can be recovered by extraction with naphtha solvent. Brown-acid sulfonates are somewhat similar to mahogany sulfonates but are more water-soluble. In the dry, oil-free state, the sodium soaps are light-colored powders

BS&W see Basic sediment and water

BTEX benzene, toluene, ethylbenzene, and the xylene isomers

Bunker fuel heavy *residual oil*, also called bunker C, bunker C fuel oil, or bunker oil. See No. 6 Fuel oil

Burner fuel oil any petroleum liquid suitable for combustion
Burning oil illuminating oil, such as kerosene (kerosine) suitable for burning in a wick lamp
Burning point see Fire point
C_1, C_2, C_3, C_4, C_5 fractions a common way of representing fractions containing a preponderance of hydrocarbons having 1, 2, 3, 4, or 5 carbon atoms, respectively, and without reference to hydrocarbon type
CAA Clean Air Act; this act is the foundation of air regulations in the United States
Calorie The amount of heat energy required to raise the temperature of 1 gram of water at 4 °C by 1 degree
Capillary forces interfacial forces between immiscible fluid phases, resulting in pressure differences between the two phases
Capillary number N_c, the ratio of viscous forces to capillary forces, and equal to viscosity times velocity divided by interfacial tension
Capillary pressure A force per area unit resulting from the surface forces to the interface between two fluids
Capillary water Water drawn through small pore spaces by capillary action and is available for plant uptake
Carbene the pentane- or heptane-insoluble material that is insoluble in benzene or toluene but which is soluble in carbon disulfide (or pyridine); a type of rifle used for hunting bison
Carboid the pentane- or heptane-insoluble material that is insoluble in benzene or toluene and which is also insoluble in carbon disulfide (or pyridine)
CAS Chemical Abstract Service
Casing the hard metal or plastic pipe that lines the well, prevents a borehole from caving in, and provides a barrier to the outside rock and groundwater; it also. It also serves to isolate fluids, such as water, gas, and oil, from the surrounding geologic formations
Catalyst a chemical agent which, when added to a reaction (process) will enhance the conversion of a feedstock without being consumed in the process
Catalyst selectivity the relative activity of a catalyst with respect to a particular compound in a mixture, or the relative rate in competing reactions of a single reactant
Catalyst stripping the introduction of steam, at a point where spent catalyst leaves the reactor, in order to strip, i.e., remove, deposits retained on the catalyst
Catalytic activity the ratio of the space velocity of the catalyst under test to the space velocity required for the standard catalyst to give the same conversion as the catalyst being tested; usually multiplied by 100 before being reported
Catalytic cracking the conversion of high-boiling feedstocks into lower boiling products by means of a catalyst that may be used in a fixed bed or fluid bed
Catchment area The area that draws surface runoff from precipitation into a stream or urban storm drain system
Cat cracking see Catalytic cracking
Catalytic reforming rearranging hydrocarbon molecules in a gasoline-boiling-range feedstock to produce other hydrocarbons having a higher antiknock quality; isomerization of paraffins, cyclization of paraffins to naphthenes, dehydrocyclization of paraffins to aromatics
Cetane index an approximation of the cetane number calculated from the density and mid-boiling point temperature; see also Diesel index
Cetane number a number indicating the ignition quality of diesel fuel; a high cetane number represents a short ignition delay time; the ignition quality of diesel fuel can also be estimated from the following formulaDiesel index = (aniline point (°F) × API gravity)100

CFR Code of Federal Regulations; Title 40 (40 CFR) contains the regulations for protection of the environment

Characterization factor the UOP characterization factor K, defined as the ratio of the cube root of the molal average boiling point, T_B, in degrees Rankine (°R = °F + 460), to the specific gravity at 60 °F/60 °F$K = (T_B)^{1/3}/\text{sp gr}$Ranges from 12.5 for paraffinic stocks to 10.0 for the highly aromatic stocks; also called the Watson characterization factor

Chemical waste Any solid, liquid, or gaseous material discharged from a process and that may pose substantial hazards to human health and environment

Chloride A chemical compound with one or more chlorine atoms bonded within the molecule; a salt of hydrochloric acid

Chromatographic adsorption selective adsorption on materials such as activated carbon, alumina, or silica gel; liquid or gaseous mixtures of hydrocarbons are passed through the adsorbent in a stream of diluent, and certain components are preferentially adsorbed

Chromatography a method of separation based on selective adsorption; see also Chromatographic adsorption

Chromatogram the resultant electrical output of sample components passing through a detection system following chromatographic separation. A chromatogram may also be called a *trace*

Cirque glacier A glacier that forms on the crests and slopes of mountains. See Valley glacier

Clay silicate minerals that also usually contain aluminum and have particle sizes are less than 0.002 micron; used in separation methods as an adsorbent and in refining as a catalyst

Cleanup a preparatory step following extraction of a sample media designed to remove components that may interfere with subsequent analytical measurements

Cloud point the temperature at which paraffin wax or other solid substances begin to crystallize or separate from the solution, imparting a cloudy appearance to the oil when the oil is chilled under prescribed conditions

Coke a gray to black solid carbonaceous material produced from petroleum during thermal processing; characterized by having a high carbon content (95%+ by weight) and a honeycomb type of appearance and is insoluble in organic solvents

Coker the processing unit in which coking takes place

Coking a process for the thermal conversion of petroleum in which gaseous, liquid, and solid (coke) products are formed

Completion A generic term used to describe the events and equipment necessary to bring a wellbore into production once drilling operations have been concluded, including but not limited to the assembly of downhole tubulars and equipment required to enable safe and efficient production from an oil or gas well. Completion quality can significantly affect production from shale reservoirs

Cone of depression A depression of the water levels in an aquifer; in confined aquifers (artesian aquifers), the cone of depression is a reduction in the pressure head surrounding the pumped well

Confined aquifer An aquifer (often an artesian aquifer) in which a confining layer prevents upward flow of groundwater

Confirmation column a secondary column in chromatography that contains a stationary phase having different affinities for components in a mixture than in the primary column. Used to confirm analyses that may not be completely resolved using the primary column

Composition the general chemical make-up of petroleum

Contaminant a substance that causes deviation from the normal composition of an environment

Coriolis force The force that dictates the direction of the movement of the sea currents which is determined by the earth rotation and which creates circular movements

Coulombic interaction The electrostatic interactions between electric charges

Cp (centipoise) a unit of viscosity

Cracking the thermal processes by which the constituents of petroleum are converted to lower molecular weight products; a process whereby the relative proportion of lighter or more volatile components of crude oil is increased by changing the chemical structure of the constituent hydrocarbons

Criteria air pollutants air pollutants or classes of pollutants regulated by the Environmental Protection Agency; the air pollutants are (including VOCs): ozone, carbon monoxide, particulate matter, nitrogen oxides, sulfur dioxide, and lead

Crude oil See *Petroleum*

Cryosphere An integral part of the global climate system with important linkages and through its influence on surface energy and moisture fluxes, clouds, precipitation, atmospheric circulation, and oceanic circulation

Cut the *distillate* obtained between two given temperatures during a distillation process

Cut point the boiling-temperature division between distillation fractions of petroleum

Cycle A collection of connected, on-going processes that circulates a common component throughout a system—such cycles are continuous with no beginning or end; examples in the Earth system include the carbon cycle, the nitrogen cycle, and the water (or hydrogeological) cycle

Cycloalkane a class of alkanes that are in the form of a ring

Cycloparaffin synonymous with Cycloalkane

Darcy a measure of the permeability of rock or sediment

Deasphaltened oil the fraction of petroleum after the asphaltene constituents has been removed

Deasphaltening removal of a solid powdery asphaltene fraction from petroleum by the addition of the low-boiling liquid hydrocarbons such as n-pentane or n-heptane under ambient conditions

Deasphalting the removal of the asphaltene fraction from petroleum by the addition of a low-boiling hydrocarbon liquid such as n-pentane or n-heptane; more correctly the removal asphalt (tacky, semi-solid) from petroleum (as occurs in a refinery asphalt plant) by the addition of liquid propane or liquid butane under pressure

Decoking removal of petroleum coke from equipment such as coking drums; hydraulic decoking uses high-velocity water streams

Delayed coking a coking process in which the thermal reaction are allowed to proceed to completion to produce gaseous, liquid, and solid (coke) products

Density the mass (or weight) of a unit volume of any substance at a specified temperature; see also Specific gravity

Desalting removal of mineral salts (mostly chlorides) from crude oils

Desorption the reverse process of adsorption whereby adsorbed matter is removed from the adsorbent; also used as the reverse of absorption

Dewaxing see Solvent dewaxing

Diesel fuel fuel used for internal combustion in diesel engines; usually that fraction which distills within the temperature range approximately 200–370 °C. A general term covering oils used as fuel in diesel and other compression ignition engines

Discharges The release of od the addition of pollutants (including animal manure or contaminated waters) to navigable waters

Distillate a product obtained by condensing the vapors evolved when a liquid is boiled and collecting the condensation in a receiver that is separate from the boiling vessel

Distillation a process for separating liquids with different boiling points

Distillation curve see Distillation profile

Distillation range the difference between the temperature at the initial boiling point and at the end point, as obtained by the distillation test

Distillation profile the distillation characteristics of petroleum or a petroleum product showing the temperature and the percent distilled

Domestic heating oil see No. 2 Fuel Oil

Drilling rig The machine used to drill a wellbore. It includes virtually everything except living quarters. Major components of the rig include the mud tanks, the mud pumps, the derrick or mast, the draw-works, the rotary table or top-drive, the drillstring, the power generation equipment and auxiliary equipment

Drinking water resource any body of water, ground or surface, that could currently, or in the future, serve as a source of drinking water for public or private water supplies

Effective permeability A relative measure of the conductivity of a porous medium for a fluid when the medium is saturated with more than one fluid. This implies that the effective permeability is a property associated with each reservoir flow, for example, gas, oil and water. A fundamental principle is that the total of the effective permeability is less than or equal to the absolute permeability

Effective porosity A fraction that is obtained by dividing the total volume of communicated pores and the total rock volume

Effluent any contaminating substance, usually a liquid that enters the environment via a domestic industrial, agricultural, or sewage plant outlet

Electric desalting- a continuous process to remove inorganic salts and other impurities from crude oil by settling out in an electrostatic field

Electrical precipitation a process using an electrical field to improve the separation of hydrocarbon reagent dispersions. May be used in chemical treating processes on a wide variety of refinery stocks

Electrostatic precipitators devices used to trap fine dust particles (usually in the size range 30-60 microns) that operate on the principle of imparting an electric charge to particles in an incoming air stream and which are then collected on an oppositely charged plate across a high voltage field

Eluate the solutes, or analytes, moved through a chromatographic column (see *elution*)

Eluent solvent used to elute sample

Elution a process whereby a solute is moved through a chromatographic column by a solvent (liquid or gas), or eluent

Emission control the use gas cleaning processes to reduce emissions

Emission standard the maximum amount of a specific pollutant permitted to be discharged from a particular source in a given environment

Enthalpy A thermodynamic quantity equivalent to the total heat content of a system; equal to the internal energy of the system plus the product of pressure and volume

Entropy A thermodynamic quantity representing the unavailability of the thermal energy of a system energy for conversion into mechanical work; often interpreted as the degree of disorder or randomness in the system

EPA Environmental Protection Agency

Epilimnion see Thermal stratification

Estuaries Coastal waters where seawater is measurably diluted with freshwater; a marine ecosystem where freshwater enters the ocean

Eutrophication The process through which a body of water becomes enriched with chemicals such as nitrates and phosphates after which algae and other aquatic plants then feed on these nutrients leading to excess growth. This leads to a reduction in the amount of dissolved oxygen available as algal blooms on the surface restrict the amount of sunlight penetrating the water limiting photosynthesis which causes the death and decomposition of plant life underwater. The lack of dissolved oxygen also kills all animal life in the water body

Evaporites A rock consisting of soluble minerals deposited as a result of evaporation of the water in which they were dissolved

Expanding clay A clay mineral that expand or swell on contact with water, such as montmorillonite

Extensive property A physical quantity where the value is (i) proportional to the size of the system that the property describes or (ii) to the quantity of matter in the system; see Intensive property

Extract the portion of a sample preferentially dissolved by the solvent and recovered by physically separating the solvent

Fabric filters filters made from fabric materials and used for removing particulate matter from gas streams (see Baghouse)

FCC fluid catalytic cracking

FCCU fluid catalytic cracking unit

Feedstock petroleum as it is fed to the refinery; a refinery product that is used as the raw material for another process; the term is also generally applied to raw materials used in other industrial processes

Ferrichrome A cyclic hexa-peptide that forms a complex with iron atoms which is composed of three glycine and three modified ornithine residues with hydroxamate groups [–N(OH) C(=O)C-]; the six oxygen atoms from the three hydroxamate groups bind the ferric iron (Fe^{3+}) in near perfect octahedral coordination

Filtration the use of an impassable barrier to collect solids but which allows liquids to pass

Fingerprint analysis a direct injection GC/FID analysis in which the detector output—the chromatogram—is compared to chromatograms of reference materials as an aid to product identification

Fire point the lowest temperature at which, under specified conditions in standardized apparatus, a petroleum product vaporizes sufficiently rapidly to form above its surface an air-vapor mixture that burns continuously when ignited by a small flame

Fischer-Tropsch process a process for synthesizing hydrocarbons and oxygenated chemicals from a mixture of hydrogen and carbon monoxide

Fixed bed a stationary bed (of catalyst) to accomplish a process (see Fluid bed)

Flame ionization detector a detector for a gas chromatograph that measures any *detector (FID)* thing that can burn

Flammability range the range of temperature over which a chemical is flammable

Flammable a substance that will burn readily

Flammable liquid a liquid having a flash point below 37.8°C (100°F)

Flammable solid a solid that can ignite from friction or from heat remaining from its manufacture, or which may cause a serious hazard if ignited

Flash point The lowest temperature to which the product must be heated under specified conditions to give off sufficient vapor to form a mixture with air that can be ignited momentarily by a flame

Flowback the portion of the injected fracturing fluid that flows back to the surface, along with oil, gas, and brine, when the well is produced

Flowback water the fracturing fluid that returns to the surface through the wellbore during and after hydraulic treatment

Flue gas gas from the combustion of fuel, the heating value of which has been substantially spent and which is, therefore, discarded to the flue or stack

Fluid catalytic cracking cracking in the presence of a fluidized bed of catalyst

Fluid coking a continuous fluidized solids process that cracks feed thermally over heated coke particles in a reactor vessel to gas, liquid products, and coke

Fly ash particulate matter produced from mineral matter in coal that is converted during combustion to finely divided inorganic material and which emerges from the combustor in the gases

Fractional composition the composition of petroleum as determined by fractionation (separation) methods

Fractional distillation the separation of the components of a liquid mixture by vaporizing and collecting the fractions, or cuts, which condense in different temperature ranges

Fractionating column a column arranged to separate various fractions of petroleum by a single distillation and which may be tapped at different points along its length to separate various fractions in the order of their boiling points

Fractionation the separation of petroleum into the constituent fractions using solvent or adsorbent methods; chemical agents such as sulfuric acid may also be used

Fracturing fluids The water and chemical additives used to hydraulically fracture the reservoir rock, and proppant (typically sand or ceramic beads) pumped into the fractures to keep them from closing once the pumping pressure is released

Freshwater Water without significant amounts of dissolved sodium chloride (salt); characteristic of rain, rivers, ponds, and most lakes

Fuel oil a general term applied to oil used for the production of power or heat. In a more restricted sense, it is applied to any petroleum product that is used as boiler fuel or in industrial furnaces. These oils are normally *residues*, but blends of distillates and *residues* are also used as fuel oil. The wider term *liquid fuel* is sometimes used, but the term *fuel oil* is preferred; also called heating oil; see also No. 1 to No. 4 Fuel oils

Fuller's earth a clay that has high adsorptive capacity for removing color from oils; attapulgus clay is a widely used fuller's earth

Functional group the portion of a molecule that is characteristic of a family of compounds and determines the properties of these compounds

Furnace oil a distillate fuel primarily intended for use in domestic heating equipment

Gas chromatography an analytical technique, employing a gaseous mobile phase, which separates mixtures into their individual components

Gas oil a petroleum distillate with a viscosity and *distillation range* intermediate between those of *kerosene* and *light lubricating oil*

Gasoline fuel for the internal combustion engine that is commonly, but improperly, referred to simply as *gas*; the terms *petrol* and *benzine* are commonly used in some countries

Gaseous pollutants gases released into the atmosphere that act as primary or secondary pollutants

Geological province A region of large dimensions characterized by similar geological and development histories

Geosphere A term often used as collective name for the atmosphere, the hydrosphere, and the lithosphere

Gibbs energy (Gibbs free energy) A thermodynamic potential that can be used to calculate the maximum of reversible work that may be performed by a thermodynamic system at constant temperature (iso thermal system) and constant pressure (isobaric system)

Gravimetric gravimetric methods weigh a residue

Grease a semisolid or solid lubricant consisting of a stabilized mixture of mineral, fatty, or synthetic oil with soaps, metal salts, or other thickeners

Greenhouse effect warming of the earth due to entrapment of the sun's energy by the atmosphere

Greenhouse gases gases that contribute to the greenhouse effect

Groundwater Water contained in porous strata below the surface of the Earth

Habitat the area in which a particular species lives; in wildlife management, the major elements of a habitat are considered to the food, water, cover, breeding space, and living space

HAP(s) hazardous air pollutant(s)

Headspace the vapor space above a sample into which volatile molecules evaporate. Certain methods sample this vapor

Heating oil see Fuel oil

Heat value The amount of heat released per unit of mass, or per unit of volume, when a substance is completely burned. The heat power of solid and liquid fuels is expressed in calories per gram or in BTU per pound. For gases, this parameter is generally expressed in kilocalories per cubic meter or in BTU per cubic foot

Heavy ends the highest boiling portion of a petroleum fraction; see also Light ends

Heavy fuel oil fuel oil having a high density and viscosity; generally residual fuel oil such as No. 5 and No 6. fuel oil

Heavy oil petroleum having an API gravity of less than 20°

Heavy petroleum see Heavy oil

Heteroatom compounds chemical compounds that contain nitrogen and/or oxygen and/or sulfur and /or metals bound within their molecular structure(s)

HF alkylation an alkylation process whereby olefins (C_3, C_4, C_5) are combined with *iso*-butane in the presence of hydrofluoric acid catalyst

Hydration unit A unit that mixes the water and chemical additives to make the hydraulic fracturing fluid; usually the blending process takes a few minutes for the water to gel to the right consistency

Hydraulic fluid a fluid supplied for use in hydraulic systems. Low viscosity and low *pour point* are desirable characteristics. Hydraulic fluids may be of petroleum or non-petroleum origin

Hydraulic head The force per unit area exerted by a column of liquid at a height above a depth (and pressure) of interest; fluids flow down a hydraulic gradient, from points of higher to lower hydraulic head

Hydrocarbons molecules that consist *only* of hydrogen and carbon atoms

Hydrocracking a catalytic high-pressure high-temperature process for the conversion of petroleum feedstocks in the presence of fresh and recycled hydrogen; carbon-carbon bonds are cleaved in addition to the removal of heteroatomic species

Hydrocracking catalyst a catalyst used for hydrocracking which typically contains separate hydrogenation and cracking functions

Hydrolysate A rock composed principally of relatively insoluble minerals produced during the weathering of the parent rock

Hygroscopic water (hygroscopic moisture) Water held in place by molecular forces during all except the driest climatic conditions

Hydrosphere also called the aquasphere; the combined mass of water found on, under, and above the surface of the Earth

Hydrothermal circulation The circulation of hot water which occurs most often in the vicinity of sources of heat within the crust of the Earth; generally occurs near volcanic activity, but can occur in the deep crust related to the intrusion of grabote or as the result of orogeny or metamorphism

Hydrotreating the removal of heteroatomic (nitrogen, oxygen, and sulfur) species by treatment of a feedstock or product at relatively low temperatures in the presence of hydrogen

Hygroscopic moisture Water held in place by molecular forces during all except the driest climatic conditions

Hypolimnion see Thermal stratification

Hypoxic A condition in which natural waters have a low concentration of dissolved oxygen (about 2 milligrams per liter, compared with a normal level of 5 to 10 milligrams per liter). Most game and commercial species of fish avoid waters that are hypoxic

Ice cap A large body of glacial ice astride a mountain, mountain range, or volcano; also called and *ice field*

Ice field See ice cap

Ignitability characteristic of liquids whose vapors are likely to ignite in the presence of ignition source; also characteristic of non-liquids that may catch fire from friction or contact with water and that burn vigorously

Illuminating oil oil used for lighting purposes

Immiscibility the inability of two or more fluids to have complete mutual solubility; they co-exist as separate phases

Immiscible two or more fluids that do not have complete mutual solubility and co-exist as separate phases

Immunoassay portable tests that take advantage of an interaction between an antibody and a specific analyte. Immunoassay tests are semi-quantitative and usually rely on color changes of varying intensities to indicate relative concentrations

Infrared spectroscopy an analytical technique that quantifies the vibration (stretching and bending) that occurs when a molecule absorbs (heat) energy in the infrared region of the electromagnetic spectrum

Intensive property A physical quantity where the value does not depend on the amount of the substance for which it is measured; see Extensive property

Interface the thin surface area separating two immiscible fluids that are in contact with each other

Interfacial film a thin layer of material at the interface between two fluids which differs in composition from the bulk fluids

Interfacial tension the strength of the film separating two immiscible fluids, e.g., oil and water or microemulsion and oil; measured in dynes (force) per centimeter or milli-dynes per centimeter

Interfacial viscosity the viscosity of the interfacial film between two immiscible liquids

Isomerization the conversion of a *normal* (straight-chain) paraffin hydrocarbon into an *iso* (branched-chain) paraffin hydrocarbon having the same atomic composition

Jet fuel fuel meeting the required properties for use in jet engines and aircraft turbine engines

Kauri butanol number A measurement of solvent strength for hydrocarbon solvents; the higher the kauri-butanol (KB) value, the stronger the solvency; the test method (ASTM D1133) is based on the principle that kauri resin is readily soluble in butyl alcohol but not in hydrocarbon solvents and the resin solution will tolerate only a certain amount of dilution and is reflected as a cloudiness when the resin starts to come out of solution; solvents such as toluene can be added in a greater amount (and thus have a higher KB value) than weaker solvents like hexane

Kerosene (kerosine) a fraction of petroleum that was initially sought as an illuminant in lamps; a precursor to diesel fuel with a *distillation* range generally falls within the limits of 150 and 300 °C; main uses are as a jet engine fuel, an illuminant, for heating purposes, and as a fuel for certain types of internal combustion engines

K-factor see Characterization factor

LAER lowest achievable emission rate; the required emission rate in non-attainment permits

Light ends the lower-boiling components of a mixture of hydrocarbons; see also Heavy ends, Light hydrocarbons

Light hydrocarbons hydrocarbons with molecular weights less than that of heptane (C_7H_{16})

Light oil the products distilled or processed from crude oil up to, but not including, the first lubricating-oil distillate

Light petroleum petroleum having an API gravity greater than 20°

Ligroine (Ligroin) a saturated petroleum naphtha boiling in the range of 20–135 °C (68–275 °F) and suitable for general use as a solvent; also called benzine or petroleum ether

Limnology The study of inland waters; see Potamology

Liquefied petroleum gas propane, butane, or mixtures thereof, gaseous at atmospheric temperature and pressure, held in the liquid state by pressure to facilitate storage, transport, and handling

Liquid chromatography a chromatographic technique that employs a liquid mobile phase

Liquid/liquid extraction an extraction technique in which one liquid is shaken with or contacted by an extraction solvent to transfer molecules of interest into the solvent phase

Lithosphere S referred to as is the *terrestrial biosphere*; the solid, outer part of the Earth, including the brittle upper portion of the mantle and the crust of the Earth

MACT maximum achievable control technology. Applies to major sources of hazardous air pollutants

Major source a source that has a potential to emit for a regulated pollutant that is at or greater than an emission threshold set by regulations

Maltenes that fraction of petroleum that is soluble in, for example, pentane or heptane; deasphaltened oil; also the term arbitrarily assigned to the pentane-soluble portion of petroleum that is relatively high boiling (>300 °C, 760 mm) (see also Petrolenes)

Mass spectrometer an analytical technique that *fractures* organic compounds into characteristic fragments based on functional groups that have a specific mass-to-charge ratio

MCL maximum contaminant level as dictated by regulations

MDL See Method detection limit

Metamorphism A process that results in the transformation of rock types

Metasomatism The chemical alteration of a rock by hydrothermal and other fluids; the replacement of one rock by another of different mineralogical and chemical composition in a process in which the minerals which compose the rocks are dissolved and new mineral formations are deposited in their place; dissolution and deposition occur simultaneously and the rock remains solid. See Mineral hydration

Method Detection Limit the smallest quantity or concentration of a substance that the instrument can measure

Microcrystalline wax wax extracted from certain petroleum residua and having a finer and less apparent crystalline structure than paraffin wax

Middle distillate one of the distillates obtained between *kerosene* and *lubricating oil* fractions in the refining processes. These include l*ight fuel oils* and *diesel fuels*

Milligrams per liter (mg/l) typically used to define the concentration of a compound dissolved in a fluid

Millidarcy (md) the customary unit of measurement of fluid permeability

equivalent to 0.001 Darcy

Mineral hydration An inorganic chemical reaction where water is added to the crystal structure pf a mineral, usually creating a new mineral (the hydrate); the process of mineral hydration is also known as *retrograde alteration* and commonly accompanies metasomatism and is often a feature of wall rock alteration around ore bodies; hydration of minerals occurs generally in concert with hydrothermal circulation which may be driven by tectonic or igneous activity. See Metasomatism

Mineral hydrocarbons petroleum hydrocarbons, considered *mineral* because they come from the earth rather than from plants or animals

Mineral oil the older term for petroleum; the term was introduced in the nineteenth century as a means of differentiating petroleum (rock oil) from whale oil or oil from plants that, at the time, were the predominant illuminant for oil lamps

Minerals naturally occurring inorganic solids with well-defined crystalline structures

Miscibility an equilibrium condition, achieved after mixing two or more fluids, which is characterized by the absence of interfaces between the fluids: (i) *first-contact miscibility:* miscibility in the usual sense, whereby two fluids can be mixed in all proportions without any interfaces forming. Example: At room temperature and pressure, ethyl alcohol and water are first -contact miscible. (ii) *multiple-contact miscibility (dynamic miscibility):* miscibility that is developed by repeated enrichment of one fluid phase with components from a second fluid phase with which it comes into contact. (iii) *minimum miscibility* pressure: the minimum pressure above which two fluids become miscible at a given temperature, or can become miscible, by dynamic processes

Mobile phase in chromatography, the phase (gaseous or liquid) responsible for moving an introduced sample through a porous medium to separate components of interest

MSDS Material safety data sheet; MSDS

Material safety data sheet a document for a chemical product or additive prepared in accordance with the Hazard Communication Standard set forth by the federal Occupational Safety and Health Administration (OSHA). MSDS must include information about the physical and chemical characteristics, physical and health hazards, and precautions for the safe use and handling (including emergency and first-aid procedures) of the hazardous chemical components contained in the specific product or additive for which the MSDS was prepared, along with the name, address, and emergency telephone number of the company that prepared it

NAAQS National Ambient Air Quality Standards; standards exist for the pollutants known as the criteria air pollutantsnitrogen oxides (NO_x), sulfur oxides (SO_x), lead, ozone, particulate matter less than 10 microns in diameter, and carbon monoxide (CO)

Naphtha a generic term applied to refined, partly refined, or unrefined petroleum products and liquid products of natural gas, the majority of which distills below 240 °C (464 °F); the volatile fraction of petroleum which is used as a solvent or as a precursor to gasoline

Naphthenes cycloparaffins

NESHAP National Emissions Standards for Hazardous Air Pollutants; emission standards for specific source categories that emit or have the potential to emit one or more hazardous air pollutants; the standards are modeled on the best practices and most effective emission reduction methodologies in use at the affected facilities

Nitrogen cycle The biogeochemical cycle by which nitrogen is converted into multiple chemical forms as it circulates among atmospheric, terrestrial, and aqueous ecosystems; the conversion of nitrogen can be carried out through both biological and physical processes, including nitrogen fixation, ammonification, nitrification, and denitrification

Non-attainment area a geographical area that does not meet NAAQS for criteria air pollutants (See also Attainment area)

NOx oxides of nitrogen

Non-aqueous phased liquids (NAPL) Organic liquids that are relatively insoluble in water and less dense than water; when mixed with water or when an aquifer is contaminated with this class of pollutant (frequently hydrocarbon in nature), these substances tend to float on the surface of the water

Nonpoint source A diffuse, unconfined discharge of water from the land to a receiving body of water; when this water contains materials that can potentially damage the receiving stream, the runoff is considered to be a source of pollutants

Non-stoichiometric compound A compound having a variable composition; see Stoichiometric compound

No. 1 Fuel oil very similar to kerosene and is used in burners where vaporization before burning is usually required and a clean flame is specified

No. 2 Fuel oil also called domestic heating oil; has properties similar to diesel fuel and heavy jet fuel; used in burners where complete vaporization is not required before burning

No. 4 Fuel oil a light industrial heating oil and is used where preheating is not required for handling or burning; there are two grades of No. 4 fuel oil, differing in safety (flash point) and flow (viscosity) properties

No. 5 Fuel oil a heavy industrial fuel oil that requires preheating before burning

No. 6 Fuel oil a heavy fuel oil and is more commonly known as Bunker C oil when it is used to fuel ocean-going vessels; preheating is always required for burning this oil

Olefin synonymous with *alkene*

Orogeny an event that leads to both structural deformation and compositional differentiation of the lithosphere of the Earth (the crust and uppermost mantle of the Earth) at convergent plate margins; an orogen or orogenic belt develops when a continental plate crumples and is pushed upwards to form one or more mountain ranges – the series of geological processes is collectively called orogenesis

Oxygenated gasoline gasoline with added ethers or alcohols, formulated according to the Federal Clean Air Act to reduce carbon monoxide emissions during winter months

Paraffin (alkane) one of a series of saturated aliphatic hydrocarbons, the lowest numbers of which are methane, ethane, and propane. The higher homologues are solid waxes

Paraffin wax the colorless, translucent, highly crystalline material obtained from the light lubricating fractions of paraffinic crude oils (wax distillates)

Particulate matter particles in the atmosphere or on a gas stream that may be organic or inorganic and originate from a wide variety of sources and processes

Partitioning in chromatography, the physical act of a solute having different affinities for the stationary and mobile phases

Partition ratios, K the ratio of total analytical concentration of a solute in the stationary phase, C_S, to its concentration in the mobile phase, C_M

Permeability A rock property for permitting a fluid pass; a factor that indicates whether a reservoir has producing characteristics or not; the capacity of a rock for transmitting a fluid which depends on the size and shape of pores in the rock, along with the size, shape, and extent of the connections between pore spaces

Petrol a term commonly used in some countries for gasoline

Petrolatum a semisolid product, ranging from white to yellow in color, produced during refining of residual stocks; see Petroleum jelly

Petroleum (Crude oil) naturally occurring mixture consisting essentially of many types of hydrocarbons, but also containing sulfur, nitrogen or oxygen derivatives. Petroleum may be of paraffinic, asphaltic or mixed base, depending on the presence of *paraffin* wax and *bitumen* in the *residue* after atmospheric distillation. Petroleum composition varies according to the geological strata of its origin

Petroleum refinery see Refinery

Petroleum refining a complex sequence of events that result in the production of a variety of products

Petrolenes the term applied to that part of the pentane-soluble or heptane-soluble material that is low boiling (<300 °C, <570 °F, 760 mm) and can be distilled without thermal decomposition (see also Maltenes)

Phase a separate fluid (or solid) that co-exists with other fluids; gas, oil, water and other stable fluids such as micro emulsions are all called phases in EOR research

Phase behavior the tendency of a fluid system to form phases as a result of changing temperature, pressure, or the bulk composition of the fluids or of individual fluid phases

Phase A quantity of matter that is homogeneous throughout

Phase Boundaries Interfaces between different phases

Phase diagram a graph of phase behavior. In chemical flooding a graph showing the relative volume of oil, brine, and sometimes one or more micro emulsion phases. In carbon dioxide flooding, conditions for formation of various liquid, vapor, and solid phases

Phase properties types of fluids, compositions, densities, viscosities, and relative amounts of oil, microemulsion, or solvent, and water formed when a micellar fluid (surfactant slug) or miscible solvent (e.g., CO2) is mixed with oil

Phase separation the formation of a separate phase that is usually the prelude to coke formation during a thermal process; the formation of a separate phase as a result of the instability/incompatibility of petroleum and petroleum products

Photoionization a gas chromatographic detection system that utilizes an *detector (PID)* ultraviolet lamp as an ionization source for analyte detection. It is usually used as a selective detector by changing the photon energy of the ionization source

Phreatic zone See Zone of saturation

PINA analysis a method of analysis for paraffins, *iso*-paraffins, naphthenes, and aromatics

PIONA analysis a method of analysis for paraffins, *iso*-paraffins, olefins, naphthenes, and aromatics

Pipe still a still in which heat is applied to the oil while being pumped through a coil or pipe arranged in a suitable firebox

Pipestill gas the most volatile fraction that contains most of the gases that are generally dissolved in the crude. Also known as *pipestill light ends*

PNA a polynuclear aromatic compound

PNA analysis a method of analysis for paraffins, naphthenes, and aromatics

Point source An identifiable and confined discharge point for one or more water pollutants, such as a pipe, channel, vessel, or ditch; the actual source need not be physically small, if its size is negligible relative to other length scales in the problem

Poisson's ratio (ν) the ratio of transverse contraction strain to longitudinal extension strain in the direction of stretching force; tensile deformation is considered positive and compressive deformation is considered negative. The definition of Poisson's ratio contains a minus sign so that normal materials have a positive ratio; also called Poisson ratio or the Poisson coefficient, or coefficient de Poisson$\nu = -\varepsilon_{trans}/\varepsilon_{longitudinal}$Strain ($\varepsilon$) is defined in elementary form as the change in length divided by the original length$E = \Delta L/L$

Pollution the introduction into the land water and air systems of a chemical or chemicals that are not indigenous to these systems or the introduction into the land water and air systems of indigenous chemicals in greater-than-natural amounts

Polynuclear aromatic compound an aromatic compound having two or more fused benzene rings, e.g. naphthalene, phenanthrene

Polycyclic aromatic hydrocarbons (PAHs) polycyclic aromatic hydrocarbons are a suite of compounds comprised of two or more condensed aromatic rings. They are found in many petroleum mixtures, and they are predominantly introduced to the //environment through natural and anthropogenic combustion processes

PONA analysis a method of analysis for paraffins (P), olefins (O), naphthenes (N), and aromatics (A)

Porosity The total volume of soil, rock, or other material that is occupied by pore spaces; a high porosity does not equate to a high permeability because the pore spaces may be poorly interconnected

Porphyrins Organometallic constituents of petroleum that contain vanadium or nickel; the degradation products of chlorophyll that became included in the protopetroleum

Positive bias A result that is incorrect and too high

Potamology The scientific study of rivers; see Limnology

Pour point The lowest temperature at which oil will pour or flow when it is chilled without disturbance under definite conditions

Precipitate A rock produced by chemical precipitation of mineral matter from aqueous solution

Produced water the naturally occurring fluid in a formation that flows to the surface through the wellbore, throughout the entire lifespan of an oil or gas well. it typically has high levels of total dissolved solids with leached out minerals from the rock

Propane deasphalting solvent deasphalting using propane as the solvent

Propane dewaxing a process for dewaxing lubricating oils in which propane serves as solvent

Proppant solid material used in hydraulic fracturing to hold open the cracks made in the reservoir rock after the high pressure of the fracturing fluids is reduced. sand, ceramic beads or miniature pellets prop open the cracks to allow for freer flow of oil or gas

Propping agent see Proppant

PSD prevention of significant deterioration

PTE potential to emit; the maximum capacity of a source to emit a pollutant, given its physical or operation design, and considering certain controls and limitations

Pure substance A substance that has a fixed chemical composition throughout

Purge and trap a chromatographic sample introduction technique in volatile components that are purged from a liquid medium by bubbling gas through it. The components are then concentrated by "trapping" them on a short intermediate column, which is subsequently heated to drive the components on to the analytical column for separation

Purge gas typically helium or nitrogen, used to remove analytes from the sample matrix in purge/trap extractions

Quality assurance/quality control a system of procedures, checks, audits, and corrective actions to ensure that all technical, operational, monitoring, and reporting activities are of high quality

RACT Reasonably Available Control Technology standards; implemented in areas of non-attainment to reduce emissions of volatile organic compounds and nitrogen oxides

Recharge A hydrologic process where water moves downward from surface water to groundwater; this process usually occurs in the vadose zone below plant roots, and is often expressed as a flux to the water table surface

Recycling the use or reuse of chemical waste as an effective substitute for a commercial products or as an ingredient or feedstock in an industrial process

Redox reaction A a type of chemical reaction that involves a transfer of electrons between two species

Reduced crude a residual product remaining after the removal, by distillation or other means, of an appreciable quantity of the more volatile components of crude oil

Refinery a series of integrated unit processes by which petroleum can be converted to a slate of useful (salable) products

Refinery gas a gas (or a gaseous mixture) produced as a result of refining operations

Refining the process(es) by which petroleum is distilled and/or converted by application of a physical and chemical processes to form a variety of products are generated

Reformate the liquid product of a reforming process

Reforming the conversion of hydrocarbons with low octane numbers into hydrocarbons having higher octane numbers; e.g. the conversion of a n-paraffin into a iso-paraffin

Reformulated gasoline (RFG) gasoline designed to mitigate smog production and to improve air quality by limiting the emission levels of certain chemical compounds such as benzene and other aromatic derivatives; often contains oxygenates

Residual fuel oil obtained by blending the residual product(s) from various refining processes with suitable diluent(s) (usually middle distillates) to obtain the required fuel oil grades

Residual oil see Residuum

Residuum (resid; *pl:*. residua) the residue obtained from petroleum after nondestructive distillation has removed all the volatile materials from crude oil, e.g. an atmospheric (345°C, 650°F+) residuum

Retention time the time it takes for an eluate to move through a chromatographic system and reach the detector. Retention times are reproducible and can therefore be compared to a standard for analyte identification

Resins that portion of the maltenes that is adsorbed by a surface-active material such as clay or alumina; the fraction of deasphaltened oil that is insoluble in liquid propane but soluble in n-heptane

Resistate A rock composed principally of residual minerals not chemically altered by the weathering of the parent rock

Rhizophere The narrow region of soil that is directly influenced by root secretions and associated soil microorganisms

River A natural, freshwater surface stream that has considerable volume compared with its smaller tributaries which are known as brooks, creeks, branches, or forks; a rivers are usually the main stems and larger tributaries of the drainage systems that convey surface runoff from the land; also a river flows from headwater areas of small tributaries to their mouths, where they may discharge into the ocean, a major lake, or a desert basin

SARA analysis a method of analysis for saturates, aromatics, resins, and asphaltenes

SARA separation see SARA analysis

Saturates paraffins and cycloparaffins (naphthenes)

Saybolt Furol viscosity the time, in seconds (Saybolt Furol Seconds, SFS), for 60 ml of fluid to flow through a capillary tube in a Saybolt Furol viscometer at specified temperatures between 70 and 210 °F; the method is appropriate for high-viscosity oils such as transmission, gear, and heavy fuel oils

Saybolt Universal viscosity the time, in seconds (Saybolt Universal Seconds, SUS), for 60 ml of fluid to flow through a capillary tube in a Saybolt Universal viscometer at a given temperature

Separatory funnel glassware shaped like a funnel with a stoppered rounded top and a valve at the tapered bottom, used for liquid/liquid separations

Shale A fine-grained sedimentary rock that formed from the compaction of finely layered silt and clay-sized minerals; a dense rock formed over millions of years from ancient sediments of decaying, organic material. Although geologists have known about the energy-potential of shale rock for generations, only within the past decade have these resources been considered economical to produce—in part due to the advances in horizontal drilling and the application of the 60-year-old technology of hydraulic fracturing. Shale is known as a *source rock* because it is the source of oil and gas deposits that are contained in sandstone and carbonate formations from which oil and gas are normally produced

Sludge A semi-solid to solid product that results from the storage instability and/or the thermal instability of petroleum and petroleum products

Soap an emulsifying agent made from sodium or potassium salts of fatty acids

Solvent a liquid in which certain kinds of molecules dissolve. While they typically are liquids with low boiling points, they may include high-boiling liquids, supercritical fluids, or gases

Solvent extraction a process for separating liquids by mixing the stream with a solvent that is immiscible with part of the waste but that will extract certain components of the waste stream

Sonication a physical technique employing ultrasound to intensely vibrate a sample media in extracting solvent and to maximize solvent/analyte interactions

Sorption The physical or chemical linkage of substances, either by absorption or by adsorption

Sour crude oil crude oil containing an abnormally large amount of sulfur compounds; see also Sweet crude oil

SOx oxides of sulfur

Soxhlet extraction an extraction technique for solids in which the sample is repeatedly contacted with solvent over several hours, increasing extraction efficiency

Specific gravity the mass (or weight) of a unit volume of any substance at a specified temperature compared to the mass of an equal volume of pure water at a standard temperature; see also Density

Specific yield The amount of water that can be extracted from an aquifer which is the percent of total volume of water in the aquifer that will drain freely from the aquifer

Spent catalyst catalyst that has lost much of its activity due to the deposition of coke and metals

Stabilization the removal of volatile constituents from a higher boiling fraction or product (q.v. stripping); the production of a product which, to all intents and purposes, does not undergo any further reaction when exposed to the air

Stationary phase in chromatography, the porous solid or liquid phase through which an introduced sample passes. The different affinities the stationary phase has for a sample allow the components in the sample to be separated, or resolved

Stoichiometric compound A compound having a definite composition; see Non-stoichiometric compound

Sulfonic acids acids obtained by of petroleum or a petroleum product with strong sulfuric acid

Sulfuric acid alkylation an alkylation process in which olefins (C_3, C_4, and C_5) combine with *iso*-butane in the presence of a catalyst (sulfuric acid) to form branched chain hydrocarbons used especially in gasoline blending stock

Supercritical fluid an extraction method where the extraction fluid is present at a pressure and temperature above its critical point

SW-846 an EPA multi-volume publication entitled *Test Methods for Evaluating Solid Waste, Physical/Chemical Methods*; the official compendium of analytical and sampling methods that have been evaluated and approved for use in complying with the RCRA regulations and that functions primarily as a guidance document setting forth acceptable, although not required, methods for the regulated and regulatory communities to use in responding to RCRA-related sampling and analysis requirements. SW-846 changes over time as new information and data are developed

Swamp A name usually applied to a wetland where trees and shrubs are an important part of the vegetative association; the term bog generally implies a water-logged area that does not have a solid foundation

Target analyte target analytes are compounds that are required analytes in U.S. EPA analytical methods. BTEX and PAHs are examples of petroleum-related compounds that are target analytes in U.S. EPA Methods

Temperature A measure of the average kinetic energy of the particles in a sample of matter, expressed in terms of units or degrees designated on a standard scale

Thermal cracking a process that decomposes, rearranges, or combines hydrocarbon molecules by the application of heat, without the aid of catalysts

Thermal pollution Occurs when the water entering the water system is warmer than the water already present in the system; one source of thermal pollution is industries such as nuclear power plants which discharge cooling water

Thermal stratification Occurs during the summer in non-flowing bodies of water (such as lakes) when a surface layer (epilimnion) is heated by solar radiation and, because of its lower density, floats upon the lower layer (hypolimnion)

Thin layer chromatography (TLC) a chromatographic technique employing a porous medium of glass coated with a stationary phase. An extract is spotted near the bottom of the medium and placed in a chamber with solvent (mobile phase). The solvent moves up the medium and separates the components of the extract, based on affinities for the medium and solvent

Total dissolved solids (TDS) The dry weight of dissolved material, organic and inorganic, contained in water and usually expressed in parts per million; non-filterable solids that pass through a filter with a pore size of 2.0 micron, after filtration the liquid is dried and residue is weighed

Total Solids (TS) The total of all solids in a water sample

Total Suspended Solids (TSS) The amount of filterable solids in a water sample, filters are dried and weighed

Total maximum daily load The maximal quantity of a particular water pollutant that can be discharged into a water body without violating a water quality standard

Total petroleum hydrocarbons (TPH) the family of several hundred chemical compounds that originally come from petroleum

TPH E gas chromatographic test for TPH extractable organic compounds

TPH V gas chromatographic test for TPH volatile organic compounds

TPH-D(DRO) gas chromatographic test for TPH diesel-range organics

TPH-G(GRO) gas chromatographic test for TPH gasoline-range organics

Trace element a chemical element that occurs at very low levels in a given system; an element used by organisms and essential to the functioning of the organism

Triple point The temperature and pressure at which the three phases (gas, liquid, solid) of that substance coexist in thermodynamic equilbrium

Ultimate analysis elemental composition

Underground source of drinking water (USDW) as defined by 40 CPR §144.3, an underground source of drinking water is an aquifer or its portion(i) which supplies any public water system or (ii) which contains a sufficient quantity of groundwater to supply a public water system; and (i) currently supplies drinking water for human consumption, or (ii) contains fewer than 10,000 mg/L total dissolved solids, and (iii) which is not an exempted aquifer

Unresolved complex the thousands of compounds that a gas chromatograph *mixture (UCM)* is unable to fully separate

Unsaturated zone see Vadose zone

Upgrading the conversion of petroleum to value-added salable products

Vacuum distillation distillation under reduced pressure

Vacuum residuum a residuum obtained by distillation of a crude oil under vacuum (reduced pressure); that portion of petroleum that boils above a selected temperature such as 510 °C (950 °F) or 565 °C (1050 °F)

Vadose water see Vadose zone

Vadose zone The area of the ground below the surface and above the region occupied by groundwater; often referred to as the unsaturated zone or zone of aeration

Valley glacier A glacier that fills a valley; also called an *alpine glacier* or *mountain glacier*. See Cirque glacier

Vapor pressure of water The pressure at which water vapor is in thermodynamic equilibrium with the condense state of water—at higher pressures water would condense; the water vapor pressure is the partial pressure water vapor in any gas mixture in equilibrium with solid or liquid water. As for other substances, water vapor pressure is a function of temperature

Vapor recovery unit (VRU) A system at a drilling site to recover vapors formed inside completely sealed crude oil or condensate tanks. The vapors are sucked through a scrubber, where the liquid trapped is returned to the liquid pipeline system or to the tanks, and the vapor recovered is pumped into gas lines

Visbreaking A process for reducing the viscosity of heavy feedstocks by controlled thermal decomposition

Viscosity A measure of the ability of a liquid to flow or a measure of its resistance to flow; the force required to move a plane surface of area 1 m^2 over another parallel plane surface 1 m away at a rate of 1 m/s when both surfaces are immersed in the fluid

VGC (viscosity-gravity constant) an index of the chemical composition of crude oil defined by the general relation between specific gravity, sg, at 60 °F and Saybolt Universal viscosity, SUV, at 100 F

VI (Viscosity index) an arbitrary scale used to show the magnitude of viscosity changes in lubricating oils with changes in temperature

Viscosity-gravity constant see VGC

Viscosity index see VI

VOC (VOCs) volatile organic compound(s); volatile organic compounds are regulated because they are precursors to ozone; carbon-containing gases and vapors from incomplete gasoline combustion and from the evaporation of solvents

Volatile Solids (VS) Those solids lost on heating to 500 °C (930 °F); an approximation of the amount of organic matter present in the solid fraction of wastewater

Vug a small space or cavity within a carbonate rock

Water-borne disease (water borne disease) Disease caused by pathogenic microorganisms that most commonly are transmitted in contaminated fresh water; infection commonly results during bathing, washing, drinking, in the preparation of food, or the consumption of food thus infected

Water quality the chemical, physical, and biological characteristics of water with respect to its suitability for a particular use

Watershed The area of land that drains into a lake or stream

Water table the subsurface level below which the pores in the soil or rock are filled with water

Watson characterization factor see Characterization factor

Wax wax of petroleum origin consists primarily of normal paraffins; wax of plant origin consists of esters of unsaturated fatty acids

Weathered crude oil crude oil which, due to natural causes during storage and handling, has lost an appreciable quantity of its more volatile components; also indicates uptake of oxygen

Wellbore (well bore) The drilled hole or borehole, including the open hole or uncased portion of the well. Borehole may refer to the inside diameter of the wellbore wall, the rock face that bounds the drilled hole

Well completion The process in which, after all the fractures are created, the downward pressure is removed from the well. Within a matter of days (typically less than a week), the release of that pressure will reverse, allowing the oil and gas to flow from the rocks and up the well

Wetlands Distinct ecosystems that is are inundated by water, either permanently or seasonally, where oxygen-free processes prevail

Wettability The relative degree to which a fluid will spread on (or coat) a solid surface in the presence of other immiscible fluids

Wobbe Index (or Wobbe Number) the calorific value of a gas divided by the specific gravity

Zeolite A crystalline aluminosilicate used as a catalyst and having a particular chemical and physical structure

Zeroth law of thermodynamics When two bodies have equality of temperature with a third body, then they have equality of temperature

Zone A rock layer identified by a characteristic microfossil species

Zone of aeration see Vadose zone

Zone of interest A segment of the formation in a single wellbore that is considered likely to produce commercial amounts of crude oil or natural gas

Zone of saturation A subsurface zone located below the zone of aeration in which the formation (soil) pores are filled with water, which is also referred to as *groundwater.*

Selected examples of ASTM standard test methods for water

1. Standard test methods for the determination of inorganic constituents in water

Designation	Title
ASTM D511	Standard Test Methods for Calcium and Magnesium in Water
ASTM D512	Standard Test Methods for Chloride Ion in Water
ASTM D513	Standard Test Methods for Total and Dissolved Carbon Dioxide in Water
ASTM D516	Standard Test Method for Sulfate Ion in Water
ASTM D857	Standard Test Method for Aluminum in Water
ASTM D858	Standard Test Methods for Manganese in Water
ASTM D859	Standard Test Method for Silica in Water
ASTM D888	Standard Test Methods for Dissolved Oxygen in Water
ASTM D1067	Standard Test Methods for Acidity or Alkalinity of Water
ASTM D1068	Standard Test Methods for Iron in Water
ASTM D1126	Standard Test Method for Hardness in Water
ASTM D1179	Standard Test Methods for Fluoride Ion in Water
ASTM D1246	Standard Test Method for Bromide Ion in Water
ASTM D1253	Standard Test Method for Residual Chlorine in Water
ASTM D1292	Standard Test Method for Odor in Water
ASTM D1426	Standard Test Methods for Ammonia Nitrogen In Water
ASTM D1429	Standard Test Methods for Specific Gravity of Water and Brine
ASTM D1687	Standard Test Methods for Chromium in Water
ASTM D1688	Standard Test Methods for Copper in Water
ASTM D1691	Standard Test Methods for Zinc in Water
ASTM D1886	Standard Test Methods for Nickel in Water
ASTM D2972	Standard Test Methods for Arsenic in Water
ASTM D3082	Standard Test Method for Boron in Water
ASTM D3223	Standard Test Method for Total Mercury in Water
ASTM D3352	Standard Test Method for Strontium Ion in Brackish Water, Seawater, and Brines
ASTM D3372	Standard Test Method for Molybdenum in Water
ASTM D3373	Standard Test Method for Vanadium in Water
ASTM D3557	Standard Test Methods for Cadmium in Water
ASTM D3558	Standard Test Methods for Cobalt in Water
ASTM D3559	Standard Test Methods for Lead in Water
ASTM D3590	Standard Test Methods for Total Kjeldahl Nitrogen in Water
ASTM D3645	Standard Test Methods for Beryllium in Water
ASTM D3651	Standard Test Method for Barium in Brackish Water, Seawater, and Brines
ASTM D3697	Standard Test Method for Antimony in Water

Designation	Title
ASTM D3859	Standard Test Methods for Selenium in Water
ASTM D3866	Standard Test Methods for Silver in Water
ASTM D3920	Standard Test Method for Strontium in Water

2. Standard test methods for analysis for organic substances in water

Designation	Title
ASTM D1252	Standard Test Methods for Chemical Oxygen Demand (Dichromate Oxygen Demand) of Water
ASTM D1783	Standard Test Methods for Phenolic Compounds in Water
ASTM D2036	Standard Test Methods for Cyanides in Water
ASTM D2580	Standard Test Method for Phenols in Water by Gas-Liquid Chromatography
ASTM D3325	Standard Practice for Preservation of Waterborne Oil Samples
ASTM D3326	Standard Practice for Preparation of Samples for Identification of Waterborne Oils
ASTM D3328	Standard Test Methods for Comparison of Waterborne Petroleum Oils by Gas Chromatography
ASTM D3415	Standard Practice for Identification of Waterborne Oils
ASTM D3973	Standard Test Method for Low-Molecular Weight Halogenated Hydrocarbons in Water
ASTM D4165	Standard Test Method for Cyanogen Chloride in Water
ASTM D4193	Standard Test Method for Thiocyanate in Water
ASTM D4489	Standard Practices for Sampling of Waterborne Oils
ASTM D4763	Standard Practice for Identification of Chemicals in Water by Fluorescence Spectroscopy
ASTM D6238	Standard Test Method for Total Oxygen Demand in Water

3. Standard test methods for quality systems, specification, and statistics

Designation	Title
ASTM D596	Standard Guide for Reporting Results of Analysis of Water
ASTM D1129	Standard Terminology Relating to Water
ASTM D4840	Standard Guide for Sample Chain-of-Custody Procedures
ASTM D5851	Standard Guide for Planning and Implementing a Water Monitoring Program
ASTM D6145	Standard Guide for Monitoring Sediment in Watersheds
ASTM D6146	Standard Guide for Monitoring Aqueous Nutrients in Watersheds

4. Standard test methods for sediments, geomorphology, and open-channel flow

Designation	Title
ASTM D3974	Standard Practices for Extraction of Trace Elements from Sediments
ASTM D3976	Standard Practice for Preparation of Sediment Samples for Chemical Analysis
ASTM D3977	Standard Test Methods for Determining Sediment Concentration in Water Samples
ASTM D4410	Terminology for Fluvial Sediment
ASTM D4411	Standard Guide for Sampling Fluvial Sediment in Motion
ASTM D4698	Standard Practice for Total Digestion of Sediment Samples for Chemical Analysis of Various Metals
ASTM D4823	Standard Guide for Core Sampling Submerged, Unconsolidated Sediments
ASTM D5073	Standard Practice for Depth Measurement of Surface Water
ASTM D5387	Standard Guide for Elements of a Complete Data Set for Non-Cohesive Sediments

Index

Note: Page numbers followed by *f* indicate figures and *t* indicate tables.

D

I

L

M

Z

Printed in the United States
By Bookmasters